Quantitative Aptitude
(Volume I)

Quantitative Aptitude
(Volume I)

Bharath Kumar Kola
TK Rama Krishna Rao

CWP

Central West Publishing

Disclaimer
Every effort has been made by the publisher and authors while preparing this book, however, no warranties are made regarding the accuracy and completeness of the content. The publisher and authors disclaim without any limitation all warranties as well as any implied warranties about sales, along with fitness of the content for a particular purpose. Citation of any website and other information sources does not mean any endorsement from the publisher and authors. For ascertaining the suitability of the contents contained herein for a particular lab or commercial use, consultation with the subject expert is needed. In addition, while using the information and methods contained herein, the practitioners and researchers need to be mindful for their own safety, along with the safety of others, including the professional parties and premises for whom they have professional responsibility. To the fullest extent of law, the publisher and authors are not liable in all circumstances (special, incidental, and consequential) for any injury and/or damage to persons and property, along with any potential loss of profit and other commercial damages due to the use of any methods, products, guidelines, procedures contained in the material herein.

A catalogue record for this book is available from the National Library of Australia

ISBN (print): 978-1-925823-86-8

About the Book

For each topic in the book, initially a brief concept has been presented, followed by the solved examples of various types including **shortcuts** in most of the cases. At the end of each topic, an assessment test has been provided for students in order to assess the knowledge. This content is mainly useful for the students who are preparing for various competitive examinations, campus recruitment training (CRT), MBA entrance tests like GMAT, MAT, CMAT, XAT, etc., along with the M. Tech. entrance examination (GATE).

Bharath Kumar Kola
TK Rama Krishna Rao

Contents

NUMBER SYSTEMS

Natural numbers:

- ✓ All counting numbers are called as natural numbers.
- ✓ Natural numbers are denoted by 'N'.
- ✓ Natural numbers are also called as 'counting numbers' N = 1, 2, 3, 4,.........

Whole numbers:

- ✓ All counting numbers including zero such as 0, 1, 2, 3,.... are called as whole numbers.
- ✓ Whole numbers are denoted by 'W'.

$$W = 0, 1, 2, 3,......$$

> **Note:** All natural numbers are whole numbers but all whole numbers are not natural numbers.

Integers:

- ✓ All counting numbers and their negatives including zero are called as integers.
- ✓ Integers are denoted by 'Z'

$$Z = -\infty,...... -3, -2, -1, 0, 1, 2, 3, ...\infty$$

- ✓ Basically integers are divided into three parts such as positive integers, negative integers and zero.

Positive integers:

- ✓ All counting numbers are called as positive integers.
- ✓ Positive integers are denoted by Z^+ and also called as natural numbers.

$$Z^+ = 1, 2, 3, 4, ...$$

Negative integers:

- ✓ Negatives of all natural numbers are called as negative integers.
- ✓ Negative integers are denoted by Z^-.

$$Z^- = -1, -2, -3, -4, ...$$

Zero:

- ✓ 'Zero' is neither positive nor negative integer.

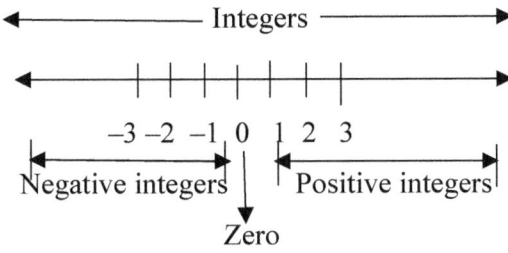

Rational numbers:

- ✓ Numbers which are able to write in the form of $\frac{a}{b}$, where both 'a' and 'b' are integers but $b \neq 0$ are called as "rational numbers".

- ✓ Rational numbers are denoted by Q.

  Examples: $\frac{2}{3}, \frac{7}{15}, \frac{-9}{8}, \frac{0}{4}, \frac{-5}{16}$ etc.

- ✓ All integers 'Z' are rational numbers.

  Since every integer can be represented in the form of $\frac{Z}{1}$.

Irrational numbers:

- ✓ Numbers which are not able to write in the form of $\frac{a}{b}$ are called as "irrational numbers" and are denoted by $\overline{Q}$.

  Examples: $\sqrt{2}, \sqrt{3}, \sqrt{5}, \sqrt{6}, \sqrt{7}, \sqrt{8}, \sqrt{10}$, …….. and also π.

- ✓ The value of π is not exactly $\frac{22}{7}$, that is only approximate value of π and hence π is also irrational number.

- ✓ In the square roots, except perfect square values remaining all values are irrational numbers.

Terminating decimals:

- ✓ Any decimal number which contains finite number of digits after the decimal point are called as "Terminating decimals".

  Examples: 0.75, 0.526, 3.7248, ……

Non – Terminating decimals:

- ✓ Any decimal number which contains infinite number of digits after the decimal point are called as "Non – Terminating decimals".

- ✓ Again these are divided into two parts.
    - i.   Recurring decimals
    - ii. Non – recurring decimals

i) Recurring decimals:

- ✓ Any decimal number in which a single digit (or) group of digits are repeated are called as "Recurring decimals".

- ✓ To write the recurring decimals in shortened form, we use the bar (–) symbol for the digits which are repeated.

Examples: $\frac{11}{3} = 3.666.... = 3.\overline{6}$ ⎤
$\frac{22}{7} = 3.142857142857...... = 3.\overline{142857}$ ⎦ Pure circulator

$\frac{41}{44} = 0.93181818 ...... = 0.93\overline{18}$ ⟶ Mixed circulator

✓ In recurring decimals, if all digits are repeated after the decimal point those are called as '**Pure Circulator**".

✓ In recurring decimals, if only few digits are repeated after decimal point those are called as "**Mixed Circulator**".

ii) Non – recurring decimals:

✓ Any decimal number in which no digit(s) cycle is/are repeated after the decimal point are called as "Non – recurring decimals".

Examples: $\sqrt{2} = 1.414213......$

$\sqrt{3} = 1.73205......$

❖ Let us look at the comparison between the terminating and non – terminating decimals.

Terminating decimals	Non – terminating decimals	
Eg: $x = 0.3275$		
$x = \frac{3275}{10000}$	Recurring decimals	Non – recurring decimals
$x = \frac{131}{400}$	$x = 3.666.... = \frac{11}{3}$	$x = 1.414213.... = \sqrt{2}$
Rational number	Rational number	Irrational number

Note:

✓ Every terminating number is a Rational number.

✓ Every non – terminating and recurring decimal is a Rational number.

✓ Every non – terminating and non- recurring decimal is an Irrational number.

Real numbers:

✓ The combination of rational numbers and irrational numbers are called as Real numbers.

✓ Real numbers are denoted by R.

Examples: $\frac{15}{17}, \frac{6}{23}, \frac{-7}{5}, \sqrt{3}, 5 + \sqrt{2}, \pi, 6 - \sqrt{3}$ etc.

✓ The sum, difference (or) product of a rational number and irrational number is always irrational.

Examples: $7 + \sqrt{2}, 5 - \sqrt{3}, 10 + 2\sqrt{5}$ all are irrational numbers.

Complex numbers:

- ✓ Numbers which can be expressed in the form of $x + iy$ are called as "Complex numbers". Where 'x' and 'y' are real numbers and 'i' is the imaginary number.

- ✓ For the complex number $x + iy$, 'x' is called as real part and 'y' is called as imaginary part.

- ✓ Basically square root of any negative real number is called as imaginary numbers.

- ✓ The concept of imaginary numbers was first introduced by great mathematician '**Euler**'.

- ✓ The value of $i = \sqrt{-1}$ was first introduced by '**Euler**'.

   Example: $5 + 3i$, $7 - 5i$, $3 + i\sqrt{5}$, $-2 + i\sqrt{3}$ etc are all complex numbers.

- ❖ Let us look at the diagrammatical representation of all types of numbers.

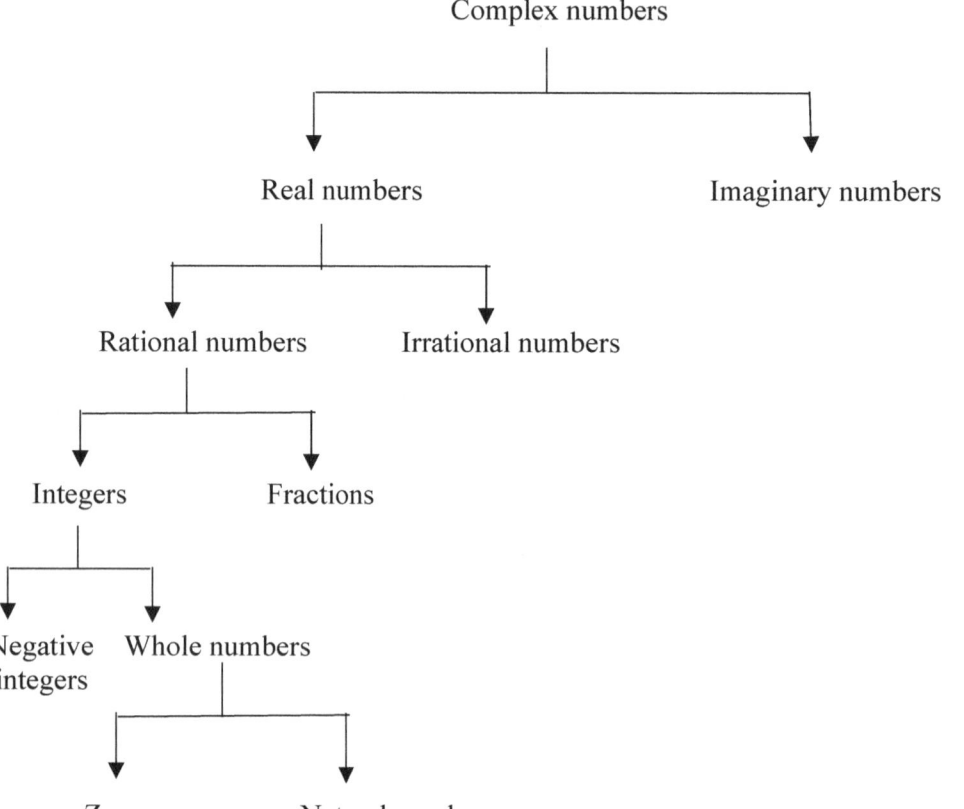

Factors:

- ✓ A number which can be expressed as a product of two (or) more numbers, then those numbers are called as factors for the given number.

  Example: $12 = 1 \times 12$ Factors of 12 are 1, 2, 3, 4, 6, 12
  $$= 2 \times 6$$
  $$= 3 \times 4$$

- ✓ In algebra terms, any equation can be expressed as a product of two (or) more independent expressions, then those independent expressions are called as factors of given equation.

  Example: $3x^2 + 17x + 10 = (3x + 2)(x + 5)$

  Factors of $3x^2 + 17x + 10$ are 1, $(3x + 2)$, $(x + 5)$, $(3x^2 + 17x + 10)$

  > **Note:**
  > - ✓ For any number 1 and itself are definitely a factors.
  > - ✓ Number of factors for any number are always finite.

Multiples:

- ✓ The product obtained by multiplying two (or) more numbers are called as multiple of the numbers being multiplied.

  Example: $5 \times 7 = 35$

  Hence 35 is a multiple of 5 and 7.

  Multiples of 2 = 2, 4, 6, 8 , 10,……

  Multiples of 3 = 3, 6, 9, 12, 15,……

  > **Note:**
  > - ✓ Zero is a multiple of all natural numbers.
  > - ✓ For any number, number of multiples are always infinite.

Prime numbers:

Numbers which have exactly two factors i.e; one and itself are called as "prime numbers".

Prime numbers upto 100:

2, 3, 5, 7, 11, 13, 17, 19, 23, 29, 31, 37, 41, 43, 47, 53, 59, 61, 67, 71, 73, 79, 83, 89, 97.

- ✓ Total 25 prime numbers from 1 to 100.
- ✓ Total 15 prime numbers from 1 to 50 and 10 prime numbers from 51 to 100.
- ✓ 2 is the only even prime number.
- ✓ 3 is the first odd prime number.

Composite numbers:

- ✓ Numbers which have more than two factors are called as "composite numbers".
- ✓ In other words; numbers greater than 1, other than prime numbers are called as "composite numbers".
- ✓ 1 is neither prime number nor composite number.

    Examples: 4, 6, 8, 9, 10, 12, 14,… are composite numbers.

Twin prime numbers:

Two prime numbers with a difference of 2 are called as "Twin prime numbers".

    Examples: (3,5), (5,7), (11,13), (17,19), (29,31), …….. are all Twin prime numbers.

Co-prime numbers:

- ✓ Two numbers doesn't have any common factor except 1, then those numbers are called as co-prime numbers.
- ✓ Here, two numbers are either prime (or) composite numbers.

    Examples: (5,7), (11,17), (13,28), (9,16), …… are all co-prime numbers.

> **Note:** Every twin prime number is a co-prime number but every co-prime number need not be a twin prime.

    Example: (5, 7) ⟶ Twin prime and also co-prime

    (9, 16) ⟶ Co-prime but not Twin prime

Prime triplet:

- ✓ Three prime numbers with a difference of 2 in between them are called as "prime triplet".
- ✓ In the entire number system, we have only one prime triplet i.e; (3, 5, 7).
- ✓ Upto 100 numbers we have only one prime triplet (3, 5, 7).
- ✓ Beyond 100 all prime numbers are ending with 1, 3, 5, 7, 9.

Now, let us look at all possible cases why there is no other prime triplet in the entire number system.

Beyond 100 prime numbers ending with:

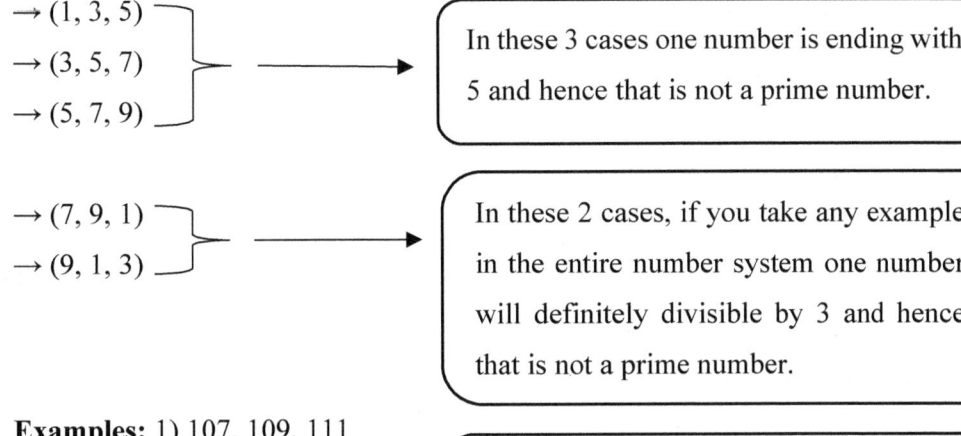

→ (1, 3, 5)
→ (3, 5, 7)
→ (5, 7, 9)

In these 3 cases one number is ending with 5 and hence that is not a prime number.

→ (7, 9, 1)
→ (9, 1, 3)

In these 2 cases, if you take any example in the entire number system one number will definitely divisible by 3 and hence that is not a prime number.

Examples: 1) 107, 109, 111

111 is divisible by 3

2) 219, 221, 223

219 is divisible by 3

Because of these reasons we have only one prime triplet in the entire number system.

Prime factors:

Numbers which can be expressed as a product of only prime numbers are called as prime factors.

Example: 72

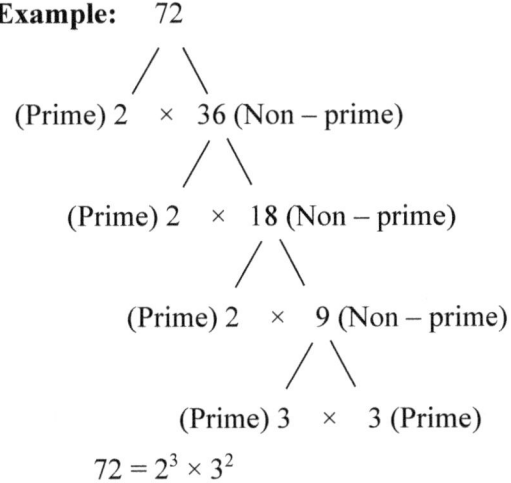

(Prime) 2 × 36 (Non – prime)

(Prime) 2 × 18 (Non – prime)

(Prime) 2 × 9 (Non – prime)

(Prime) 3 × 3 (Prime)

$72 = 2^3 \times 3^2$

How to identify whether a number is prime (or) not:

Let us consider the given number is '*x*', then

❖ Take the nearest whole number 'W' which is greater than square root of given number.

$(W > \sqrt{x})$

❖ Take all the prime numbers less than or equal to 'W'.

❖ If no prime number divides the given number '*x*', then '*x*' is a prime number, if not '*x*' is not a prime number.

Example: $x = 211 \Rightarrow 15 > \sqrt{211}$

Now take all prime numbers up to 15

2, 3, 5, 7, 11, 13

Here, no prime number divides the number 211 exactly. Hence 211 is a prime number.

Arithmetical Operations:

In mathematics, there are four arithmetical operations to be performed on numbers. Those are addition, subtraction, multiplication and division.

Addition:

- ✓ To find the total (or) sum by combining two (or) more numbers are called as "Addition".
- ✓ Addition is denoted by '+' sign.

Example: $35 + 62 + 87 = 184$.

Subtraction:

- ✓ When one (or) more numbers are taken out from a larger number, then it is called as "Subtraction".
- ✓ Subtraction is denoted by '–' sign.

Example: $641 - 328 = 313$.

Multiplication:

- ✓ Multiplication means repeated addition.
- ✓ When 'a' is multiplied by 'b', then it means 'a' is added 'b' times (or) 'b' is added 'a' times.
- ✓ Multiplication is denoted by '×' sign.

$$a \times b = ab$$

Where a = Multiplicand, b = Multiplier, ab = Product

Example: $4 \times 5 = 20$ $[4 + 4 + 4 + 4 + 4 = 20$ (or) $5 + 5 + 5 + 5 = 20]$

Division:

- ✓ Division means repeated subtraction.
- ✓ Basically in division problems, we have four elements. They are

  i) **Dividend** – Number which is to be divided.

  ii) **Divisor** – Number which divides the other number.

  iii) **Quotient** – The result obtained after the division.

  iv) **Remainder** – The value which is left undivided.

- ✓ Division is denoted by '÷' sign.

Divisor) Dividend (Quotient

xxxxx
———
Remainder
———

The relationship between these four elements is
Dividend = Divisor × Quotient + Remainder

Example: $17 \div 5$

It means 5 is subtracted repeated

3 times from 17 and the remainder left is 2

```
5) 17 (3
   15
   ──
    2
```

DIVISIBILITY RULES:

Divisibility by 2:

A number is divisible by 2, if it is ends with either 0, 2, 4, 6 (or) 8.

Example: 30, 242, 674, 1096, 4238….

Divisibility by 3:

A number is divisible by 3, if the sum of the digits of a number is divisible by 3.

Example: 32247 is divisible by 3. Since sum of the digits

$3 + 2 + 2 + 4 + 7 = 18$ is divisible by 3.

Divisibility by 4:

A number is divisible by 4, if the last two digits of the number are divisible by 4.

Example: 52472 is divisible by 4. Since the last two digits 72 is divisible by 4.

Divisibility by 5:

A number is divisible by 5, if it is ends with either 0 (or) 5.

Example: 250, 325, 560, 785, ……..

Divisibility by 6:

A number is divisible by 6, if it is divisible by both 2 and 3.

Example: 9504 is divisible by 6. Since it is ending with 4 so divisible by 2 and sum of the digits $9 + 5 + 0 + 4 = 18$ is divisible by 3.

Divisibility by 7:

A number is divisible by 7, if the difference between number of tens and twice the unit's place is divisible by 7.

Example: 665

No. of ten's = 66, twice of unit's place = $2 \times 5 = 10$, difference = $66 - 10 = 56$

Since 56 is divisible by 7, hence 665 is divisible by 7.

➤ Sometimes difference is higher digit number (3-digit, 4-digit, ……) in that case, we have to follow the same rule till we get the single-digit (or) 2-digit number.

Example: 3794

No. of tens = 379, twice the units place = $2 \times 4 = 8$,

Difference = $379 - 8 = 371$

Again, follow the same rule for 371

No. of tens = 37, twice the unit's place = $2 \times 1 = 2$

Difference = 37 − 2 = 35

Since 35 is divisible by 7, hence 3794 is divisible by 7.

Divisibility by 8:

A number is divisible by 8, if the last three digits of the number is divisible by 8.

Example: 34568 is divisible by 8, since the last 3 digits is 568 is divisible by 8.

Divisibility by 9:

A number is divisible by 9, if the sum of the digits of the number is divisible by 9.

Example: 96678 is divisible by 9.

Since the sum of the digits 9 + 6 + 6 + 7 + 8 = 36 is divisible by 9.

Divisibility by 10:

A number is divisible by 10, if the number is ends with zero.

Example: 3640 is divisible by 10, since it is ends with '0'.

Divisibility by 11:

A number is divisible by 11, if the difference between the sum of odd place digits and the sum of even place digits is either zero (or) multiple of 11.

Example: 1) 26543

Sum of odd place digits = 2 + 5 + 3 = 10

Sum of even place digits = 6 + 4 = 10

Difference = 10 − 10 = 0

Hence, the number 26543 is divisible by 11.

2) 62128

Sum of odd place digits = 6 + 1 + 8 = 15

Sum of even place digits = 2 + 2 = 4

Difference = 15 − 4 = 11

Hence, the number 62128 is divisible by 11.

Divisibility by 12:

A number is divisible by 12, if the number is divisible by both 3 and 4.

Example: 82104 is divisible by 12, since the sum of the digits 8 + 2 + 1 + 0 + 4 = 15 is divisible by 3 and the last two digits of the number is divisible by 4, so the number is divisible by 12.

Divisibility by 13:

A number is divisible by 13, if the sum of number of tens and four times of unit's place is divisible by 13.

Example: 312

Number of tens = 31, Four times of unit's place = $4 \times 2 = 8$

Sum = $31 + 8 = 39$

Since 39 is divisible by 13, so 312 is divisible by 13.

Divisibility by 14:

A number is divisible by 14, if the number is divisible by both 2 and 7.

Example: 364 is divisible by 14, since it is divisible by both 2 and 7.

Divisibility by 15:

A number is divisible by 15, if the number is divisible by both 3 and 5.

Example: 1335 is divisible by 15, since sum of the digits $1 + 3 + 3 + 5 = 12$ is divisible by 3 so the number is divisible by 3 and also the number is ending with 5 so it is divisible by 5.

Divisibility by 16:

A number is divisible by 16, if the last 4 digits of the number is divisible by 16.

Example: 3256432 is divisible by 16, since last 4 digits 6432 is divisible by 16.

Divisibility by 17:

A number is divisible by 17, if the difference between the number of tens and 5 times of unit's place is divisible by 17.

Example: 986

No. of tens = 98, Five times of unit's place = $5 \times 6 = 30$

Difference = $98 - 30 = 68$

Since 68 is divisible by 17, so 986 is also divisible by 17.

Divisibility by 18:

A number is divisible by 18, if an even number which satisfies the divisibility rule of 9.

Example: 1242 is divisible by 18, since it is an even number and also the sum of the digits $1 + 2 + 4 + 2 = 9$ is divisible by 9.

Divisibility by 19:

A number is divisible by 19, if the sum of number of tens and twice the units place is divisible by 19.

Example: 703

No. of tens = 70, twice the unit's place = $2 \times 3 = 6$

Sum = $70 + 6 = 76$ Since 76 is divisible by 19, so 703 is also divisible by 19.

Divisibility by 20:

A number is divisible by 20, if the number is divisible by both 4 and 5.

Example: 720 is divisible by 20, since the number is ending with '0' so it is divisible by 5 and the last two digits 20 is divisible by 4.

Divisibility by 24:

A number is divisible by 24, if it is divisible by both 3 and 8.

Divisibility by 25:

A number is divisible by 25, if the last two digits of the number is divisible by 25 (or) last two digits are zeros.

Example: 37575 is divisible by 25, since the last two digits 75 is divisible by 25.

Divisibility by 36:

A number is divisible by 36, if it is divisible by both 4 and 9.

Divisibility by 45:

A number is divisible by 45, if it is divisible by both 5 and 9.

Divisibility by 77:

A number is divisible by 77, if it is divisible by both 7 and 11.

Divisibility by 125:

A number is divisible by 125, if the last three digits of the number is divisible by 125 (or) the last three digits are zeros.

Example: 7564250 is divisible by 125, since the last three digits 250 is divisible by 125.

Divisibility rules of $x^n - y^n$:

Here 'n' is a natural number, to get the conclusions for $x^n - y^n$ we have to take few examples.

$$n = 1 \quad \Rightarrow \quad x - y$$

$$n = 2 \quad \Rightarrow \quad x^2 - y^2 = (x + y)(x - y)$$

$$n = 3 \quad \Rightarrow \quad x^3 - y^3 = (x - y)(x^2 + xy + y^2)$$

$$n = 4 \quad \Rightarrow \quad x^4 - y^4 = (x^2)^2 - (y^2)^2 = (x^2 + y^2)(x^2 - y^2) = (x^2 + y^2)(x + y)(x - y)$$

Conclusion 1: $x^n - y^n$ is always divisible by $(x - y)$.

Conclusion 2: $x^n - y^n$ is divisible by $(x + y)$, if 'n' is an even number.

Conclusion 3: $x^n - y^n$ is never divisible by $(x + y)$, if 'n' is an odd number.

Divisibility rules of $x^n + y^n$:

Here 'n' is a natural number, to get the conclusions for $x^n + y^n$ we have to take few examples.

n = 1	$\Rightarrow$	$x + y$
n = 2	$\Rightarrow$	$x^2 + y^2$
n = 3	$\Rightarrow$	$x^3 + y^3 = (x + y)(x^2 - xy + y^2)$
n = 4	$\Rightarrow$	$x^4 + y^4$

Conclusion 1: $x^n + y^n$ is divisible by $(x + y)$, if 'n' is an odd number.

Conclusion 2: $x^n + y^n$ is never divisible by $(x + y)$, if 'n' is an even number.

Conclusion 3: $x^n + y^n$ is never divisible by $(x - y)$ for all values of 'n'.

How to find number of factors of a given number:

✓ We all know that what are factors, for example factors of the number 24 are 1, 2, 3, 4, 6, 8, 12, 24

∴ Number of factors = 8.

✓ For small numbers we can write the factors and then we can easily count the number of factors but for large numbers it is very difficult to write all the factors.

✓ In order to avoid that difficulty, we have one simplest method to find the number of factors of any number easily.

Shortcut method:

✓ Let us consider a given number N, write that number interms of prime factors.

$$N = P_1^a \times P_2^b \times P_3^c \times \ldots\ldots$$

Where $P_1, P_2, P_3, \ldots..$ are prime numbers.

a, b, c $\ldots\ldots$ are natural numbers.

No. of factors = (a + 1)(b + 1)(c + 1) $\ldots\ldots$

Example: 1) $72 = 2^3 \times 3^2$

No. of factors = (3 + 1)(2 + 1) = 4 × 3 = 12 factors

2) $196 = 2^2 \times 7^2$

No. of factors = (2 + 1)(2 + 1) = 3 × 3 = 9 factors

Note:

1) For any perfect square, number of factors are always odd number because the power of prime number is always even number for perfect squares.

2) For non – perfect squares, number of factors are always even number.

Number of ways of expressing a given number as a product of two factors:

✓ For small numbers, we can easily write the number of ways.

Example: $24 = 1 \times 24$

$\qquad\qquad\quad 2 \times 12 \qquad$ Total 4 ways for the number 24.

$\qquad\qquad\quad 3 \times 8$

$\qquad\qquad\quad 4 \times 6$

✓ But for large numbers it is difficult to write all the ways. So in order to avoid that difficulty we have one simplest method to find the number of ways.

Shortcut method:

✓ Number of ways are always depends on number of factors. There are two cases to find the number of ways.

Case 1: If number of factors are even, then

$$\text{Number of ways} = \frac{Number\ of\ factors}{2}$$

Example: 24 factors are 1, 2, 3, 4, 6, 8, 12, 24.

No. of factors = 8 (even number)

∴ No. of ways $= \frac{8}{2} = 4$ ways.

Case 2: If number of factors are odd, then again we have two conditions.

i) As a product of two factors

$$\text{Number of ways} = \frac{No.of\ factors + 1}{2}$$

1×36
2×18
3×12
4×9
6×6

Example: 36 factors are 1, 2, 3, 4, 6, 9, 12, 18, 36.

No. of factors = 9 (odd number)

∴ No. of ways $= \frac{9+1}{2} = 5$ ways.

ii) As a Product of two different factors

$$\text{Number of ways} = \frac{No.of\ factors - 1}{2}$$

1×36
2×18
3×12
4×9

Example: 36 factors are 1, 2, 3, 4, 6, 9, 12, 18, 36.

No. of factors = 9 (odd number)

∴ No. of ways $= \frac{9-1}{2} = 4$ ways.

Sum of factors:

- ✓ Calculating sum of all factors is easy for small numbers.

    Example: 12 factors are 1, 2, 3, 4, 6, 12

    Sum of factors = $1 + 2 + 3 + 4 + 6 + 12 = 28$

- ✓ However, it is some what difficult to calculate sum of factors for large numbers.

- ✓ In order to avoid that difficulty, we have one simplest method to calculate sum of all factors.

Shortcut method:

- ✓ Let us consider a given number 'N'. Write that number interms of prime factors.

    $$N = P_1{}^a \times P_2{}^b \times P_3{}^c \times \dots\dots\dots$$

    Where $P_1, P_2, P_3, \dots\dots$ are prime numbers & a, b, c $\dots\dots$ are natural numbers.

    $$\boxed{\text{Sum of factors} = \left[\frac{P_1{}^{a+1}-1}{P_1-1}\right]\left[\frac{P_2{}^{b+1}-1}{P_2-1}\right]\left[\frac{P_3{}^{c+1}-1}{P_3-1}\right]\dots\dots}$$

    Example: $12 = 2^2 \times 3^1$

    $$\text{Sum of factors} = \frac{2^{2+1}-1}{2-1} \times \frac{3^{1+1}-1}{3-1} = 7 \times \frac{8}{2} = 28.$$

How to find unit's place digit:

i) Product form:

- ✓ To find the unit's place digit of any product, first take unit's place digit of each number and then multiply these digits.

- ✓ If there is any ten's place digit, ignore that digit.

    Example: 346×257

    Unit's place digit in $6 \times 7 = 42$

    $\therefore$ Unit's place digit in $346 \times 257 = 2$.

ii) Index form (Interms of powers):

- ✓ Basically in the number system, we have ten digits 0, 1, 2, 3, 4, 5, 6, 7, 8, 9.

- ✓ For clear understanding of concept, we have to divide these ten digits into three categories.

    1) 0, 1, 5, 6   2) 4, 9   3) 2, 3, 7, 8

    Category 1:

    If a number ends with either 0, 1, 5 (or) 6 to any power, then the unit's place digit is always the same digit.

    Example: Unit's place digit of $(240)^{163} = 0$

    Unit's place digit of $(651)^{327} = 1$

    Unit's place digit of $(425)^{648} = 5$

    Unit's place digit of $(586)^{732} = 6$

Category 2:

- ✓ First let us look at the powers of 4 and 9.
- ✓ If you clearly observes the unit's place, you will get one conclusion regarding unit's place.
- ✓ **If a number ends with 4:**

  → For odd power of 4, unit's place digit is '4'.

  → For even power of 4, unit's place digit is '6'.
- ✓ **If a number ends with 9:**

  → For odd power of 9, unit's place digit is '9'.

  → For even power of 9, unit's place digit is '1'.

$4^1 = 4$	$9^1 = 9$
$4^2 = 16$	$9^2 = 81$
$4^3 = 64$	$9^3 = 729$
$4^4 = 256$	$9^4 = 6561$

Example: Unit's place digit of $(364)^{147} = 4$

Unit's place digit of $(724)^{686} = 6$

Unit's place digit of $(529)^{371} = 9$

Unit's place digit of $(279)^{548} = 1$

Category 3:

- ✓ First let us look at the powers of 2, 3, 7, 8.

$2^1 = 2$	$3^1 = 3$	$7^1 = 7$	$8^1 = 8$
$2^2 = 4$	$3^2 = 9$	$7^2 = 49$	$8^2 = 64$
$2^3 = 8$	$3^3 = 27$	$7^3 = 343$	$8^3 = 512$
$2^4 = 16$	$3^4 = 81$	$7^4 = 2401$	$8^4 = 4096$
$2^5 = 32$	$3^5 = 243$	$7^5 = 16807$	$8^5 = 32768$

- ✓ In third category digits the concept of cycle length is very important.

Cycle length:

After how many cycles unit's place digit is repeated that is called as cycle length.

Cycle-length of the digits 0, 1, 5, 6 is 1

Cycle-length of the digits 4, 9 is 2

Cycle-length of the digits 2, 3, 7, 8 is 4

- ✓ If a number ends with 2, then the unit's place digit is $2^{\text{Remainder}}$.
- ✓ If a number ends with 3, then the unit's place digit is $3^{\text{Remainder}}$.
- ✓ If a number ends with 7, then the unit's place digit is $7^{\text{Remainder}}$.
- ✓ If a number ends with 8, then the unit's place digit is $8^{\text{Remainder}}$.

Note:

- ✓ We are getting remainder by dividing the given power value with cycle length (4).
- ✓ If remainder is '0' take that value as '4' instead of '0' because remainder '0' means it is in fourth cycle. So we have to take the unit's place digit at fourth cycle.

Examples: 1) Unit's place digit in $(562)^{327} = 2^{\text{Remainder}}$.

$$\frac{327}{4} \quad \Rightarrow \quad \text{Remainder} = 3$$

∴ Unit's place digit in $(562)^{327} = 2^3 = 8$.

2) Unit's place digit in $(347)^{652} = 7^{\text{Remainder}}$.

$$\frac{652}{4} \quad \Rightarrow \quad \text{Remainder} = 0 \text{ means we have to take } 4$$

∴ Unit's place digit in $(347)^{652} = 7^4 = 2401 = 1$.

Highest power of a prime number in N!:

- ✓ N! means it is a continuous product of 'n' natural numbers.
- ✓ To solve the problems in this concept, first we have to check whether the given number is prime (or) not. If it is not a prime number, then convert that number into prime factors.
- ✓ Here 'N' is any natural number.

Example: Highest power of 2 in 10!

Sol: Here '2' is a prime number and

$10! = 1 \times 2 \times 3 \times 4 \times 5 \times 6 \times 7 \times 8 \times 9 \times 10$

Write every possible number in terms of 2

$10! = 1 \times 2^1 \times 3 \times 2^2 \times 5 \times (2^1 \times 3^1) \times 7 \times 2^3 \times 9 \times (2^1 \times 5^1)$

$10! = 2^8 \times 3^2 \times 5^2 \times 7 \times 9$

∴ Highest power of 2 in 10! is 8.

- ✓ In the above example, 10! is a small number. So we can solve the problem easily, but for large numbers like 20!, 50!, 100!, 120! etc it is not easy to write all the values.
- ✓ In order to avoid that difficulty, we have one simplest technic to solve these kind of problems.

$$\text{Highest power of a prime number in N!} = \frac{Sum\ of\ all\ quotients}{Power\ of\ prime\ number}$$

Example 1: Find the highest power of 2 in 50!

Sol: **Step 1:** Check whether the number is prime (or) not.

Here '2' is a prime number.

Step 2: Divide the given factorial number with the prime number to get the quotients and leave the remainder every time while division.

$\therefore$ Highest Power of 2^1 in 50!

$= \dfrac{Sum\ of\ all\ quotients}{Power\ of\ 2}$

$= \dfrac{25 + 12 + 6 + 3 + 1}{1} = 47$

$\therefore$ Highest power of 2 in 50! is 47

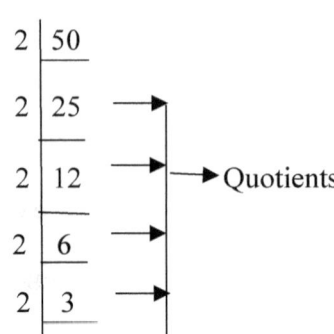

Example 2: Find the highest power of 6 in 100!

Sol: **Step 1:** Here '6' is not a prime number, so we have to convert that into prime factors.

$6 = 2 \times 3$

Step 2: Here '6' is a combination of both 2 and 3. So we have to divide 100 by both 2 and 3.

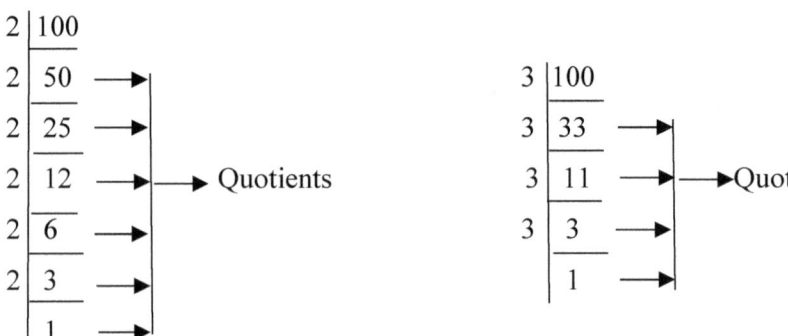

Highest power of 2 in 100! $= \dfrac{50 + 25 + 12 + 6 + 3 + 1}{1} = 97$

Highest power of 3 in 100! $= \dfrac{33 + 11 + 3 + 1}{1} = 48$

Since '6' is a combination of both 2 and 3, so we have to take combined value of both 2 and 3 which is 48.

Note:

✓ If a number is a combination of 2 (or) more than 2 prime numbers and their powers are also equal, then we can just divide with highest prime number.

✓ For this kind of problems, the required answer is always ''**Least value**''.

Example 3: Find the highest power of 24 in 100!

Sol: **Step 1:** Here '24' is not a prime number, so we have to convert that into prime factors.

$$24 = 2^3 \times 3^1$$

Step 2: Here, we have two prime numbers but their powers are not equal. So we have to divide with both the prime numbers.

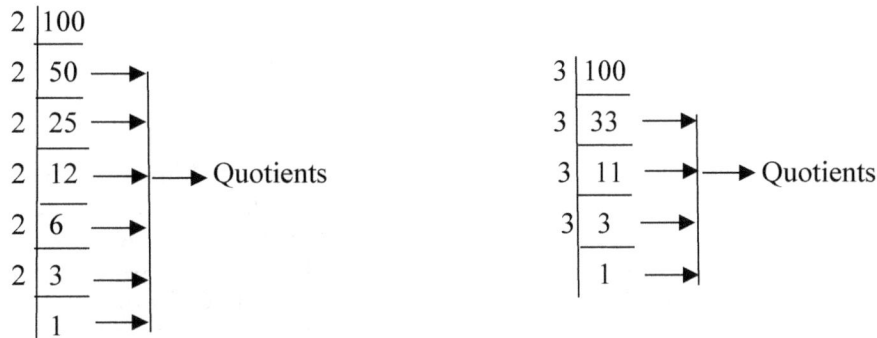

Highest power of 2^3 in 100! = $\dfrac{50 + 25 + 12 + 6 + 3 + 1}{3} = \dfrac{97}{3} = 32$ (leave remainder)

Highest power of 3 in 100! = $\dfrac{33 + 11 + 3 + 1}{1} = 48$

∴ Highest power of 24 ($2^3 \times 3$) in 100! is 32 because combined value in both the cases is 32.

Types of number systems:

❖ Basically number systems are divided into 4 types.

1) Decimal number system

2) Binary number system

3) Octal number system

4) Hexa decimal number system

✓ Decimal number system consisting of 10 digits from 0 to 9. It is denoted by base 10. In general we are using decimal number system.

✓ Binary number system consisting of only 2 digits, those are 0 and 1. It is denoted by base 2.

✓ Octal number system consisting of 8 digits from 0 to 7. It is denoted by base 8.

✓ Hexa decimal number system consisting of 16 digits from 0–9, A–F. It is denoted by base 16. In this A = 10, B = 11, C = 12, D = 13, E = 14 and F = 15.

Octal numbers interms of Binary:

- ✓ We are writing every octal number interms of binary as three digits.
- ✓ To convert any octal number into binary number

  Step 1: First divide the given octal number with 2 successively.

  Step 2: After completion of division, write all the remainders from bottom to top.

  Example: $(6)_8 = $ 2 | 6

                 2 | 3 – 0 ↑ $(6)_8 = (110)_2$

                    | 1 – 1 |

- ✓ Following table shows the octal numbers interms of binary numbers.

Octal number	0	1	2	3	4	5	6	7
Binary number	000	001	010	011	100	101	110	111

Hexa decimal numbers interms of Binary:

- ✓ We are writing every hexa decimal number interms of binary as four digits.
- ✓ To convert any hexa decimal number into binary number

  Step 1: First divide the given hexa decimal number with 2 successively.

  Step 2: After completion of division, write all the remainders from bottom to top.

  Example: $(9)_{16} = $ 2 | 9

                 2 | 4 – 1 ↑ $(9)_{16} = (1001)_2$

                 2 | 2 – 0

                    | 1 – 0 |

- ✓ Following table shows the hexa decimal numbers interms of binary numbers

Hexa decimal number	Binary number	Hexa decimal number	Binary number
0	0000	8	1000
1	0001	9	1001
2	0010	A	1010
3	0011	B	1011
4	0100	C	1100
5	0101	D	1101
6	0110	E	1110
7	0111	F	1111

Conversion of number system:

1) Decimal to Binary conversion:

    ✓ To convert any decimal number into binary number first divide the given decimal number with 2 successively and then write all the remainders from bottom to top.

Example: $(74)_{10} =$

```
2 | 74
2 | 37 − 0 ↑      (74)₁₀ = (1001010)₂
2 | 18 − 1
2 | 9  − 0
2 | 4  − 1
2 | 2  − 0
  | 1  − 0
```

$(74)_{10} = (1001010)_2$

2) Decimal to Octal conversion:

    ✓ To convert any decimal number into octal number, first divide the given decimal number with 8 successively and then write all the remainders from bottom to top.

Example: $(254)_{10} =$

```
8 | 254
8 | 31 − 6 ↑      (254)₁₀ = (376)₈
  | 3  − 7
```

$(254)_{10} = (376)_8$

3) Decimal to Hexa decimal conversion:

    ✓ To convert any decimal number into hexa decimal number, first divide the given decimal number with 16 successively and then write all the remainders from bottom to top.

Example: $(2148)_{10} =$

```
16 | 2148
16 | 134 − 4 ↑      (2148)₁₀ = (864)₁₆
   | 8   − 6
```

$(2148)_{10} = (864)_{16}$

4) Binary to Decimal conversion:

    ✓ To convert any binary number into decimal number it all depends on place values of binary number system.

    ✓ In binary number system, places are starts from 2^0 and the power values are increases from right hand side to left hand side.

Example: $(1011011)_2 =$ 1 0 1 1 0 1 1

$$2^6 \ 2^5 \ 2^4 \ 2^3 \ 2^2 \ 2^1 \ 2^0 \quad \longleftarrow \text{ places}$$

$$= 2^6 \times 1 + 2^5 \times 0 + 2^4 \times 1 + 2^3 \times 1 + 2^2 \times 0 + 2^1 \times 1 + 2^0 \times 1$$

$$= 64 + 16 + 8 + 2 + 1 = 91$$

$$\therefore (1011011)_2 = (91)_{10}$$

5) <u>Octal to Decimal conversion:</u>

- ✓ To convert any octal number into decimal number, it all depends on place values of octal number system.

- ✓ In octal number system, places are starts from 8^0 and the power values are increases from right hand side to left hand side.

    Example: $(247)_8 =$ 2   4   7

    8^2 8^1 8^0 ◄——— places

    $= 8^2 \times 2 + 8^1 \times 4 + 8^0 \times 7 = 128 + 32 + 7 = 167$

    $\therefore (247)_8 = (167)_{10}$

6) <u>Hexa decimal to Decimal conversion:</u>

- ✓ To convert any hexa decimal number into decimal number, it all depends on place values of hexa decimal number system.

- ✓ In hexa decimal number system, places are starts from 16^0 and the power values are increases from right hand side to left hand side.

    Example: $(13A9)_{16} =$ 1   3   A   9

    16^3 16^2 16^1 16^0 ◄——— places

    $= 16^3 \times 1 + 16^2 \times 3 + 16^1 \times 10 + 16^0 \times 9$ $[A = 10]$

    $= 4096 + 768 + 160 + 9 = 5033$

    $\therefore (13A9)_{16} = (5033)_{10}.$

7) <u>Octal to Binary conversion</u>:

- ✓ We know that, we will write every octal number interms of binary as three digits.

- ✓ Hence, write all the given octal digits as three digits of binary.

    Example: $(36475)_8$

    $\therefore (36475)_8 = (011110100111101)_2.$

Octal digits	3	6	4	7	5
Binary digits	011	110	100	111	101

8) <u>Hexa decimal to Binary conversion</u>:

- ✓ We know that, we will write every hexa decimal number interms of binary as four digits.

- ✓ Hence, write all the given hexa decimal digits as four digits of binary.

    Example: $(946A3D)_{16}$

Hexa decimal digits	9	4	6	A	3	D
Binary digits	1001	0100	0110	1010	0011	1101

$\therefore (946A3D)_{16} = (100101000110101000111101)_2.$

9) Binary to Octal conversion:

- ✓ To convert any binary number into octal number, split every three digits of the given binary number from right hand side to left hand side.
- ✓ After that write down their respective octal numbers.

Example: $(100110101110011)_2$

100	110	101	110	011

$$= (46563)_8$$

10) Binary to Hexa decimal conversion:

- ✓ To convert any binary number into hexa decimal number, split every four digits of the given binary number from right hand side to left hand side.
- ✓ After that write down their respective hexa decimal numbers.

Example: $(100011010110101001011001)_2$

1000	1101	0110	1010	0101	1001

$$= (8D6A59)_{16}$$

Arithmetic Progression:

- ✓ Arithmetic Progression is a series of numbers in which the difference between any two consecutive terms is always same.

  Example: 3, 7, 11, 15, 19, 23, ……

  The above series of numbers are in A.P because those are having a common difference of 4.

- ✓ The general form of an arithmetic series is

$$a, a + d, a + 2d, a + 3d, ….$$

Where a = First term, d = Common difference

d = Second term – First term

$$\boxed{n^{th} \text{ term in A.P, } T_n = a + (n - 1)d}$$

Sum of 'n' terms, $S_n = \dfrac{n}{2}[\text{First term} + \text{Last term}]$

$$S_n = \dfrac{n}{2}[a + a + (n - 1)d]$$

$$\boxed{S_n = \dfrac{n}{2}[2a + (n - 1)d]}$$

Geometric Progression:

✓ Geometric Progression is a series of numbers in which there is a common ratio between any two consecutive terms.

Example: 2, 4, 8, 16, 32, 64,…..

The above series of numbers are in geometric progression because those are having a common ratio of 2.

✓ The general form of geometric series is

$$a, ar, ar^2, ar^3, ar^4, \ldots\ldots$$

Where a = First term, r = Common ratio

r = Second term ÷ First term

n^{th} term in G.P, $T_n = ar^{n-1}$

Sum of 'n' terms, $S_n = \dfrac{a(1 - r^n)}{1 - r}$, if r < 1

$S_n = \dfrac{a(r^n - 1)}{r - 1}$, if r > 1

Sum of infinite terms $= \dfrac{a}{1 - r}$

Some important points to be remember:

✓ Sum of 'n' natural numbers $= \dfrac{n(n + 1)}{2}$.

✓ Sum of squares of 'n' natural numbers $= \dfrac{n(n + 1)(2n + 1)}{6}$.

✓ Sum of cubes of 'n' natural numbers $= \dfrac{n^2(n + 1)^2}{4}$ (or) $[\dfrac{n(n + 1)}{2}]^2$.

✓ Sum of first 'n' odd numbers $= n^2$.

✓ Sum of first 'n' even numbers $= n(n + 1)$.

✓ Numbers of prime factors of $a^P \times b^q \times c^r \times d^s$……. is p + q + r + s + ……. where a, b, c, d are prime numbers.

✓ The product of 3 consecutive natural numbers is always divisible by 6.

✓ For any natural number 'n', $(n^3 - n)$ is divisible by 6.

✓ If the difference between the squares of two consecutive numbers is x, then the numbers are $\dfrac{x - 1}{2}$ and $\dfrac{x + 1}{2}$.

✓ If the ratio of the sum and difference of two numbers is $x : y$, then the ratio of these two numbers is $\dfrac{x + y}{x - y}$ $(x > y)$

✓ If 'a' divides b and c, then 'a' divides their sum and difference also.

Example: 5 divides 25 and 40, then '5' also divides 25 + 40 = 65 and 40 – 25 = 15.

✓ When two numbers are divided by a third number, leaves the same remainder, the difference of those two numbers must be divisible by third number.

✓ If the sum of a number and its square is x, then the number is $\frac{\sqrt{1+4x}-1}{2}$.

✓ $\frac{(x+1)^n}{x}$ always gives the remainder as 1.

✓ $\frac{x^n}{x+1}$ gives the remainder as 1, if 'n' is an even number and gives the remainder as 'x' itself, if 'n' is an odd number.

Some important algebraic formulae:

✓ $(a+b)^2 = a^2 + 2ab + b^2$

✓ $(a-b)^2 = a^2 - 2ab + b^2$

✓ $(a+b+c)^2 = a^2 + b^2 + c^2 + 2(ab+bc+ca)$

✓ $(a+b)^3 = a^3 + b^3 + 3ab\,(a+b)$

✓ $(a-b)^3 = a^3 - b^3 - 3ab\,(a-b)$

✓ $(a+b+c)^3 = a^3 + b^3 + c^3 + 3\,(a+b)(b+c)(c+a)$

✓ $a^2 - b^2 = (a+b)(a-b)$

✓ $(a+b)^2 - (a-b)^2 = 4ab$

✓ $a^3 + b^3 = (a+b)(a^2 - ab + b^2)$

✓ $a^3 - b^3 = (a-b)(a^2 + ab + b^2)$

✓ $a^3 + b^3 + c^3 - 3abc = (a+b+c)(a^2 + b^2 + c^2 - ab - bc - ca)$

SOLVED EXAMPLES

Q - 1   **Two-fifth of three-seventh of a number is 54. Find the number.**

Sol: Consider, a number is 'N'.

$\frac{2}{5}$ of $\frac{3}{7}$ of N = 54 $\Rightarrow$ $\frac{2}{5} \times \frac{3}{7} \times N = 54$ $\Rightarrow$ N = 315

∴ Required number is 315.

Q - 2   **One-fourth of a number when subtracted from half of it gives 18. Find the number.**

Sol: Consider a number is 'x'.

According to question,

$\frac{x}{2} - \frac{x}{4} = 18$ $\rightarrow$ $\frac{x}{4} = 18$ $\Rightarrow$ $x = 72$.

Q - 3   **How many numbers up to 200 are divisible by 3?**

Sol:
```
3) 200 (66
   198          ∴ 66 numbers are divisible by 3 upto 200.
   ───
     2
```

Q - 4   **How many numbers up to 300 are divisible by both 2 and 7?**

Sol: If a number is divisible by both 2 and 7,

then that number must be divisible by 14 (LCM of 2, 7)

```
14) 300 (21
    294          ∴ 21 numbers are divisible by both 2 and 7 i.e; 14 upto 300.
    ───
      6
```

Q - 5   **Find the number of factors for the numbers**

a) 504   **b) 1156**

Sol:

✓ To find the number of factors, first write down the given number interms of prime factors.

$$N = P_1{}^a \times P_2{}^b \times P_3{}^c \times \ldots\ldots$$

Where $P_1, P_2, P_3, \ldots$ are prime numbers.

a, b, c ...... are natural numbers.

No. of factors $= (a + 1)(b + 1)(c + 1)\ldots\ldots$

a) $504 = 9 \times 56 = 3^2 \times 7 \times 8 = 2^3 \times 3^2 \times 7^1$

∴ No. of factors $= (3 + 1)(2 + 1)(1 + 1) = 4 \times 3 \times 2 = 24$.

b) $1156 = 34 \times 34 = 2^2 \times 17^2$

$\therefore$ No. of factors $= (2 + 1)(2 + 1) = 3 \times 3 = 9$.

Q - 6 | **Find the number of factors for the numbers**

a) 848   **b) 1764**

Sol: a) $848 = 8 \times 106 = 2^4 \times 53^1$

$\therefore$ No. of factors $= (4 + 1)(1 + 1) = 5 \times 2 = 10$.

b) $1764 = 42 \times 42 = (2 \times 3 \times 7)^2 = 2^2 \times 3^2 \times 7^2$

$\therefore$ No. of factors $= (2 + 1)(2 + 1)(2 + 1) = 3 \times 3 \times 3 = 27$.

Q - 7 | **In how many ways the following numbers can be expressed as a product of two factors?**

a) 742   **b) 1296**

Sol:

✓   No. of ways is always depends on no. of factors, so first we have to calculate no. of factors.

a) $742 = 7 \times 106 = 2^1 \times 7^1 \times 53^1$

No. of factors $= (1 + 1)(1 + 1)(1 + 1) = 2 \times 2 \times 2 = 8$

> If no. of factors are even, then
>
> No. of ways $= \dfrac{No.of\ factors}{2}$

$\therefore$ No. of ways $= \dfrac{8}{2} = 4$ ways.

b) $1296 = 36 \times 36 = 2^4 \times 3^4$

No. of factors $= (4 + 1)(4 + 1) = 5 \times 5 = 25$

> If no. of factors are odd, then
>
> i) As a product of 2 factors   ii) As a product of 2 different factors
>
> No. of ways $= \dfrac{No.of\ factors + 1}{2}$ No. of ways $= \dfrac{No.of\ factors - 1}{2}$

Here, in the given question he asked about as a product of 2 factors.

$\therefore$ No. of ways $= \dfrac{25 + 1}{2} = 13$ ways.

Q - 8   **In how many ways the following numbers can be expressed as a product of two different factors?**

a) **1444**   b) **3249**

Sol: a) $1444 = 38 \times 38 = 2^2 \times 19^2$

No. of factors $= (2 + 1)(2 + 1) = 3 \times 3 = 9$

If no. of factors are odd and also as a product of 2 different factors, then

$\therefore$ No. of ways $= \dfrac{No.of\ factors - 1}{2} = \dfrac{9 - 1}{2} = 4$ ways.

b) $3249 = 57 \times 57 = 3^2 \times 19^2$

No. of factors $= (2 + 1)(2 + 1) = 3 \times 3 = 9$

$\therefore$ No. of ways $= \dfrac{No.of\ factors - 1}{2} = \dfrac{9 - 1}{2} = 4$ ways.

Q - 9   **Find the sum of factors for the numbers**

a) **628**   b) **576**

Sol: Consider, given number is 'N' write that number interms of Prime factors i.e;

$N = P_1^a \times P_2^b \times P_3^c \times \ldots\ldots$

Where $P_1, P_2, P_3\ldots\ldots$are prime numbers.

a, b, c ,$\ldots\ldots$ are natural numbers.

$$\text{Sum of factors} = \frac{P_1^{a+1} - 1}{P_1 - 1} \times \frac{P_2^{b+1} - 1}{P_2 - 1} \times \frac{P_3^{c+1} - 1}{P_3 - 1} \times \ldots\ldots$$

a) $628 = 2^2 \times 157^1$

$\therefore$ Sum of factors $= \dfrac{2^{2+1} - 1}{2 - 1} \times \dfrac{157^{1+1} - 1}{157 - 1} = 1106.$

b) $576 = 24 \times 24 = 2^6 \times 3^2$

$\therefore$ Sum of factors $= \dfrac{2^{6+1} - 1}{2 - 1} \times \dfrac{3^{2+1} - 1}{3 - 1} = 1651.$

Q - 10   **What is the unit's place digit in 1482 × 327 × 568?**

Sol: Unit's place digit in $1482 \times 327 \times 568 = 2 \times 7 \times 8 = 112 = 2$

$\therefore$ Required unit's place digit is 2.

Q - 11 **Find the unit's place digit in**

a) $(13)^{726}$ **b)** $(27)^{843}$

Sol: a) $(13)^{726}$

✓ If a number ends with 3, then the unit's place digit is $3^{\text{remainder}}$.

✓ To get the remainder value, we have to divide the given power value with cycle length of 3 which is equal to 4.

$$\frac{726}{4} \quad \text{remainder} = 2$$

∴ Unit's place digit in $(13)^{726}$ is $3^2 = 9$.

b) $(27)^{843}$

✓ If a number ends with 7, then the unit's place digit is $7^{\text{remainder}}$.

$$\frac{843}{4} \quad \text{remainder} = 3$$

∴ Unit's place digit in $(27)^{843}$ is $7^3 = 343 = 3$.

Q - 12 **Find the unit's place digit in**

a) $(32)^{124}$ **b)** $(48)^{362}$

Sol: a) $(32)^{124}$

✓ If a number ends with 2, then the unit's place digit is $2^{\text{remainder}}$.

$$\frac{124}{4} \quad \text{remainder} = 0 \qquad \boxed{\textbf{Note:} \text{ If remainder is '0', then take that value as 4.}}$$

∴ Unit's place digit in $(32)^{124} = 2^4 = 16 = 6$.

b) $(48)^{362}$

✓ If a number ends with 8, then the unit's place digit is $8^{\text{remainder}}$.

$$\frac{362}{4} \quad \text{remainder} = 2$$

∴ Unit's place digit in $(48)^{362} = 8^2 = 64 = 4$.

Q - 13 **What is the unit's place digit in $(345)^{312} + (219)^{679} + (953)^{247}$?**

Sol: $(345)^{312} + (219)^{679} + (953)^{247}$

✓ If a number ends with 5 to any power, then its unit's place is 5.

✓ If a number ends with 9, for odd power of 9 unit's place is 9.

✓ If a number ends with 3, then the unit's place digit is $3^{\text{remainder}}$.

$$(953)^{247} \quad \Rightarrow \quad \frac{247}{4} \quad \Rightarrow \quad \text{remainder} = 3$$

Unit's place digit in $(953)^{247} = 3^3 = 27 = 7$

∴ Required unit's digit $= 5 + 9 + 7 = 21 = 1$.

Q - 14 | **Find the unit's place digit in $(786)^{824} - (642)^{329}$.**

Sol: $(786)^{824} - (642)^{329}$

✓ If a number ends with 6 to any power, then its unit's place is 6

✓ If a number ends with 2, then the unit's place is $2^{remainder}$.

$(642)^{329}$ $\Rightarrow$ $\dfrac{329}{4}$ $\Rightarrow$ remainder = 1

Unit's place digit in $(642)^{329}$ is $2^1 = 2$

∴ Required unit's digit = 6 – 2 = 4.

Q - 15 | **What is the units place digit in $(869)^{284} - (763)^{159}$?**

Sol: $(869)^{284} - (763)^{159}$

✓ If a number ends with 9, for even power of 9 unit's place is 1.

✓ If a number ends with 3, then the unit's place is $3^{remainder}$.

$(763)^{159}$ $\Rightarrow$ $\dfrac{159}{4}$ $\Rightarrow$ remainder = 3

Unit's place in $(763)^{159}$ is $3^3 = 27 = 7$

∴ Required unit's place is 1 – 7

Take borrow 10, then 11 – 7 = 4.

Q - 16 | **Find the units place digit in $(874)^{126} \times (329)^{832} \times (623)^{257}$.**

Sol:

✓ If a number ends with 4: For even power of 4, then its unit's place is 6.

✓ If a number ends with 9: For even power of 9, then its unit's place is 1.

✓ If a number ends with 3, then the unit's place is $3^{remainder}$.

$(623)^{257}$ $\Rightarrow$ $\dfrac{257}{4}$ $\Rightarrow$ remainder = 1

Unit's place in $(623)^{257}$ is $3^1 = 3$

∴ Required unit's place = 6 × 1 × 3 = 18 = 8.

Q - 17 | **Find the highest power of**

a) 3 in 120!   b) 7 in 160!

Sol:

$$\boxed{\text{Highest power of a prime number in N!} = \frac{Sum\ of\ all\ Quotients}{Powers\ of\ Prime\ number}}$$

a) Highest power of 3 in 120!

Step 1: Check whether the number is prime (or) not.

Here 3 is a prime number.

Step 2: Divide the given factorial number with prime number to get the quotients and leave the remainders every time while division.

```
3 | 120
3 | 40  ┐
3 | 13  ├──→ Quotients
3 | 4   ┘
    | 1
```

∴ Highest power of 3^1 in 120! $= \dfrac{40 + 13 + 4 + 1}{1} = 58$.

b) Highest power of 7 in 160!

Step 1: Here 7 is a prime number.

Step 2: Divide 160 with 7 and leave the remainders every time.

```
7 | 160
7 | 22  ┐
    | 3  ┘──→ Quotients
```

∴ Highest Power of 7 in 160! $= \dfrac{Sum\ of\ all\ quotients}{Power\ of\ 7} = \dfrac{22 + 3}{1} = 25$.

Q - 18   **Find the highest power of**

a) 8 in 110!   **b) 9 in 170!**

Sol: a) Highest power of 8 in 110!

Step 1: Here 8 is not a prime number, so we have to convert that into prime factors.

$$8 = 2^3$$

Step 2: Divide 110 with 2 and leave the remainders every time.

```
2 | 110
2 | 55  ┐
2 | 27  │
2 | 13  ├──→ Quotients
2 | 6   │
2 | 3   ┘
    | 1
```

∴ Highest power of $2^3(8)$ in 110! $= \dfrac{Sum\ of\ all\ quotients}{Power\ of\ 2} = \dfrac{55 + 27 + 13 + 6 + 3 + 1}{3} = 35$.

b) Highest power of 9 in 170!

Step 1: Here 9 is not a prime number, so we have to convert that into prime factors.

$$9 = 3^2$$

Step 2: Divide 170 with 3 and leave the remainders every time.

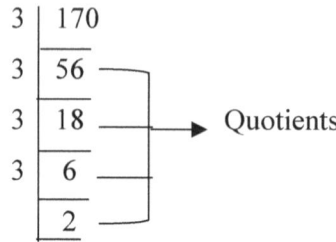

∴ Highest power of 3^2 (9) in 170! $= \dfrac{Sum\ of\ all\ quotients}{Power\ of\ 3} = \dfrac{56 + 18 + 6 + 2}{2} = 41.$

| Q - 19 |

Find the highest power of

a) 24 in 200!   **b) 36 in 220!**

Sol: a) Highest power of 24 in 200!

Step 1: Here, 24 is not a prime number. So we have to convert that into prime factors.

$$24 = 2^3 \times 3^1$$

Step 2: Here, we have two prime numbers. So we have to divide 200 with both the prime Numbers.

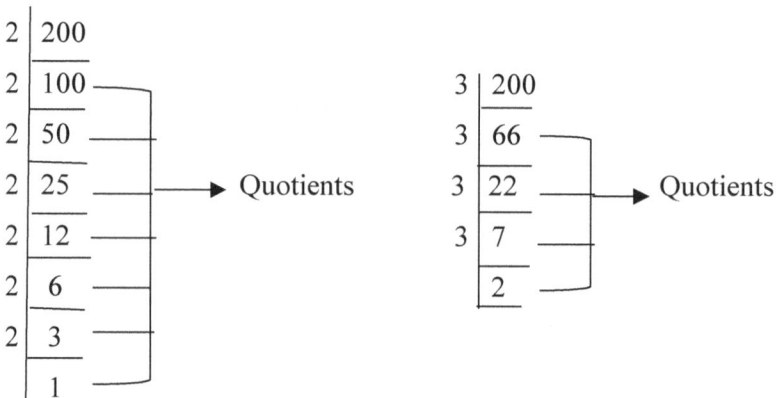

∴ Highest power of 2^3 in 200! $= \dfrac{100 + 50 + 25 + 12 + 6 + 3 + 1}{3} = \dfrac{197}{3} = 65$ (leave remainder)

Highest power of 3^1 in 200! $= \dfrac{66 + 22 + 7 + 2}{1} = 97$

Since 24 is a combination of both 2^3 and 3^1, so we have to take the combined value of both 2^3 and 3^1 which is 65.

b) Highest power of 36 in 220!

Step 1: Here, 36 is not a prime number. So we have to convert that into prime factors.

$$36 = 2^2 \times 3^2$$

Step 2: Here, we have two prime numbers and their powers are also equal, So we can just divide 220 with only highest prime number 3 to get the answer.

Highest power of 3^2 in 220!

$$= \frac{73 + 24 + 8 + 2}{2} = 53 \text{ (leave remainder)}$$

∴ Highest power of 36 in 220! is 53.

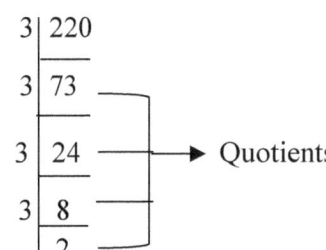

| Q - 20 | **Find the number of zeros at the end of 140!** |

Sol:

No. of zeros at the end of 140! Means highest power of 10 in 140!

Step 1: Here, 10 is not a prime number. So we have to convert that into prime factors.

$$10 = 2^1 \times 5^1$$

Step 2: Here, we have 2 prime numbers and their powers are also equal. So we can just divide 140 with only highest prime number 5 to get the answer.

Highest power of 5^1 in 140! $= \frac{28 + 5 + 1}{1} = 34$

Highest power of 10 in 140! is 34

∴ No. of zeros at the end of 140! is 34.

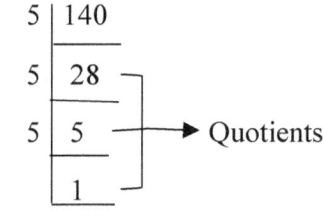

| Q - 21 | **Convert the decimal number $(224)_{10}$ into binary number.** |

Sol:

To convert any decimal number into binary, first divide the given decimal number with 2 successively and then write all the remainders from bottom to top.

```
2 | 224
2 | 112 – 0  ▲
2 | 56  – 0  |
2 | 28  – 0  |
2 | 14  – 0  |
2 | 7   – 0  |
2 | 3   – 1  |
  | 1   – 1  |
```

∴ $(224)_{10} = (11100000)_2$

Q - 22 | **Convert the decimal number $(3452)_{10}$ into hexadecimal number.**

Sol: To convert any decimal number into hexa decimal number, first divide the given decimal number with 16 successively and then write all the remainders from bottom to top.

$$16 \underline{\,|\,3452}$$
$$16 \underline{\,|\,215 - 12(C)}$$
$$13(D) - 7$$

$\therefore (3452)_{10} = (D7C)_{16}$

Q - 23 | **Convert the binary number $(111011010110100110)_2$ into octal number.**

Sol:

✓   To convert any binary number into octal number, split every three-digit parts of the given binary number from right hand side to left hand side.

✓   After that write down their respective octal numbers.

111	011	010	110	100	110

$= (732646)_8$

Q - 24 | **Convert the binary number $(11000111010110001010)_2$ into hexadecimal number.**

Sol:

✓   To convert any binary number into hexa decimal number, split every four digit parts of the given binary number from right hand side to left hand side.

✓   After that write down their respective hexa decimal numbers.

1100	0111	0101	1000	1010

$= (C758A)_{16}$

Q - 25 | **Convert the binary number $(101101011)_2$ into decimal number.**

Sol:

✓   To convert any binary number into decimal number, it all depends on place values of binary number system.

✓   In binary number system, places are starts from 2^0 and the power values are increases from right hand side to left hand side.

$(101101011)_2 = 1 \quad 0 \quad 1 \quad 1 \quad 0 \quad 1 \quad 0 \quad 1 \quad 1$

$ 2^8 \ 2^7 \ 2^6 \ 2^5 \ 2^4 \ 2^3 \ 2^2 \ 2^1 \ 2^0 \longleftarrow$ places

$= 2^8 \times 1 + 2^7 \times 0 + 2^6 \times 1 + 2^5 \times 1 + 2^4 \times 0 + 2^3 \times 1 + 2^2 \times 0 + 2^1 \times 1 + 2^0 \times 1$

$= 256 + 64 + 32 + 8 + 2 + 1 = 363$

$\therefore (101101011)_2 = (363)_{10}.$

Q - 26   **Convert the hexadecimal number (2C4)₁₆ into decimal number.**

Sol:

- ✓ To convert any hexa decimal number into decimal number, it all depends on place values of hexa decimal number system.
- ✓ In hexa decimal number system, places are starts from 16^0 and the power values are increases from right hand side to left hand side.

$$(2C4)_{16} = \begin{matrix} 2 & C & 4 \\ 16^2 & 16^1 & 16^0 \end{matrix} \longleftarrow \text{places}$$

$$= 16^2 \times 2 + 16^1 \times 12 + 16^0 \times 4 \qquad \because C = 12 \text{ in hexa decimal number system}$$

$$= 512 + 192 + 4 = 708$$

$\therefore (2C4)_{16} = (708)_{10}$

Q - 27   **The difference between the squares of two consecutive numbers is 49. Find the numbers.**

Sol: Consider, two consecutive numbers are 'x' and '$x + 1$'

According to question,

$$(x + 1)^2 - x^2 = 49 \qquad \Rightarrow \qquad x^2 + 2x + 1 - x^2 = 49$$

$$2x + 1 = 49 \qquad \Rightarrow \qquad x = 24$$

$\therefore$ Two numbers are 24 and 25.

Shortcut:

> If the difference between 2 consecutive numbers is x, then the numbers are $\dfrac{x - 1}{2}$ and $\dfrac{x + 1}{2}$.

Here, $x = 49$

$\therefore$ Two numbers are $\dfrac{49 - 1}{2} = 24$ and $\dfrac{49 + 1}{2} = 25$.

Q - 28   **Divide 90 into two parts, so that the sum of their reciprocals is $\dfrac{1}{20}$. Find the two parts.**

Sol: Consider two parts are 'x' and '$90 - x$' $\because$ Total = 90

According to question,

$$\frac{1}{x} + \frac{1}{90 - x} = \frac{1}{20} \qquad \Rightarrow \qquad \frac{90 - x + x}{x(90 - x)} = \frac{1}{20} \qquad \Rightarrow \qquad 90 \times 20 = x(90 - x)$$

$$x^2 - 90x + 1800 = 0 \qquad \Rightarrow \qquad x^2 - 60x - 30x + 1800 = 0$$

$$x(x - 60) - 30(x - 60) = 0 \qquad \Rightarrow \qquad (x - 60)(x - 30) = 0 \qquad \Rightarrow \qquad x = 60, 30$$

$\therefore$ Two parts are 30 and 60.

> **Note:** This problem can be solved easily through option verification method.

Q - 29 A number when divided by 426 leaves the remainder as 38 and the value of quotient is 12. Find the number.

Sol:

> Dividend = Divisor × Quotient + Remainder

Here, Divisor = 426, Quotient = 12 and Remainder = 38

∴ Required number = 426 × 12 + 38 = 5150.

Q - 30 The quotient arising from a division of a number by 54 is 135 and the remainder is 28. What is the number?

Sol:

> Dividend = Divisor × Quotient + Remainder

Here, divisor = 54, quotient = 135 and remainder = 28

∴ Required number = 54 × 135 + 28 = 7318.

Q - 31 What is the least value of k, so that the number 34861k2 is divisible by 9?

Sol:

A number is divisible by 9, if its sum of the digits are divisible by 9

Sum of digits = 3 + 4 + 8 + 6 + 1 + k + 2 = 24 + k

(24 + k) is divisible by 9, so the least value of k is 3 in order to satisfies the given condition.

Q - 32 What is the least value of *x*, so that the number 452*x*73 is divisible by 11?

Sol:

A number is divisible by 11, if the difference between sum of odd place digits and sum of even place digits is either 0 (or) multiple of 11.

$(4 + 2 + 7) - (5 + x + 3) = 0 \Rightarrow 13 - (8 + x) = 0$

∴ Least value of $x = 5$.

Q - 33 Find the number of prime factors and number of distinct prime factors in the expression $12^6 \times 5^{10} \times 7^{13}$

Sol:

- ✓ To find the no. of prime factors, first all base values must be prime numbers.
- ✓ If all bases are not prime numbers, then convert them into prime numbers.

  $12^6 \times 5^{10} \times 7^{13} = (3 \times 2^2)^6 \times 5^{10} \times 7^{13} = 2^{12} \times 3^6 \times 5^{10} \times 7^{13}$

  No. of prime factors in $a^p \times b^q \times c^r \times d^s \times \ldots\ldots\ldots = p + q + r + s + \ldots..$

  where a, b, c, d ...... are prime numbers

∴ No. of prime factors in $2^{12} \times 3^6 \times 5^{10} \times 7^{13} = 12 + 6 + 10 + 13 = 41$ and

No. of distinct prime factors = 4 (2, 3, 5, 7).

Q - 34 | **How many natural numbers less than 1200 have exactly three factors?**

Sol:

> **Note:** The square of any prime number is having exactly 3 factors.

Less than 1200, last perfect square is $34^2 (1156)$

∴ Required no. of natural numbers which are having exactly 3 factors is 11

$(2^2, 3^2, 5^2, 7^2, 11^2, 13^2, 17^2, 19^2, 23^2, 29^2, 31^2)$.

Q - 35 | **How many natural numbers between 100 and 700 have odd number of factors?**

Sol:

> **Note:** Every perfect square is having odd no. of factors.

After 100 first perfect square $= 11^2 = 121$

Before 700 last perfect square $= 26^2 = 676$

∴ Required no. of natural numbers which are having odd no. of factors between 100 and

700 is 16 (11^2 to 26^2).

Q - 36 | **A number is decreased by 7 and divided by 4, the result is 11. What would be the result, if 9 is subtracted from the number and divided by 6?**

Sol: Consider, a number be 'x'.

According to question, $\frac{x-7}{4} = 11$ $\Rightarrow$ $x = 11 \times 4 + 7 = 51$

∴ Required result $= \frac{51-9}{6} = 7$.

Q - 37 | **The sum of squares of two numbers is 92 and the square of their difference is 64. Find the product of two numbers.**

Sol: Consider, two numbers are x and y.

According to question,

$x^2 + y^2 = 92$ and $(x-y)^2 = 64$

$(x-y)^2 = x^2 + y^2 - 2xy$ $\Rightarrow$ $xy = \frac{x^2 + y^2 - (x-y)^2}{2}$

∴ Product of two numbers, $xy = \frac{92-64}{2} = 14$.

Q - 38 | **The product of two numbers is 196. The sum of their squares is 284. Find the sum of the numbers.**

Sol: Consider, two numbers are x and y.

Given that, $xy = 196$ and $x^2 + y^2 = 284$

$(x+y)^2 = x^2 + y^2 + 2xy$ $\Rightarrow$ $x + y = \sqrt{x^2 + y^2 + 2xy}$

∴ Sum of the numbers, $x + y = \sqrt{284 + 2 \times 196} = 26$.

Q - 39 | The sum of three numbers is 154. Second number is thrice the first and third number is half of the second. Find the second number.

Sol: Given that, F + S + T = 154 $\because$ F – First number

$S = 3F$ and $T = \dfrac{S}{2} = \dfrac{3F}{2}$ S – Second number

$F + 3F + \dfrac{3F}{2} = 154$ T – Third number

$\dfrac{11F}{2} = 154 \qquad \Rightarrow \qquad F = 28$

$\therefore$ Second number $S = 3F = 3 \times 28 = 84$.

Q - 40 | The denominator of a rational number is 2 more than its numerator. If the numerator is increased by 15 and denominator is decreased by 5, we obtain 4. Find the rational number.

Sol: Consider, a rational number is $\dfrac{a}{b}$

Given that, b = a + 2 $\Rightarrow$ Rational number $= \dfrac{a}{a+2}$

Now, $\dfrac{a+15}{(a+2)-5} = 4$ $\Rightarrow$ $\dfrac{a+15}{a-3} = 4$ $\Rightarrow$ $a + 15 = 4a - 12$

$3a = 27$ $\Rightarrow$ $a = 9$

$b = a + 2 = 9 + 2 = 11$

$\therefore$ Required rational number $= \dfrac{a}{b} = \dfrac{9}{11}$.

Q - 41 | Find the sum of first 42 odd numbers.

Sol:

| Sum of first 'n' odd numbers = n^2 |

$\therefore$ Sum of first 42 odd numbers $= 42^2 = 1764$.

Q - 42 | Find the sum of first 38 even numbers.

Sol:

| Sum of first 'n' even numbers = n(n + 1) |

$\therefore$ Sum of first 38 even numbers $= 38(38 + 1) = 1482$.

Q - 43 | What is the least number that should be added to 2072 to get a number is exactly divisible by 31?

Sol:
```
31) 2072 (66
    2046
    ----
      26  → Remainder
```

$\therefore$ Required least number added = Divisor – Remainder = 31 – 26 = 5.

Q - 44 | What is the least number to be subtracted from 2167 to get a number is exactly divisible by 43?

Sol:

$$43) \, 2167 \, (50$$
$$\underline{2150}$$
$$17 \longrightarrow \text{Remainder}$$

$\therefore$ Required least number subtracted = Remainder = 17.

Q - 45 | The sum of five consecutive odd numbers A, B, C, D and E is 375. Find the product of B and D.

Sol: Consider, 5 consecutive odd numbers A, B, C, D and E are x, $x + 2$, $x + 4$, $x + 6$, $x + 8$.

$\because$ Gap between any two consecutive odd numbers is 2.

Given that, $x + (x + 2) + (x + 4) + (x + 6) + (x + 8) = 375$

$\qquad 5x + 20 = 375 \qquad \Rightarrow \qquad x = 71$

5 odd numbers A, B, C, D and E are 71, 73, 75, 77, 79

$\therefore$ Product of B and D = $73 \times 77 = 5621$.

Alternate method:

Consider, 5 consecutive odd numbers A, B, C, D and E are $x - 4$, $x - 2$, x, $x + 2$, $x + 4$.

Given that, $(x - 4) + (x - 2) + x + (x + 2) + (x + 4) = 375$

$\qquad 5x = 375 \qquad \Rightarrow \qquad x = 75$

5 odd numbers A, B, C, D and E are 71, 73, 75, 77, 79

$\therefore$ Product of B and D = $73 \times 77 = 5621$.

Q - 46 | The sum of four consecutive even numbers P, Q, R and S is 348. What is the product of P and R?

Sol: Consider, 4 consecutive even numbers P, Q, R, S are x, $x + 2$, $x + 4$, $x + 6$.

$\because$ Gap between any two consecutive even numbers is 2.

Given that, $x + (x + 2) + (x + 4) + (x + 6) = 348$

$\qquad 4x + 12 = 348 \qquad \Rightarrow \qquad x = 84$

4 consecutive even numbers P, Q, R, S are 84, 86, 88, 90

$\therefore$ Product of P and R = $84 \times 88 = 7392$.

Alternate method:

Consider, 4 consecutive even numbers P, Q, R, S are $(x - 3)$, $(x - 1)$, $(x + 1)$, $(x + 3)$

Given that, $(x - 3) + (x - 1) + (x + 1) + (x + 3) = 348$

$\qquad 4x = 348 \qquad \Rightarrow \qquad x = 87$

4 consecutive even numbers P, Q, R, S are 84, 86, 88, 90

$\therefore$ Product of P and R = $84 \times 88 = 7392$.

Q - 47   **A number when divided by 72 gives 48 as remainder. If the same remainder is divided by 9, then what will be the remainder?**

Sol:

$$\boxed{\text{Dividend} = \text{Divisor} \times \text{Quotient} + \text{Remainder}}$$

Consider, a number is N

$N = 72Q + 48$ $\because$ Divisor = 72, Remainder = 48

Now, $\dfrac{N}{9} = \dfrac{72Q + 48}{9} = \dfrac{(72Q + 45) + 3}{9} = \dfrac{9(8Q + 5) + 3}{9}$

$\therefore$ Remainder obtained by dividing the same number with 9 is 3.

Shortcut:

Since, 72 is a multiple of 9, so we can directly divide given remainder with 9.

Therefore, required remainder when 48 divided by 9 is 3.

Q - 48   **A number when divided by 323 gives 82 as remainder. If the same number is divided by 19, then what will be the remainder?**

Sol:

$$\boxed{\text{Dividend} = \text{Divisor} \times \text{Quotient} + \text{Remainder}}$$

Consider, a number is N.

$N = 323Q + 82$ $\because$ Divisor = 323, Remainder = 82

Now, $\dfrac{N}{19} = \dfrac{323Q + 82}{19} = \dfrac{19(17Q + 4) + 6}{19}$

$\therefore$ Remainder obtained by dividing the same number with 19 is 6.

Shortcut:

Since, 323 is a multiple of 19, so we can directly divide given remainder with 19.

$\therefore$ Required remainder when 82 divided by 19 is 6.

Q - 49   **A number when divided by 6, leaves 5 as remainder. What will be the remainder, when square of the same number is divided by 6?**

Sol:

$$\boxed{\text{Dividend} = \text{Divisor} \times \text{Quotient} + \text{Remainder}}$$

Let us consider, a number is N.

$N = 6Q + 5$ $\because$ Divisor = 6, Remainder = 5

$N^2 = (6Q + 5)^2 = 36Q^2 + 60Q + 25 = 36Q^2 + 60Q + 24 + 1$

$N^2 = 6(6Q^2 + 10Q + 4) + 1$

Now, $\dfrac{N^2}{6}$ $\Rightarrow$ $\dfrac{6(6Q^2 + 10Q + 4) + 1}{6}$

$\therefore$ Remainder obtained by dividing the square of the number with 6 is 1.

Q - 50 | If the sum of a natural number and its square is 210, then what is the number?

Sol: Consider, a number be 'x'.

Given that, $x + x^2 = 210$ $\Rightarrow$ $x^2 + x - 210 = 0$

$x^2 + 15x - 14x - 210 = 0$ $\Rightarrow$ $(x + 15)(x - 14) = 0$ $\Rightarrow$ $x = 14 \ \& \ x \neq -15$

$\therefore$ Required number = 14.

Shortcut:

If the sum of a number and its square is x, then number is $\dfrac{\sqrt{1 + 4x} - 1}{2}$

Here $x = 210$

$\therefore$ Required number = $\dfrac{\sqrt{1 + 4 \times 210} - 1}{2} = 14$.

Q - 51 | The ratio of the sum and the difference of the two numbers is 8 : 3. Find the ratio of those two numbers.

Sol: Consider, two numbers are x and y.

Given that, $(x + y) : (x - y) = 8 : 3$

$\dfrac{x + y}{x - y} = \dfrac{8}{3}$ $\Rightarrow$ $3x + 3y = 8x - 8y$ $\Rightarrow$ $5x = 11y$

$\therefore x : y = 11 : 5$.

Shortcut:

Given that, $\dfrac{x + y}{x - y} = \dfrac{8}{3}$

By componendo and dividendo method,

$\therefore \dfrac{(x + y) + (x - y)}{(x + y) - (x - y)} = \dfrac{8 + 3}{8 - 3}$ $\Rightarrow$ $x : y = 11 : 5$.

Q - 52 | Find the number of zeros at the end of the product $15 \times 20 \times 36 \times 24 \times 25 \times 35$.

Sol: No. of zeros at the end of the product can be formed by the combination of 2 and 5.

First write down the given product interms of prime factors.

$\Rightarrow$ $(3 \times 5) \times (2^2 \times 5) \times (2^2 \times 3^2) \times (2^3 \times 3) \times 5^2 \times (5 \times 7)$

$\Rightarrow$ $(2^7 \times 5^5) \times 3^4 \times 7$

$\therefore$ No. of zeros at the end of the given product is 5,

since the combined value of 2 and 5 is 5.

Q - 53 | **Find the number which when multiplied by 27 is increased by 754.**

Sol: Let the number be 'x'.

According to question,

$$27x - x = 754 \qquad \Rightarrow \qquad 26x = 754 \qquad \Rightarrow \qquad x = 29$$

∴ Required number $x = 29$.

Q - 54 | **The difference between the squares of two consecutive numbers is 43. Find the numbers.**

Sol: Consider, two consecutive numbers are 'x' and '$x + 1$'.

According to question,

$$(x + 1)^2 - x^2 = 43 \qquad \Rightarrow \qquad x^2 + 2x + 1 - x^2 = 43$$

$$2x + 1 = 43 \qquad \Rightarrow \qquad x = 21$$

∴ Two consecutive numbers are 21 and 22.

Shortcut:

> If the difference between two consecutive numbers is 'x', then
> the numbers are $\dfrac{x - 1}{2}$ and $\dfrac{x + 1}{2}$.

Here, $x = 43$

∴ Two consecutive numbers are $\dfrac{43 - 1}{2} = 21$ and $\dfrac{43 + 1}{2} = 22$.

Q - 55 | **Calculate** $\dfrac{762 \times 762 \times 762 + 938 \times 938 \times 938}{762 \times 762 - 762 \times 938 + 938 \times 938}$

Sol: The given question is in the form of $\dfrac{a^3 + b^3}{a^2 - ab + b^2}$

$$a^3 + b^3 = (a + b)(a^2 - ab + b^2)$$

$$\frac{a^3 + b^3}{a^2 - ab + b^2} = \frac{(a + b)(a^2 - ab + b^2)}{a^2 - ab + b^2} = a + b$$

Here, $a = 762$ and $b = 938$

∴ Required number $= a + b = 762 + 938 = 1700$.

Q - 56 | **Calculate** $\dfrac{834 \times 834 \times 834 - 375 \times 375 \times 375}{834 \times 834 + 834 \times 375 + 375 \times 375}$

Sol: The given question is in the form of $\dfrac{a^3 - b^3}{a^2 + ab + b^2}$

$$a^3 - b^3 = (a - b)(a^2 + ab + b^2)$$

$$\frac{a^3 - b^3}{a^2 + ab + b^2} = \frac{(a - b)(a^2 + ab + b^2)}{a^2 + ab + b^2} = a - b$$

Here, $a = 834$ and $b = 375$

∴ Required number $= a - b = 834 - 375 = 459$.

Q - 57 On dividing 12662 by a certain number, we get 29 as quotient and 18 as remainder. Find the divisor.

Sol:

$$\boxed{\text{Dividend} = \text{Divisor} \times \text{Quotient} + \text{Remainder}}$$

$\text{Divisor} = \frac{Dividend - Remainder}{Quotient}$ $\quad \because$ Dividend = 12662, Remainder = 18, Quotient = 29

$\therefore$ Divisor = $\frac{12662 - 18}{29}$ = 436.

Q - 58 If the product 1976 × 6x5 is divisible by 24, then find the least value of x.

Sol: Given that, 1976 × 6x5 is divisible by 24

If a number is divisible by 24, then it must be divisible by both 3 and 8

Here 1976 is divisible by 8 but not divisible by 3, so 6x5 must be divisible by 3

According to divisibility rule of 3,

Sum of digits of $6x5 = 6 + x + 5 = 11 + x$ is divisible by 3.

$\therefore$ The least value of x is 1 in order to divisible by 3.

Q - 59 The difference of two numbers is 1985. By dividing the larger number with smaller number, we got a quotient of 7 and remainder is 35. Find the larger number.

Sol: According to question,

$L - S = 1985 \quad \Rightarrow \quad L = S + 1985$ $\because$ L – Larger number & S – Smaller number

$$\boxed{\text{Dividend} = \text{Divisor} \times \text{Quotient} + \text{Remainder}}$$

$L = S \times 7 + 35 \quad \Rightarrow \quad S + 1985 = 7S + 35$

$6S = 1950 \quad \Rightarrow \quad S = 325$

$\therefore$ Larger number L = S + 1985 = 325 + 1985 = 2310.

Q - 60 What will be the remainder when $(38^{43} + 38)$ is divided by 39?

Sol: $38^{43} + 38 = (38^{43} + 1^{43}) + 37$

$x^n + y^n$ is divisible by $x + y$, if 'n' is an odd number.

Here, $x = 38, y = 1$ and n = 43

$38^{43} + 1^{43}$ is divisible by 38 + 1 = 39

Now, $\frac{(38^{43} + 1^{43}) + 37}{39}$

$\therefore$ Required remainder when $38^{43} + 38$ is divided by 39 is 37.

Q - 61 A 3-digit number 5a7 is added to another 3-digit number 786 to give a 4-digit number 1b53, which is divisible by 11. Then, find the value of a – b.

Sol: Given that, 1b53 is divisible by 11 means

Sum of odd place digits = Sum of even place digits

$$1 + 5 = b + 3 \quad \Rightarrow \quad b = 3$$

$$\begin{array}{r} 5a7 \\ 786 \\ \hline 1b53 \end{array}$$

$5a7 + 786 = 1353 \Rightarrow 5a7 = 1353 - 786 = 567 \Rightarrow a = 6$

∴ Required value of a – b = 6 – 3 = 3.

Q - 62 The sum of three consecutive odd numbers is 25 more than the two-third of 48. What are the numbers?

Sol: Consider, 3 consecutive numbers are $x - 2, x, x + 2$.

According to question,

$$(x - 2) + x + (x + 2) = \frac{2}{3} \times 48 + 25 \qquad \Rightarrow \qquad 3x = 57 \qquad \Rightarrow \qquad x = 19$$

∴ Required numbers are 17, 19, 21.

Q - 63 Find the number of times the keys of a type writer must be pressed to type the first 1025 natural numbers.

Sol: Upto 1025 natural numbers, we have 1-digit numbers, 2-digit numbers, 3-digit numbers and 4-digit numbers.

To type 1-digit no's, type writer press the key for 1 time

To type 2-digit no's, type writer press the key for 2 times

To type 3-digit no's, type writer press the key for 3 times

To type 4-digit no's, type writer press the key for 4 times

1 – digit no's	2 – digit no's	3 – digit no's	4 – digit no's
1 to 9	10 to 99	100 to 999	1000 to 1025
$9 \times 1 = 9$	$90 \times 2 = 180$	$900 \times 3 = 2700$	$26 \times 4 = 104$

∴ Type writer press the key for 9 + 180 + 2700 + 104 = 2993 times.

Q - 64 How many digits are required to write the numbers on a book contains 500 pages?

Sol: Upto 500 pages, we have 1-digit no's, 2-digit no's and 3-digit no's.

1 – digit no's	2 – digit no's	3 – digit no's
1 to 9	10 to 99	100 to 500
$9 \times 1 = 9$	$90 \times 2 = 180$	$401 \times 3 = 1203$

∴ No. of digits required to write 500 pages = 9 + 180 + 1203 = 1392.

Q - 65 **In a division sum, the divisor is 15 times the quotient and 7 times the remainder. If the remainder is 60, then find the value of dividend.**

Sol: Given that, divisor = 15Q, divisor = 7R and R = 60

Divisor = 7 × 60 = 420   Q – Quotient

420 = 15Q ⇒ Q = 28   R – Remainder

Dividend = Divisor × Quotient + Remainder

∴ Dividend = 420 × 28 + 60 = 11820.

Q - 66 **The sum of the digits of a 2-digit number is 9. If the digits are reversed the number is decreased by 45. Find the number.**

Sol: Consider, a 2-digit number is '*xy*'.

Given that, $x + y = 9$ .... (1)

$yx = xy - 45$ ⇒ $10y + x = 10x + y - 45$

$9(x - y) = 45$ ⇒ $x - y = 5$ .... (2)

(1) + (2) ⇒ $2x = 9 + 5 = 14$ ⇒ $x = 7$

$x + y = 9$ ⇒ $7 + y = 9$ ⇒ $y = 2$

∴ Required 2-digit number is 72.

> **Note:** This problem can also be calculated by option verification method.

Alternate method:

Consider, a 2-digit number is '*xy*'.

Given that, $x + y = 9$

By trial and error method, all possibilities are

Original number	Reversed number	Change in number
18	81	increased by 63
27	72	increased by 45
36	63	increased by 27
45	54	increased by 9
63	36	decreased by 27
72	27	decreased by 45
81	18	decreased by 63

The given condition is, after reversing the number it is decreased by 45. This satisfies when the original number is 72.

∴ Required 2-digit number is 72.

Q - 67   **If the number 48x365y2 is divisible by 36, then find the minimum value of x − y.**

Sol: If a number is divisible by 36, it must be divisible by both 4 and 9.

Divisibility rule of 4 → Last two digits are divisible by 4.

Possible values of 'y', in order to divisible by 4 are $y = 1, 3, 5, 7, 9$

Divisibility rule of 9 → Sum of the digits divisible by 9.

$48x365y2 = 4 + 8 + x + 3 + 6 + 5 + y + 2 = 28 + (x + y)$

Possible values of $(x + y)$ are $x + y = 8, x + y = 17$

∵ Maximum sum of 2 single digit no's are 18.

To find the minimum value of $(x − y)$, the value of 'y' must be maximum and 'x' is minimum.

So, maximum value of $y = 9$.

If $y = 9$, then $x + y \neq 8$. So, $x + y = 17$ ⇒ $x + 9 = 17$ ⇒ $x = 8$

∴ Minimum value of $x − y = 8 − 9 = −1$.

Q - 68   **The sum of the digits of a two-digit number is 13 and the difference between the two digits of that number is 1. What is the product of the two digits of that two-digit number?**

Sol: Consider, a 2-digit number is 'xy'.

Given that, $x + y = 13$ ……. (1) $x − y = 1$ ……. (2)

(1) + (2)   ⇒ $2x = 14$ ⇒ $x = 7$ ……… substitute in (1)

$x + y = 13$ ⇒ $7 + y = 13$ ⇒ $y = 6$

∴ Product of two digits $= x \times y = 7 \times 6 = 42$.

Q - 69   **A number when divided by a divisor leaves the remainder as 16. When twice the original number is divided by the same divisor, the remainder is 14. What is the value of the divisor?**

Sol: Consider, a divisor is D and quotient is Q.

Number $N = D \times Q + R = DQ + 16$

Twice the number $= 2N = 2DQ + 32$

Now, 2DQ is completely divisible by D.

On dividing 32 by D, remainder is 14

∴ Required divisor $D = 32 − 14 = 18$.

Q - 70 Find the common factor of $(26^{53} + 28^{53})$ and $(26^{55} + 28^{55})$.

Sol: $x^n + y^n$ is always divisible by $x + y$, if 'n' is odd.

$26^{53} + 28^{53}$ is always divisible by $26 + 28 = 54$

$26^{55} + 28^{55}$ is always divisible by $26 + 28 = 54$

∴ Required common factor = 54.

Q - 71 When 7571 and 6667 are divided by a 3-digit number N, leaves the same remainder in both the cases. Find all the possible 3-digit divisors.

Sol:

> **Note:** When x and y are divided by z, leaves the same remainder, then the difference of x and y is exactly divisible by z.

$7571 - 6667 = 904$ is exactly divisible by N

Now, $904 = 1 \times 904 = 2 \times 452 = 4 \times 226 = 8 \times 113$

∴ Required 3-digit divisors are 904, 452, 226, 113.

Q - 72 On division with zero remainder, a student took 18 as divisor instead of 24. The quotient obtained by him was 36. Find the correct quotient.

Sol:

Dividend = Divisor × Quotient + Remainder

According to question,

Wrongly taken divisor as 18, then Quotient = 36 and Remainder = 0

Dividend = $18 \times 36 + 0 = 648$

Now, if he takes the correct divisor which is 24, then

$648 = 24 \times Q \implies Q = 27$

∴ Correct quotient Q = 27.

Q - 73 Find the remainder of $\dfrac{15^{12^{13^{14}.......\infty}}}{7}$

Sol: Remainder of $\dfrac{15^{12^{13^{14}......\infty}}}{7}$ = Remainder of $\dfrac{(7 \times 2 + 1)^{12^{13^{14}......\infty}}}{7}$

∴ Remainder of $\dfrac{(1)^{12^{13^{14}.......\infty}}}{7} = 1$.

Q - 74 The ratio between a 2-digit number and the sum of the digits of that number is 5 : 1. If the unit's digit is 1 more than ten's place digit, then find the number.

Sol: Consider, a 2-digit number is 'xy'

According to question,

$$\frac{xy}{x+y} = \frac{5}{1} \qquad \Rightarrow \qquad \frac{10x+y}{x+y} = \frac{5}{1} \qquad \Rightarrow \qquad 10x+y = 5x+5y$$

$5x = 4y \ \dots\dots (1)$

Also given that, $y = x + 1$ .... substitute in (1)

$5x = 4(x + 1) \qquad \Rightarrow \qquad 5x = 4x + 4 \qquad \Rightarrow \qquad x = 4$

$y = x + 1 = 4 + 1 = 5$

∴ Required 2-digit number = xy = 45.

> **Note:** This problem can be solved easily through options.

Q - 75 If the digits of a 3-digit number is reversed, then the newly formed number is 396 less than the original number. Find the difference between unit's place and hundred's place digit.

Sol: Consider, a 3-digit number is 'xyz'.

According to question,

$zyx = xyz - 396 \qquad \Rightarrow \qquad 100z + 10y + x = 100x + 10y + x - 396$

$99 (x - z) = 396 \qquad \Rightarrow \qquad x - z = 4$

∴ Required difference $x - z = 4$.

Q - 76 In a 2-digit number, the digit at unit's place is 3 more than twice the digit at ten's place. If the digits are interchanged, then the number is increased by 45. What is the original number?

Sol: Consider, a 2-digit number is 'xy'.

According to question,

$y = 2x + 3$ and $yx = xy + 45$

$10y + x = 10x + y + 45 \qquad \Rightarrow \qquad 9 (y - x) = 45 \qquad \Rightarrow \qquad y - x = 5$

$(2x + 3) - x = 5 \qquad \Rightarrow \qquad x = 2$

So, $y = 2x + 3 = 2 \times 2 + 3 = 7$

∴ Required 2-digit number is 27.

Q - 77 | A number when successively divided by 6, 4 and 7 leaves the remainders 3, 1 and 6 respectively. Find the number.

Sol: Consider, a number 'x'.

$$\boxed{\text{Dividend} = \text{Divisor} \times \text{Quotient} + \text{Remainder}}$$

$$
\begin{array}{c|l}
6 & x \\ \hline
4 & y-3 \\ \hline
7 & z-1 \\ \hline
& 1-6
\end{array}
$$

$x = 6y + 3, \; y = 4z + 1$

$z = 7 \times 1 + 6 = 13$

$y = 4 \times 13 + 1 = 53$

∴ Required number $x = 6 \times 53 + 3 = 321$.

Q - 78 | A number when successively divided by 5, 3 and 8 leaves the remainders 4, 2 and 7 respectively. Find the respective remainders, if the order of divisors are reversed.

Sol: Consider, a number be 'x'.

$$
\begin{array}{c|l}
5 & x \\ \hline
3 & y-4 \\ \hline
8 & z-2 \\ \hline
& 1-7
\end{array}
$$

$x = 5y + 4, \; y = 3z + 2$

$z = 8 \times 1 + 7 = 15$

$y = 3 \times 15 + 2 = 47$

Number $x = 5 \times 47 + 4 = 239$.

If the order of divisors are reversed i.e; 8, 3 and 5, then

$$
\begin{array}{c|l}
8 & 239 \\ \hline
3 & 29-7 \\ \hline
5 & 9-2 \\ \hline
& 1-4
\end{array}
$$

∴ Required remainders are 7, 2, 4 respectively.

Q - 79 | Find the sum of all even natural numbers less than 83.

Sol: All even natural numbers less than 83 are 2, 4, 6……., 80, 82

The above series of numbers are in arithmetic progression (A.P)

Sum of 'n' terms in A. P, $S_n = \dfrac{n}{2}[\text{First term} + \text{Last term}]$

Where 'n' is no. of observations.

Here, n = 41

∴ $S_n = \dfrac{41}{2}[2 + 82] = 1722$.

Alternate method:

Sum of even numbers = $2 + 4 + 6 + \ldots + 80 + 82 = 2 \,(1 + 2 + \ldots\ldots 40 + 41)$

Sum of 'n' natural numbers = $\dfrac{n(n+1)}{2}$

∴ Required sum = $2 \times \left[\dfrac{41(41+1)}{2}\right] = 1722$.

Q - 80   **Find the sum of 8, 17, 26, ..., 134.**

Sol: Given series 8, 17, 26, ..., 134 is in A.P.

Since the gap between two consecutive numbers is same i.e; 9.

Sum of 'n' terms in A.P, $S_n = \frac{n}{2}$[First term + Last term]

n^{th} term = a + (n – 1)d

a = 8, d = 17 – 8 = 9, n^{th} term = 134

134 = 8 + (n – 1)9   ⇒   n = 15

∴ $S_n = \frac{15}{2}$[8 + 134] = 1065.

Q - 81   **Find the sum of 3, 3^2, 3^3, ..., 3^8.**

Sol: Given series 3, 3^2, 3^3.........3^8 is in geometric progression (G.P).

Sum of 'n' terms in G. P, $S_n = \frac{a(r^n - 1)}{r - 1}$, if r > 1

Here, r = $\frac{r_2}{r_1} = \frac{3^2}{3}$ = 3, a = 3 and n = 8

∴ Required sum, $S_n = \frac{3(3^8 - 1)}{3 - 1}$ = 9840.

Q - 82   **On dividing a number by 231, a candidate used the method of short division. He divided the number successively by 3, 7 and 11 (factors of 231) and got the remainders 2, 5 and 8 respectively. If he had divided the number by 231, then what is the remainder?**

Sol: Consider, a number is '*x*'

| 3 | *x* | x = 3y + 2, y = 7z + 5
|----|---------|
| 7 | *y* – 2 | z = 11 × 1 + 8 = 19
| 11 | *z* – 5 | y = 7 × 19 + 5 = 138
| | 1 – 8 |

Required number *x* = 3 × 138 + 2 = 416

Now, 416 ÷ 231, then remainder = 185

∴ If the candidate divides the number with 231, then the remainder is 185.

Q - 83 | The sum of how many terms of the series 4, 12, 20, 28, …….. is 1764?

Sol: The given series 4, 12, 20, 28, …… is in A.P

$S_n = \frac{n}{2}$[First term + Last term] $= \frac{n}{2}$ [a + a + (n – 1)d]

$S_n = \frac{n}{2}$[2a + (n – 1)d]

Here, a = 4, d = 12 – 4 = 8 and S_n = 1764

$1764 = \frac{n}{2}$[2 × 4 + (n – 1)8] $\Rightarrow$ 3528 = n(8 + 8n – 8)

$8n^2 = 3528$ $\Rightarrow$ n = 21

∴ No. of terms of the series are 21.

Q - 84 | How many terms are there in G.P. 4, 12, 36, 108, ….., 8748?

Sol: n^{th} term in G.P, $T_n = a \times r^{n-1}$

Here, T_n = 8748, a = 4 and $r = \frac{r_2}{r_1} = \frac{12}{4} = 3$

$8748 = 4 \times 3^{n-1}$ $\Rightarrow$ $3^{n-1} = 2187$

$3^{n-1} = 3^7$ $\Rightarrow$ n – 1 = 7 $\Rightarrow$ n = 8

∴ No. of terms in G.P are 8.

Q - 85 | A number when divided by 13, gives the remainder as 11. When the same number is divided by 17, the remainder is 8. Find the number.

Sol:

| Dividend = Divisor × Quotient + Remainder |

Consider, a number is 'N'.

According to question,

N = 13x + 11 and N = 17y + 8

13x + 11 = 17y + 8 $\Rightarrow$ $y = \frac{13x + 3}{17}$

The least value of x for which $y = \frac{13x + 3}{17}$ is a whole number is x = 56

∴ N = 13 × 56 + 11 = 739.

Q - 86   **If the sum of first 15 terms of an arithmetic progression equal to that of first 19 terms, then what is the sum of first 34 terms?**

Sol: According to question,

$$S_{15} = S_{19} \qquad\qquad \because S = \frac{n}{2}[2a + (n-1)d]$$

$$\frac{15}{2}[2a + 14d] = \frac{19}{2}[2a + 18d]$$

$$30a + 210d = 38a + 342d \qquad \Rightarrow \qquad 8a + 132d = 0$$

$$2a + 33d = 0$$

∴ Sum of first 34 terms $S_{34} = \frac{34}{2}[2a + 33d] = 14 \times 0 = 0$.

Q - 87   **Find the remainder when $(15)^{3129}$ is divided by 14.**

Sol:
$$\boxed{\frac{(x+1)^n}{x} \text{ always gives the remainder as } 1.}$$

∴ $\dfrac{(14+1)^{3129}}{14}$ will always gives the remainder as 1.

Q - 88   **Find the remainder when $(2)^{348}$ is divided by 9.**

Sol:
$$\boxed{\frac{x^n}{x+1} \text{ gives the remainder as } 1, \text{ if 'n' is even.}}$$

$$\frac{(2)^{348}}{9} = \frac{(2^3)^{116}}{9} = \frac{8^{116}}{8+1}$$

It is in the form of $\dfrac{x^n}{x+1}$. Here, $x = 8$ and $n = 116$ (even number)

∴ $\dfrac{8^{116}}{8+1}$ gives the remainder as 1.

Q - 89   **Find the remainder when $(23)^{1469}$ is divided by 24.**

Sol:
$$\boxed{\frac{x^n}{x+1} \text{ gives the remainder as '}x\text{' itself, if 'n' is odd.}}$$

$$\frac{(23)^{1469}}{24} = \frac{(23)^{1469}}{23+1}$$

It is in the form of $\dfrac{x^n}{x+1}$. Here, $x = 23$ and $n = 1469$ (odd number)

∴ $\dfrac{(23)^{1469}}{23+1}$ gives the remainder as 23.

Q - 90 **Find the remainder when $(42)^{37!}$ is divided by 43.**

Sol:

$\dfrac{x^n}{x+1}$ gives the remainder as 1, if 'n' is even.

$$\frac{(42)^{37!}}{43} = \frac{(42)^{37!}}{42+1}$$

It is in the form of $\dfrac{x^n}{x+1}$. Here, $x = 42$ and $n = 37!$ (even number)

$\therefore \dfrac{(42)^{37!}}{42+1}$ gives the remainder as 1.

ASSESSMENT TEST

1. $(\frac{5}{8})^{th}$ of a number is 135. What is one-third of that number?

2. Find the value of x in $\frac{x}{15} \times \frac{x}{240} = 1$.

3. How many numbers upto 250 are divisible by 7?

4. The difference between the squares of two consecutive numbers is 63. Find the greater number.

5. Divide 72 into two parts, so that the sum of their reciprocals is $\frac{1}{16}$. Find the two parts.

6. A number when divided by 548 leaves the remainder as 121 and the value of quotient is 23. Find the number.

7. Calculate $\frac{547 \times 547 \times 547 + 653 \times 653 \times 653}{547 \times 547 - 547 \times 653 + 653 \times 653}$.

8. Calculate $\frac{621 \times 621 \times 621 - 471 \times 471 \times 471}{621 \times 621 + 621 \times 471 + 471 \times 471}$.

9. Find the number of factors for the numbers

   a) 1536   b) 2304

10. In how many ways the following numbers can be expressed as a product of two factors.

    a) 2160   b) 3025

11. In how many ways the following numbers can be expressed as a product of two different factors.

    a) 2704   b) 1764

12. Find the sum of factors for the numbers.

    a) 484   b) 764

13. What is the unit's place digit in $638 \times 492 \times 259$?

14. Find the unit's place digit in $(164)^{379} + (721)^{542} + (327)^{217}$.

15. Find the unit's place digit in $(438)^{672} - (357)^{585}$.

16. What is the unit's place digit in $(348)^{561} \times (104)^{213}$?

17. Find the highest power of

    a) 2 in 150!   b) 5 in 150!

18. Find the highest power of

    a) 16 in 130!   b) 25 in 190!

19. Find the highest power of

    a) 45 in 140!   b) 48 in 250!

20. Find the number of zeros at the end of 160!

21. Convert the decimal number $(1096)_{10}$ into octal number.

22. Convert the octal number $(643)_8$ into decimal number.

23. Convert the decimal number $(5364)_{10}$ into hexa decimal number.

24. Convert the binary number $(111010110)_2$ into decimal number.

25. Convert the hexa decimal number $(1B9)_{16}$ into decimal number.

26. Convert the binary number $(110110110110111000111001)_2$ into hexa decimal number.

27. What is the least value of x, so that the number $61x3568$ is divisible by 9?

28. Find the least value of k, so that the number $9745k3$ is divisible by 11.

29. Find the number of prime factors and number of distinct prime factors in the expression $18^7 \times 11^{15} \times 17^{12}$.

30. How many natural numbers less than 1500 have exactly 3 factors?

31. How many natural numbers between 150 and 1000 have odd number of factors?

32. A number is decreased by 9 and divided by 6, the result is 13. What would be the result, if 7 is subtracted from the number and divided by 8?

33. The sum of two numbers is 48 and their difference is 25. Find their difference of squares.

34. The sum of squares of two numbers is 154 and the square of their difference is 36. Find the product of two numbers.

35. The product of two numbers is 169 and sum of their squares is 446. Find the sum of the numbers.

36. If $x + y = 8$ and $xy = 15$, then find the value of $x^3 + y^3$.

37. Sum of three numbers is 255. First number is twice the second and third number is one-fifth of first. Find the third number.

38. If the sum of a natural number and its square is 462, then what is the number?

39. The ratio of the sum and difference of two numbers is 9 : 2. Find the ratio of these two numbers.

40. The ratio between a two-digit number and the sum of the digits of that number is 3 : 1. If the unit's digit is 5 more than ten's place digit, then find the number.

41. Find the number which when multiplied by 33 is increased by 512.

42. The difference between the squares of two consecutive numbers is 67. Find the numbers.

43. On dividing 12339 by a certain number, we get 38 as quotient and 27 as remainder. Find the divisor.

44. The difference of two numbers is 1645. By dividing the larger number with smaller number, we get a quotient of 7 and remainder is 157. Find the smaller number.

45. What will be the remainder when $(47^{51} + 47)$ is divided by 48?

46. A 3-digit number $5a8$ is added to another 3-digit number 898 to give a 4-digit number $14b6$, which is divisible by 11. Then, find the value of a + b.

47. A number when divided by 11 gives the remainder as 9. When the same number is divided by 15, the remainder is 1. Find the number.

48. The sum of three consecutive even numbers is 12 less than the four-fifth of 90. Find the greatest number.

49. How many digits are required to write first 1120 natural numbers?

50. In a division sum, the divisor is 13 times the quotient and 6 times the remainder. If the remainder is 52, then find the value of dividend.

51. The sum of the digits of a 2-digit number is 12. If the digits are reversed the number is increased by 36. What is the number?

52. The denominator of a rational number is 5 more than its numerator. If the numerator is increased by 9 and denominator is decreased by 4, we obtain 2. What is the rational number?

53. If the number $12x547y6$ is divisible by 72, then find the maximum value of $x - y$.

54. Find the sum of first 53 odd numbers.

55. Find the sum of first 26 even numbers.

56. What is the least number must be added to 1316 to get a number is exactly divisible by 29?

57. The sum of five consecutive odd numbers A, B, C, D and E is 485. Find the product of B and C.

58. When one-fifth of a number is subtracted from the number itself, it gives the same value as the sum of all the angles of a rectangle. What is the number?

59. The sum of the digits of 2-digit number is 11 and the difference between the two digits of that number is 5. What is the product of the two digits of 2-digit number?

60. A number when divided by 156 gives 79 as remainder. If the same number is divided by 12, then what will be the remainder?

61. A number when divided by 8, leaves 3 as remainder. What will be the remainder, when square of the same number is divided by 8?

62. A number when divided by a divisor leaves the remainder of 27. When twice the original number is divided by the same divisor, the remainder is 23. What is the value of divisor?

63. Find the common factor of $(31^{47} + 35^{47})$ and $(31^{49} + 35^{49})$.

64. On division with zero remainder, a student took 15 as divisor instead of 27. The quotient obtained by him was 54. Find the correct quotient.

65. When 7269 and 5985 are divided by a 3-digit number N leaves the same remainder in both the cases. Find all the possible 3-digit divisors.

66. Find the remainder of $\dfrac{19^{16^{17^{18\ldots\ldots\infty}}}}{9}$.

67. The sum of the digits of a 2-digit number is $\dfrac{1}{5}$ of the difference between the number and the number obtained by interchanging the positions of the digits. What definitely is the difference between the digits of that number?

68. If the digits of a 3-digit number are reversed, then newly formed number is 594 more than the original number. Find the difference between unit's place and hundred's place digit.

69. In a 2-digit number, the digit at ten's place is one more than twice of the unit's place. If the digits are interchanged, then the number is decreased by 45. Find the original number.

70. A number when successively divided by 4, 7 and 9 leaves the remainders 1, 3 and 6 respectively. Find the respective remainders, if the order of divisors are reversed.

71. Find the sum of 7, 13, 19, ...115.

72. Find the sum of $2, 2^2, 2^3, ... 2^{12}$.

73. If the product $3725 \times 53x2$ is divisible by 45, then find the least value of x.

74. The sum of how many terms of the series 3, 8, 13, 18 ...... is 1452?

75. On dividing a number by 455, a candidate used the method of short division. He divided the number successively by 5, 7 and 13 (factors of 455) and got the remainders 3, 4 and 12 respectively. If he had divided the number by 455, then what is the remainder?

76. If the sum of first 17 terms of arithmetic progression equal to that of first 25 terms, then what is the sum of first 42 terms?

77. Find the remainder when $(18)^{2368}$ is divided by 17.

78. Find the remainder when $(12)^{564}$ is divided by 13.

79. Find the remainder when $(7)^{237}$ is divided by 8.

80. Find the remainder when $(51)^{64!}$ is divided by 52.

KEY

1. 72
2. 60
3. 35
4. 32
5. 24 and 48
6. 12725
7. 1200
8. 150
9. a) 20   b) 27
10. a) 20   b) 5
11. a) 7   b) 13
12. a) 931   b) 1344
13. 4
14. 2
15. 9
16. 2
17. a) 146   b) 37
18. a) 32   b) 23
19. a) 33   b) 61
20. 39
21. $(2110)_8$
22. $(419)_{10}$
23. $(14F4)_{16}$
24. $(470)_{10}$
25. $(441)_{10}$
26. $(DB6E39)_{16}$
27. 7
28. 2
29. 48 and 4
30. 12
31. 19
32. 10

33. 1200

34. 59

35. 28

36. 152

37. 30

38. 21

39. 11 : 7

40. 27

41. 16

42. 33 and 34

43. 324

44. 248

45. 46

46. 18

47. 526

48. 22

49. 3373

50. 7540

51. 48

52. $\frac{7}{12}$

53. 5

54. 2809

55. 702

56. 18

57. 9215

58. 450

59. 24

60. 7

61. 1

62. 31

63. 66

64. 30

65. 642, 428, 321, 214, 107

66. 1

67. 5

68. 6

69. 94

70. 1, 6, 2

71. 1159

72. 8190

73. 8

74. 24

75. 443

76. 0

77. 1

78. 1

79. 7

80. 1

LCM AND HCF

MULTIPLES: The numbers which are exactly divisible by the given number.

Example: Multiples of 2 = {2, 4, 6, 8, 10,…………}

FACTORS: The numbers which are exactly divides the given number.

Example: Factors of 12 = {1, 2, 3, 4, 6, 12}

> **Note:** For any number factors are finite and multiples are infinite.

Least Common Multiple (LCM):

LCM of two (or) more numbers is the least number which is exactly divisible by each of the given numbers.

Example: LCM of 2, 3, 6.

Multiples of 2 = 2, 4, 6, 8, 10, 12, 14, 16, 18, 20, 22, 24,……

Multiples of 3 = 3, 6, 9, 12, 15, 18, 21, 24,……

Multiples of 6 = 6, 12, 18, 24, 30, 36,……

Common multiples of 2, 3 and 6 = 6, 12, 18, 24,……

∴ LCM of 2, 3 and 6 = 6.

Highest Common Factor (HCF):

HCF of two (or) more numbers is the greatest number which divides each of the given numbers exactly without leaving any remainder.

HCF is also called as GCD (Greatest Common Divisor).

Example: HCF of 8, 12.

Factors of 8 = 1, 2, 4, 8

Factors of 12 = 1, 2, 3, 4, 6, 12

Common factors of 8 and 12 = 1, 2, 4

∴ HCF of 8 and 12 = 4.

❖ LCM and HCF both are calculated by two methods.

  1) Long division method

  2) Factorisation method

1) Long division method:

a) LCM:

According to long division method in LCM,

First – Divide all the given numbers by a prime number which exactly divides atleast any 2 of the given numbers and write down the quotients and the undivided numbers in the below line.

Second – Repeat this process of division till we get the numbers which are prime to one another.

Third – The product of all divisors and the numbers in the last line will be the required LCM.

Example: Find the LCM of 12, 18, 20.

Sol:

$$
\begin{array}{r|ccc}
2 & 12 & , & 18 & , & 20 \\
\hline
2 & 6 & , & 9 & , & 10 \\
\hline
3 & 3 & , & 9 & , & 5 \\
\hline
& 1 & , & 3 & , & 5 \\
\end{array}
$$

∴ Required LCM = 2 × 2 × 3 × 3 × 5 = 180.

b) HCF:

According to long dividion method in HCF,

First – Divide the greater number by the smaller number.

Second – Divide the divisor by the remainder.

Third – Repeat this division process till the remainder is zero. The last divisor is the required HCF.

> **Note:** To calculate HCF of more than two numbers, calculate the HCF of first 2 numbers, then the third number and HCF of first 2 numbers and calculate their HCF and so on. The resulting HCF is the required HCF of numbers.

Example: Find the HCF of 32, 424.

Sol:

```
        32) 424 (13
             32
            ____
            104
             96
            ____
          8) 32 (4
             32
            ____
              ×
```

∴ Required HCF = 8.

2) Factorisation method:

a) LCM:

According to factorisation method in LCM,

First – Write down the given numbers into their prime factors.

Second – The product of all prime factors with their highest powers is the required LCM.

Example: Find the LCM of 12, 16, 21.

Sol: $12 = 2^2 \times 3^1$ $16 = 2^4$ $21 = 3^1 \times 7^1$

$\therefore$ Required LCM = Product of all prime factors with their highest powers.

$\Rightarrow \quad 2^4 \times 3^1 \times 7^1 = 336.$

b) HCF:

According to factorisation method in HCF,

First – Write down the given numbers into their prime factors.

Second – The product of common prime factors with their lowest powers is the required HCF.

Example: Find the HCF of 24, 30, 36.

Sol: $24 = 2^3 \times 3^1$ $30 = 2^1 \times 3^1 \times 5^1$ $36 = 2^2 \times 3^2$

$\therefore$ Required HCF = Product of common prime factors with their least powers.

$\Rightarrow \quad 2^1 \times 3^1 = 6.$

Some important points to be remember:

✓ HCF of any two prime numbers, any two consecutive numbers (or) any two consecutive odd numbers is always 1.

✓ LCM of fractions $= \dfrac{\text{LCM of numerators}}{\text{HCF of denominators}}$

✓ HCF of fractions $= \dfrac{\text{HCF of numerators}}{\text{LCM of denominators}}$

✓ Let a,b are two numbers. If their LCM and HCF both are given, then

$$\boxed{\text{Product of two numbers } (a \times b) = \text{LCM} \times \text{HCF}}$$

✓ LCM of numbers is always divisible by HCF of given numbers.

✓ **Co-prime numbers:** Two numbers doesn't have any common factor except 1, then those numbers are called as co-prime numbers. Two numbers might be prime (or) composite numbers.

Examples: (3, 7), (4, 9), (5, 16).

✓ LCM of any two co-prime numbers is product of those two numbers & their HCF is always 1.

✓ If the HCF of each pair of 'n' given numbers is H and their LCM is L, then the product of these numbers is

$$\boxed{(\text{HCF})^{n-1} \times \text{LCM} = (\text{H})^{n-1} \times \text{L}}$$

❖ Basically there are three models of problems in both LCM and HCF.

LCM → Model – 1: (Same remainder)

The least number which when divided by x, y, z leaves the same remainder (R) in each case, then

$$\boxed{\text{Required number = LCM } (x, y, z) + \text{Remainder (R)}}$$

Example: Find the least number which when divided by 4, 9, 12 leaves the same remainder 3 in each case.

Sol: Required least number = LCM (4, 9, 12) + 3 = 36 + 3 = 39.

LCM → Model – 2: (Different remainders)

The least number which when divided by x, y, z leaves the remainders R_1, R_2 and R_3 respectively, then

$$\boxed{\text{Required number = LCM } (x, y, z) - \text{Common difference}}$$

Where, common difference $= x - R_1 = y - R_2 = z - R_3$

> **Note:** Common difference value must be equal for all the values. If common difference values are not equal, then the answer is calculated through options.

Example: Find the least number which when divided by 10, 15, 20 leaves the remainders 6, 11 and 16 respectively.

Sol: Here, common difference = 10 – 6 = 15 – 11 = 20 – 16 = 4 (Equal for all values)

Required least number = LCM (10, 15, 20) – Common difference = 60 – 4 = 56.

LCM → Model – 3: (Based on number of digits)

Again in LCM third model we have 3 cases.

Case 1: The least number of n–digits which is exactly divisible by x, y, z. Then,

$$\boxed{\text{Required number = Least n–digit number + LCM – Remainder (R)}}$$

Where 'R' is the remainder obtained by dividing least n–digit number with LCM of x, y, z.

Example: Find the least number of 4–digits which is exactly divisible by 6, 8, 12.

Sol: Required number = Least 4–digit number + LCM (6, 8, 12) – R

LCM of 6, 8, 12 = 24 & Least 4–digit number = 1000

```
24) 1000 (41
     96
    ────
     40        ∴ Required number = 1000 + 24 – 16 = 1008.
     24
    ────
     16 → R
```

Case 2: The least number of n–digits which when divided by x, y, z so as to leave the same remainder R_1 in each case. Then,

> Required number = (Least n–digit number + LCM – R) + R_1

Where 'R' is the remainder obtained by dividing least n–digit number with LCM of x, y, z and 'R_1' is the remainder mentioned in the given problem.

Example: Find the least number of 4 digits which when divided by 10, 14, 16 leaves the same remainder 7 in each case.

Sol: Required number = Least 4–digit number + LCM (10, 14, 16) – R + 7

LCM of 10, 14, 16 = 560 & Least 4–digit number = 1000

$$
\begin{array}{r}
560)\,1000\,(1 \\
\underline{560} \\
440 \rightarrow R
\end{array}
$$

∴ Required number = 1000 + 560 – 440 + 7 = 1127.

Case 3: The least number of n–digits which when divided by x, y, z so as to leave the remainders R_1, R_2 and R_3 respectively. Then,

> Required number = (Least n–digit number + LCM – R) – Common difference

Where 'R' is the remainder obtained by dividing Least n–digit number with LCM of x, y, z and common difference = $x - R_1 = y - R_2 = z - R_3$

Example: Find the least number of 4–digits which when divided by 4, 9, 12 leaves the remainders 2, 7, 10 respectively.

Sol: Required number = Least 4–digit number + LCM (4, 9, 12) – R – Common difference

LCM of 4, 9, 12 = 36, Least 4–digit number = 1000 and

Common difference = 4 – 2 = 9 – 7 = 12 – 10 = 2

$$
\begin{array}{r}
36)\,1000\,(27 \\
\underline{72} \\
280 \\
\underline{252} \\
28 \rightarrow R
\end{array}
$$

∴ Required number = 1000 + 36 – 28 – 2 = 1006.

HCF →Model – 1: (Same remainder)

Again in HCF first model we have 2 cases.

Case 1: (Remainder mentioned)

The greatest number which will divide x, y, z so as to leave the same remainder R in each case. Then,

$$\boxed{\text{Required number} = \text{HCF } [(x - R), (y - R), (z - R)]}$$

Example: Find the greatest number which will divide 14, 24, 34 so as to leave the same remainder 4 in each case.

Sol: Required number = HCF [(14 – 4), (24 – 4), (34 – 4)] = HCF [10, 20, 30] = 10.

Case 2: (Remainder not mentioned)

The greatest number which on dividing x, y, z so as to leave the same remainder in each case. Then,

$$\boxed{\text{Required number} = \text{HCF } [|x - y|, |y - z|, |z - x|]}$$

Example: Find the greatest number which on dividing 34, 52, 76 leves the same remainder in each case.

Sol: Required number = HCF [|34 – 52|, |52 – 76|, |76 – 34|] = HCF (18, 24, 42) = 6.

HCF →Model – 2: (Different remainders)

The greatest number which will divide x, y, z so as to leave the remainders R_1, R_2 and R_3 respectively. Then,

$$\boxed{\text{Required number} = \text{HCF } [(x - R_1), (y - R_2), (z - R_3)]}$$

Example: Find the greatest number which will divide 126 and 234 so as to leave the remainders 6 and 2 respectively.

Sol: Required number = HCF [(126 – 6), (234 – 2)] = HCF (120, 232) = 8.

HCF →Model – 3: (Based on number of digits)

Again in HCF third model we have 3 cases.

Case 1: The greatest number of n–digits which when divided by x, y, z leaves no remainder. Then,

$$\boxed{\text{Required number} = \text{Greatest n–digit number} - \text{Remainder (R)}}$$

Where 'R' is the remainder obtained by dividing greatest n–digit number with LCM.

Example: Find the greatest 4–digit number which is exactly divisible by 8, 14, 26.

Sol: Required number = Greatest 4–digit number – Remainder (R)

Greatest 4–digit number = 9999 & LCM of 8, 14, 26 = 728

$$728) \overline{9999} (13$$
$$\underline{728}$$
$$2719$$
$$\underline{2184}$$
$$535 \rightarrow R$$

∴ Required number = 9999 – 535 = 9464.

Case 2: The greatest number of n–digits which when divided by x, y, z leaves the same remainder R_1 in each case.

$$\boxed{\text{Required number} = (\text{Greatest n–digit number} – \text{Remainder R}) + R_1}$$

Where 'R' is the remainder obtained by dividing greatest n–digit number with LCM and R_1 is the remainder mentioned in the given problem.

Example: Find the greatest number of 4–digits which when divided by 12, 16 and 18 leaves the same remainder 9 in each case.

Sol: Required number = (Greatest 4–digit number – Remainder R) + R_1

Greatest 4–digit number = 9999, R_1 = 9 & LCM of 12, 16, 18 = 144

$$144) \overline{9999} (69$$
$$\underline{864}$$
$$1359$$
$$\underline{1296}$$
$$63 \rightarrow R$$

∴ Required number = 9999 – 63 + 9 = 9945.

Case 3: The gretest number of n–digits which when divided by x, y, z leaves the remainders R_1, R_2 and R_3 respectively.Then,

$$\boxed{\text{Required number} = (\text{Greatest n–digit number} – \text{Remainder R}) – \text{Common difference}}$$

Where 'R' is the remainder obtained by dividing greatest n–digit number with LCM.

Common difference = $x – R_1 = y – R_2 = z – R_3$

Example: Find the greatest 4–digit number which when divided by 24, 32, 48 leaves the remainders 15, 23, 39 respectively.

Sol: Required number = (Greatest 4–digit number – R) – Common difference

Common difference = 24 – 15 = 32 – 23 = 48 – 39 = 9 & LCM of 24, 32, 48 = 96

$$
\begin{array}{r}
96)\,\overline{9999}\,(104 \\
\underline{96} \\
399 \\
\underline{384} \\
15 \rightarrow R
\end{array}
$$

∴ Required number = 9999 – 15 – 9 = 9975.

SOLVED EXAMPLES

Q - 1 **Find the LCM of 24, 30, 35 and 42 by long division method.**

Sol: LCM of 24, 30, 35 and 42 by long division method.

2	24 , 30 , 35 , 42
3	12 , 15 , 35 , 21
5	4 , 5 , 35 , 7
7	4 , 1 , 7 , 7
	4 , 1 , 1 , 1

∴ Required LCM = $2 \times 3 \times 5 \times 7 \times 4 = 840$.

Q - 2 **Find the LCM of 18, 27, 45 and 120 by factorisation method.**

Sol: LCM of 18, 27, 45 and 120 by factorisation method.

First write the given numbers interms of prime factors

$$18 = 2 \times 3^2$$
$$27 = 3^3$$
$$45 = 3^2 \times 5$$
$$120 = 2^3 \times 3 \times 5$$

To get the LCM, calculate the product of prime factors with their highest powers.

∴ Required LCM = $2^3 \times 3^3 \times 5^1 = 1080$.

Q - 3 **Find the HCF of 48, 60 and 84 by long division method.**

Sol: HCF of 48, 60 and 84 by long division method.

```
48) 60 (1              HCF  ⟵  12) 84 (7
    48                               84
    12) 48 (4                         ×
        48                HCF of 12 and 84 is 12
         ×
HCF of 48 and 60 is 12
```

∴ HCF of 48, 60 and 84 is 12.

| Q - 4 | **Find the HCF of 168 and 720 by factorisation method.** |

Sol: Write the given numbers interms of prime factors.

$$168 = 2^3 \times 3 \times 7$$

$$720 = 2^4 \times 3^2 \times 5$$

To get the HCF, calculate the product of common prime factors with their least powers.

∴ Required HCF $= 2^3 \times 3^1 = 24$.

| Q - 5 | **Find the greatest number which divides 896, 1064 and 1260 exactly.** |

Sol: HCF of 896, 1064 and 1260

```
896) 1064 (1                        56) 1260 (22
     896                                 1232
     168) 896 (5           HCF  ◄——  28) 56 (2
          840                             56
          56) 168 (3                       ×
             168                   HCF of 56 and 1260 is 28
              ×
HCF of 896 and 1064 is 56
```

∴ HCF of 896, 1064 and 1260 is 28.

| Q - 6 | **Find the LCM of 2.6, 0.52 and 13.** |

Sol: First multiply the given numbers with 100, then calculate the LCM of 260, 52, 1300.

2	260 , 52 , 1300
2	130 , 26 , 650
5	65 , 13 , 325
13	13 , 13 , 65
	1 , 1 , 5

LCM of 260, 52 and 1300 $= 2 \times 2 \times 5 \times 13 \times 5 = 1300$

∴ LCM of 2.6, 0.52 and 13 $= \dfrac{1300}{100} = 13$.

| Q – 7 | **Find the HCF of 3.45 and 15.** |

Sol: First multiply the given numbers with 100, then calculate the HCF of 345 and 1500.

$$
\begin{array}{r}
345)\ 1500\ (4 \\
\underline{1380} \\
120)\ 345\ (2 \\
\underline{240} \\
105)\ 120\ (1 \\
\underline{105} \\
\text{HCF} \longleftarrow \quad 15)\ 105\ (7 \\
\underline{105} \\
\times
\end{array}
$$

HCF of 345 and 1500 is 15

$\therefore$ HCF of 3.45 and 15 $= \dfrac{15}{100} = 0.15$.

| Q - 8 | **Find the LCM of $\dfrac{27}{120}, \dfrac{81}{65}$ and $\dfrac{117}{80}$.** |

Sol:

$$
\text{LCM of fractions} = \frac{LCM\ of\ numerators}{HCF\ of\ denominators}
$$

$\therefore$ Required LCM $= \dfrac{LCM\ of\ 27,\ 81,\ 117}{HCF\ of\ 120,\ 65,\ 80} = \dfrac{1053}{5} = 210\dfrac{3}{5}$.

| Q - 9 | **Find the HCF of $\dfrac{33}{57}$ and $2\dfrac{6}{19}$.** |

Sol: HCF of $\dfrac{33}{57}$ and $\left(2\dfrac{6}{19}\right) = \dfrac{44}{19}$

$$
\text{HCF of fractions} = \frac{HCF\ of\ numerators}{LCM\ of\ denominators}
$$

$\therefore$ Required HCF $= \dfrac{HCF\ of\ 33,\ 44}{LCM\ of\ 57,\ 19} = \dfrac{11}{57}$.

| Q -10 | **The HCF of two numbers is 26 and their LCM is 884. If one of the numbers is 104, then find the second number.** |

Sol:

$$
\text{Product of two numbers } (a \times b) = \text{LCM} \times \text{HCF}
$$

Given that, LCM = 884, HCF = 26 and one number a = 104

$104 \times b = 884 \times 26 \quad \Rightarrow \quad b = 221$

$\therefore$ Second number b = 221.

Q-11 Find the LCM of $x^2 - 5x$, $x^2 - 2x - 15$ and $x^3 + 6x^2 + 9x$.

Sol: $x^2 - 5x = x(x - 5)$

$x^2 - 2x - 15 = (x + 3)(x - 5)$

$x^3 + 6x^2 + 9x = x(x^2 + 6x + 9) = x(x + 3)^2$

To calculate LCM, we are taking product of all factors with their highest powers.

∴ Required LCM = $x(x - 5)(x + 3)^2$.

Q-12 Find the HCF of $27(16x^3 + 52x^2 + 12x)$, $18(20x^4 + 105x^3 + 25x^2)$ and $24(28x^3 - 49x^2 - 14x)$.

Sol: $27(16x^3 + 52x^2 + 12x) = 27 \times 4x(4x^2 + 13x + 3) = 108x(4x^2 + 12x + x + 3)$

$27(16x^3 + 52x^2 + 12x) = 2^2 \times 3^3 \times x \times (x + 3)(4x + 1)$

$18(20x^4 + 105x^3 + 25x^2) = 18 \times 5x^2(4x^2 + 21x + 5) = 2 \times 3^2 \times 5 \times x^2(x + 5)(4x + 1)$

$24(28x^3 - 49x^2 - 14x) = 24 \times 7x(4x^2 - 7x - 2) = 2^3 \times 3 \times 7 \times x(x - 2)(4x + 1)$

To calculate HCF, we are taking product of common factors with their lowest powers.

∴ Required HCF = $2 \times 3 \times x(4x + 1) = 6x(4x + 1)$.

Q-13 What is the LCM of $(2^4 \times 3^2 \times 5)$, $(2^3 \times 3^3 \times 7)$ and $(2^2 \times 5^2 \times 7^2 \times 13)$?

Sol:

> **Note:** LCM of given numbers by factorisation method is the product of prime factors with their highest powers.

Given numbers are $(2^4 \times 3^2 \times 5)$, $(2^3 \times 3^3 \times 7)$ and $(2^2 \times 5^2 \times 7^2 \times 13)$

∴ Required LCM = $2^4 \times 3^3 \times 5^2 \times 7^2 \times 13$.

Q-14 Find the HCF of $(2^2 \times 3 \times 5^3)$, $(2^3 \times 3^2 \times 7^2 \times 11)$ and $(2^4 \times 3^3 \times 5 \times 13)$.

Sol:

> **Note:** HCF of given numbers by factorisation method is the product of common prime factors with their least powers.

Given numbers are $(2^2 \times 3 \times 5^3)$, $(2^3 \times 3^2 \times 7^2 \times 11)$ and $(2^4 \times 3^3 \times 5 \times 13)$

∴ Required HCF = $2^2 \times 3^1 = 12$.

Q-15 What is the product of LCM and HCF of 24 and 36?

Sol: LCM of 24, 36 = 72

HCF of 24, 36 = 12

∴ Product of LCM and HCF = $72 \times 12 = 864$.

| Q -16 | The difference of two numbers is $\frac{1}{8}$ of their sum. The sum of two numbers is 64. Find their LCM. |

Sol: Consider, two numbers are x and y.

Given that, $x - y = \frac{1}{8} \times (x + y)$ and $x + y = 64$ .... (1)

$x - y = \frac{1}{8} \times 64 = 8$ .... (2)

$(1) + (2) \qquad \Rightarrow \qquad 2x = 72 \qquad \Rightarrow \qquad x = 36$ .... sub in (1)

$36 + y = 64 \qquad \Rightarrow \qquad y = 64 - 36 = 28$

$\therefore$ LCM of 36 and 28 = 252.

| Q -17 | The ratio of two numbers is 7 : 8. If their LCM is 840, then what is their HCF? |

Sol: Consider, two numbers are $7x$ and $8x$.

Where 'x' is a common factor

LCM of $7x$ and $8x = 840 \qquad \Rightarrow \qquad 56x = 840 \qquad \Rightarrow \qquad x = 15$

$\therefore$ HCF of $7x$ and $8x = x = 15$.

| Q -18 | The ratio of three numbers is 4 : 6 : 7. If their LCM is 2100, then find their HCF. |

Sol: Consider three numbers are $4x$, $6x$ and $7x$

LCM of $4x$, $6x$ and $7x = 2100$

$84x = 2100 \qquad \Rightarrow \qquad x = 25$

$\therefore$ HCF of $4x$, $6x$ and $7x = x = 25$.

| Q -19 | The LCM of two numbers is 144 and their HCF is 12. If sum of the two numbers is 84, then find the sum of their reciprocals. |

Sol: If HCF of two numbers is 12, then two numbers must be a multiples of 12.

Consider two numbers are $12x$ and $12y$

LCM of $12x$ and $12y = 12xy \qquad \Rightarrow \qquad 12xy = 144 \qquad \Rightarrow \qquad xy = 12$

Also, $12x + 12y = 84 \qquad \Rightarrow \qquad x + y = 7$

$\therefore$ Sum of their reciprocals $= \frac{1}{12x} + \frac{1}{12y} = \frac{y + x}{12xy} = \frac{7}{12 \times 12} = \frac{7}{144}$.

Q-20   The LCM of two numbers is 480 and their HCF is 12. If the sum of the numbers is 156, then find their difference.

Sol: Consider, two numbers are $12x$ and $12y$.

LCM of $12x$ and $12y = 480$ $\Rightarrow$ $12xy = 480$ $\Rightarrow$ $xy = 40$

Also, given that $12x + 12y = 156$ $\Rightarrow$ $x + y = 13$

$(x + y)^2 - (x - y)^2 = 4xy$ $\Rightarrow$ $x - y = \sqrt{(x + y)^2 - 4xy}$

$x - y = \sqrt{13^2 - 4 \times 40} = 3$

$\therefore$ Difference of two numbers $= 12(x - y) = 12 \times 3 = 36$.

Q-21   Find the least number which is to be added to 3788, so that the number is exactly divisible by 6, 8, 9 and 12.

Sol: LCM of 6, 8, 9, 12 = 72

```
72) 3788 (52
    360
    ‾‾‾‾
    188
    144
    ‾‾‾‾
     44 → R
```

$\therefore$ Required least number added to 3788 = 72 − 44 = 28.

Q-22   Find the least number which when diminished by 21, is divisible by each one of the numbers 28, 36, 42 and 45.

Sol: LCM of 28, 36, 42, 45

2	28, 36, 42, 45
2	14, 18, 21, 45
3	7, 9, 21, 45
3	7, 3, 7, 15
7	7, 1, 7, 5
	1, 1, 1, 5

LCM = 2 × 2 × 3 × 3 × 7 × 5 = 1260

According to question,

After diminishing of 21 the number is divisible by 28, 36, 42 and 45

$\therefore$ Required number = 1260 + 21 = 1281.

Q-23 | What is the least number which when increased by 18, is divisible by each one of the numbers 15, 20, 26, 34?

Sol: LCM of 15, 20, 26, 34

$$
\begin{array}{c|cccc}
2 & 15, & 20, & 26, & 34 \\
\hline
5 & 15, & 10, & 13, & 17 \\
\hline
  & 3, & 2, & 13, & 17
\end{array}
$$

LCM = $2 \times 5 \times 3 \times 2 \times 13 \times 17 = 13260$.

According to question,

After increasing of 18 the number is divisible by 15, 20, 26, 34

∴ Required number = 13260 – 18 = 13242.

Q-24 | The LCM and HCF of two numbers are 3024 and 24 respectively. If one of the numbers lies between 400 and 500, then find that number.

Sol: Consider, two numbers are $24x$ and $24y$.

LCM of $24x, 24y = 3024$ $\Rightarrow$ $24xy = 3024$ $\Rightarrow$ $xy = 126$

The possible co-prime pairs of x and y, whose product is 126 are

(1, 126), (7, 18), (9, 14)

Among the three pairs (7, 18) satisfies the given condition

i.e; one number lies between 400 and 500

∴ Required number lies between 400 and 500 = $24 \times 18 = 432$.

Q-25 | The ratio of two numbers is 6 : 7 and their HCF is 8. What is the LCM of those two numbers?

Sol: Consider, two numbers are $6x, 7x$ respectively.

HCF of $6x$ and $7x = x = 8$

Two numbers are $6 \times 8 = 48$ and $7 \times 8 = 56$

∴ Required LCM of 48 and 56 = 336.

Q-26 | The LCM and HCF of two numbers are 140 and 14 respectively. If the ratio of two numbers is 2 : 5, then find the difference of two numbers.

Sol: Consider, two numbers are $2x$ and $5x$.

HCF of $2x$ and $5x = x = 14$

Two numbers are $2 \times 14 = 28$, $5 \times 14 = 70$

∴ Difference of two numbers = 70 – 28 = 42.

Q-27 The LCM and HCF of two numbers are 1323 and 21 respectively. If the first number is divided by 7, the quotient is 21. Then, find the second number.

Sol: Given that, LCM = 1323 and HCF = 21

First number = Divisor × Quotient + Remainder = 7 × 21 + 0 = 147

> Product of two numbers a × b = LCM × HCF

147 × b = 1323 × 21   ⇒   b = 189

∴ Second number b is 189.

Q-28 Three numbers which are co-prime to each other are such that the product of first two is 713 and that of last two is 1457, then find the sum of three numbers.

Sol: In the first 2 numbers product and last 2 numbers product, middle number is common.

Middle number = HCF of 713 and 1457

$$
\begin{array}{r}
713)\,1457\,(2 \\
\underline{1426} \\
31)\,713\,(23 \\
\underline{713} \\
\times
\end{array}
$$

HCF ←

Middle number = 31

First number = $\dfrac{713}{31}$ = 23 and Third number = $\dfrac{1457}{31}$ = 47

∴ Sum of three numbers = 23 + 31 + 47 = 101.

Q-29 LCM of two prime numbers 'a' and 'b' (a < b) is 437. What is the value of 4a – 3b?

Sol: Given that, LCM of 2 prime numbers a and b = 437

LCM of any two prime numbers = Product of those 2 prime numbers a × b = 437

Possible values of a and b are 19 and 23 which gets the product as 437.

∴ 4a – 3b = 4 × 19 – 3 × 23 = 76 – 69 = 7.

Q-30 What is the least number which when doubled, will be exactly divisible by 15, 24, 27 and 36?

Sol: LCM of 15, 24, 27 and 36

3	15, 24, 27, 36
3	5, 8, 9, 12
2	5, 8, 3, 4
2	5, 4, 3, 2
	5, 2, 3, 1

LCM of 15, 24, 27, 36 = 3 × 3 × 2 × 2 × 5 × 2 × 3 = 1080.

According to question,

After double the number it is divisible by 15, 24, 27, 36.

∴ Required number = $\dfrac{1080}{2}$ = 540.

Q-31 There are four numbers. HCF of each possible pair is 6 and LCM of all the four numbers is 1250. Find the product of all four numbers.

Sol:

> **Note:** If HCF of each pair of 'n' given numbers is H and their LCM is L, then the product of these numbers is $(HCF)^{n-1} \times LCM$

Given that, n = 4, HCF = 6 and LCM = 1250

∴ Product of all four numbers = $(6)^{4-1} \times 1250 = 270000$.

Q-32 There are five numbers. HCF of each possible pair is 5 and LCM of all five numbers is 2364. Find the product of all five numbers.

Sol:

> **Note:** If HCF of each pair of 'n' given numbers is H and their LCM is L, then the product of these numbers is $(HCF)^{n-1} \times LCM$

Given that, n = 5, HCF = 5 and LCM = 2364

∴ Product of all five numbers = $(5)^{5-1} \times 2364 = 1477500$.

Q-33 The sum of LCM and HCF of two numbers is 1472 and their LCM is 63 times their HCF. If one of the numbers is 161, then find the other number.

Sol: Given that, LCM + HCF = 1471 and LCM = 63 × HCF

63HCF + HCF =1472 ⇒ 64HCF =1472

HCF = 23 ⇒ LCM = 63 × 23 = 1449

> Product of two numbers a × b = LCM × HCF

Here, LCM = 1449 and HCF = 23 and a = 161

∴ Second number b = $\dfrac{1449 \times 23}{161} = 207$.

Q-34 The sum of LCM and HCF is 3293 and their LCM is 88 times their HCF. If one of the numbers is 296, then find the four times of second number.

Sol: Given that, LCM + HCF = 3293 and LCM = 88 × HCF

88HCF + HCF = 3293 ⇒ 89HCF = 3293 ⇒ HCF = 37

LCM = 88 × 37 = 3256

> Product of two numbers a × b = LCM × HCF

Here, LCM = 3256, HCF = 37 and a = 296

Second number b = $\dfrac{3256 \times 37}{296} = 407$

∴ Required number = 4 × 407 = 1628.

Q-35 Four bells toll at regular intervals of 6 minutes, 8 minutes, 9 minutes and 12 minutes respectively. If they toll together at 11 AM, then at what time will they toll together for the first time after 11 AM?

Sol: LCM of (6, 8, 9 and 12) minutes = 72 minutes.

All four bells toll together after 72 minutes for the first time after 11 AM

∴ Required time = 11 AM + 72 minutes = 12:12 PM.

Q-36 A, B and C start running around a circular stadium. They complete their revolutions in 54, 63 and 72 seconds respectively. After how many seconds will they be together at the starting point?

Sol: Given that A, B and C complete their revolutions in 54, 63 and 72 seconds.

LCM of (54, 63 and 72) seconds = 1512 seconds.

∴ A, B and C will meet for the first time at the starting point after 1512 seconds.

Q-37 Find the least number which when divided by 4, 9, 10, 12 and 15 leaves the same remainder 3 in each case.

Sol: The least number which when divided by x, y, z so as to leave the same remainder R in each case. Then,

$$\boxed{\text{Required number} = \text{LCM } (x, y, z) + \text{Remainder (R)}}$$

∴ Required number = LCM (4, 9, 10, 12, 15) + 3 = 180 + 3 = 183.

Q-38 Find the least number which when divided by 16, 20, 25 and 32 leaves the same remainder 12 in each case.

Sol: The least number which when divided by x, y, z leaves the same remainder R in each case. Then,

$$\boxed{\text{Required number} = \text{LCM } (x, y, z) + \text{Remainder (R)}}$$

∴ Required number = LCM (16, 20, 25, 32) + 12 = 800 + 12 = 812.

Q-39 Find the least number which when divided by 12, 24, 45 and 54 leaves remainders 4, 16, 37 and 46 respectively.

Sol: The least number which when divided by x, y, z leaves the remainders R_1, R_2 and R_3 respectively. Then,

$$\boxed{\text{Required number} = \text{LCM } (x, y, z) - \text{Common difference}}$$

Where, common difference = $x - R_1 = y - R_2 = z - R_3$

Common difference = 12 – 4 = 24 – 16 = 45 – 37 = 54 – 46 = 8

∴ Required number = LCM (12, 24, 45, 54) – 8 = 1080 – 8 = 1072.

Q-40 **Find the least number of 5 digits which is exactly divisible by 16, 18, 24 and 32.**

Sol: The least number of n-digits which is exactly divisible by x, y, z. Then,

> Required number = Least n-digit number + LCM – Remainder (R)

Where 'R' is the remainder obtained by dividing least n-digit number with LCM of x, y, z.

Required number = Least 5-digit number + LCM (16, 18, 24, 32) – Remainder (R)

Least 5-digit number = 10000 & LCM (16, 18, 24, 32) = 288

$$
\begin{array}{r}
288) \overline{10000} (34 \\
\underline{864} \\
1360 \\
\underline{1152} \\
208 \rightarrow R
\end{array}
$$

∴ Required number = 10000 + 288 – 208 = 10080.

Q-41 **Find the smallest 6-digit number which when divided by 6, 7, 8, 9, 10 leaves the remainder 5 in each case.**

Sol: The least number of n-digits which when divided by x, y, z so as to leave the same remainder R_1 in each case. Then,

> Required number = (Least n-digit number + LCM – R) + R_1

Where 'R' is the remainder obtained by dividing least n-digit number with LCM of x, y, z and 'R_1' is the remainder mentioned in the given problem.

Required number = (Least 6-digit number + LCM – R) + R_1

LCM (6, 7, 8, 9, 10) = 2520, $R_1 = 5$ and least 6-digit number = 100000.

$$
\begin{array}{r}
2520) \overline{100000} (39 \\
\underline{7560} \\
24400 \\
\underline{22680} \\
1720 \rightarrow R
\end{array}
$$

∴ Required number = (100000 + 2520 – 1720) + 5 = 100805.

Q -42 **Find the smallest 5-digit number which when divided by 12, 16, 20 and 28 leaves the remainders 6, 10, 14 and 22 respectively.**

Sol: The least number of n-digits which when divided by x, y, z so as to leave the remainders R_1, R_2 and R_3 in each case. Then,

Required number = (Least n-digit number + LCM – R) – Common difference

Where 'R' is the remainder obtained by dividing least n-digit number with LCM and

Common difference = $x – R_1 = y – R_2 = z – R_3$

LCM (12, 16, 20, 28) = 1680, least 5-digit number = 10000

Common difference = $12 – 6 = 16 – 10 = 20 – 14 = 28 – 22 = 6$

Required number = (Least 5-digit number + LCM – R) – Common difference

$$1680) 10000 (5$$
$$\underline{8400}$$
$$1600 \rightarrow R$$

∴ Required number = $10000 + 1680 – 1600 – 6 = 10074$.

Q -43 **Find the greatest number which will divide 973 and 1269 so as to leave the same remainder 11 in each case.**

Sol: The greatest number which will divide x, y, z so as to leave the same remainder R in each case. Then,

Required number = HCF $[(x – R), (y – R), (z – R)]$

Required number = HCF $[(973 – 11), (1269 – 11)] =$ HCF $[962, 1258]$

$$962) 1258 (1$$
$$\underline{962}$$
$$296) 962 (3$$
$$\underline{888}$$
HCF ← $74) 296 (4$
$$\underline{296}$$
$$\times$$

∴ Required greatest number = 74 (Last divisor).

Q -44   **Find the greatest number which on dividing 46, 115 and 161 leaves the same remainder.**

Sol: The greatest number which on dividing x, y, z so as to leave the same remainder in each case, then

$$\boxed{\text{Required number} = \text{HCF} \left[|x - y|, |y - z|, |z - x| \right]}$$

Required number = HCF $[|46 - 115|, |115 - 161|, |161 - 46|]$ = HCF $[69, 46, 115]$

```
46) 69 (1              HCF  ←  23) 115 (5
    46                             115
   ───                             ───
   23) 46 (2                        ×
       46
      ───
       ×
```

HCF of 46 and 69 = 23   HCF of 46, 69 and 115 = 23

∴ Required greatest number = 23.

Q -45   **Find the greatest number that will divide 867, 1036, 1167 leaves the remainders 9, 13, 12 respectively.**

Sol: The greatest number which will divide x, y, z so as to leave the remainders R_1, R_2 and R_3 respectively. Then,

$$\boxed{\text{Required number} = \text{HCF} \left[(x - R_1), (y - R_2), (z - R_3) \right]}$$

Required number = HCF $[(867 - 9), (1036 - 13), (1167 - 12)]$ = HCF $[858, 1023, 1155]$

```
858) 1023 (1           HCF  ←  33) 1155 (35
     858                            1155
    ────                            ────
    165) 858 (5                       ×
         825
        ────
         33) 165 (5
             165
            ────
             ×
```

HCF of 858 and 1023 = 33   HCF of 858, 1023 and 1155 = 33

∴ Required greatest number = 33.

Q -46   **Find the greatest number of 5-digits which is exactly divisible by 15, 18, 20, 24.**

Sol: The greatest number of n-digits which when divided by x, y, z leaves no remainder. Then,

$$\boxed{\text{Required number} = \text{Greatest n-digit number} - \text{R}}$$

Where 'R' is the remainder obtained by dividing greatest n-digit number with LCM.

LCM (15, 18, 20, 24) = 360 & Greatest 5-digit number = 99999

$$360) \overline{99999} (277$$
$$\underline{720}$$
$$2799$$
$$\underline{2520}$$
$$2799$$
$$\underline{2520}$$
$$279 \rightarrow R$$

∴ Required number = Greatest 5-digit number – R = 99999 – 279 = 99720.

Q -47   **Find the greatest number of 5-digits which when divided by 5, 10, 25 and 30 leaves the same remainder 4 in each case.**

Sol: The greatest number of n-digits which when divided by x, y, z leaves the same remainder R_1 in each case.

Required number = (Greatest n-digit number – R) + R_1

Where 'R' is the remainder obtained by dividing greatest n-digit number with LCM and 'R_1' is the remainder mentioned in the given problem.

LCM (5, 10, 25, 30) = 150, Greatest 5-digit number = 99999 and R_1 = 4

$$150) \overline{99999} (666$$
$$\underline{900}$$
$$999$$
$$\underline{900}$$
$$999$$
$$\underline{900}$$
$$99 \rightarrow R$$

∴ Required number = (Greatest 5-digit number – R) + R_1 = 99999 – 99 + 4 = 99904.

Q -48   **Find the greatest number of 4-digits which when divided by 4, 5, 6 and 7 leaves the remainder 2, 3, 4 and 5 respectively.**

Sol: The greatest number of n-digits which when divided by x, y, z leaves the remainders R_1, R_2, R_3 respectively. Then,

Required number = (Greatest n-digit number – R) – Common difference

Where 'R' is the remainder obtained by dividing greatest n-digit number with LCM.

Common difference = $x - R_1 = y - R_2 = z - R_3$

LCM (4, 5, 6, 7) = 420, Greatest 4-digit number = 9999

Common difference = 4 – 2 = 5 – 3 = 6 – 4 = 7 – 5 = 2

$$420) \, 9999 \, (23$$
$$\underline{840}$$
$$1599$$
$$\underline{1260}$$
$$\overline{339} \to \text{R}$$

Required number = (Greatest 4-digit number – R) – Common difference

∴ Required number = 9999 – 339 – 2 = 9658.

Q -49 | **The sum of two numbers is 576 and their HCF is 32. Find the number of such pairs which satisfies the above condition.**

Sol: If HCF is 32, then the two numbers must be a multiples of 32.

Consider two numbers are $32x$ and $32y$.

According to question,

$32x + 32y = 576$ $\Rightarrow$ $x + y = 18$

Possible pairs of $x + y = 18$

(1, 17), (2, 16), (3, 15), (4, 14), (5, 13), (6, 12), (7, 11), (8, 10), (9, 9)

Among all the possible pairs we have to take only **"co-prime"** numbers in order to satisfy the given condition.

∴ Required number of pairs which are co-primes are (1, 17), (5, 13), (7, 11) = 3 pairs.

Q -50 | **The product of two numbers is 4332 and their HCF is 19. Find the number of such pairs which satisfies the above condition.**

Sol: Consider, two numbers are $19x$ and $19y$.

According to question,

$19x \times 19y = 4332$ $\Rightarrow$ $xy = 12$

Possible pairs of $xy = 12$ are (1, 12), (2, 6), (3, 4)

Among all the possible pairs we have to take only **"co-prime"** numbers in order to satisfy the given condition.

∴ Required number of pairs which are "co-primes" (1, 12), (3, 4) = 2 pairs.

Q -51 | **Five bells ring at an intervals of 8, 10, 12, 15 and 20 seconds respectively. They start ringing simultaneously. How many times will they ring together in 54 minutes?**

Sol: LCM of 8, 10, 12, 15 and 20 seconds = 120 seconds = 2 minutes

All 5 bells ring simultaneously for every 2 minutes

In 54 minutes, they ring together for $\frac{54}{2} + 1 = 28$ times (including the ring at the starting).

Q -52 **Five bells toll at regular intervals of 4, 6, 9, 10 and 12 seconds respectively. How many times will they toll together in a span of $1\frac{1}{2}$ hour (excluding the toll at the start)?**

Sol: LCM of 4, 6, 9, 10 and 12 seconds = 180 seconds = 3 minutes

All 5 bells toll together for every 3 minutes

∴ In $1\frac{1}{2}$ hour (90 minutes), they toll together (excluding the toll at the start)

for $\frac{90}{3}$ = 30 times.

Q -53 **Five members are participating in a shooting competition. They hit the target once in a every 4, 5, 6, 7 and 8 seconds respectively. If all of them hit the target at 10 AM, then at what time will they hit the target together for the first time after 10 AM?**

Sol: LCM of 4, 5, 6, 7, 8 seconds = 840 seconds = 14 minutes

All 5 members hit the target together after 14 minutes

∴ Required time = 10 AM + 14 minutes = 10:14 AM.

Q -54 **In a petrol bunk, there are 338L, 494L and 650L of petrol is available in three different tanks. What will be the capacity of the largest container to measure the above mentioned petrol in three tanks?**

Sol: The largest capacity of the container to measure the given quantities means HCF of given values.

Required number = HCF [338, 494, 650]

```
338) 494 (1                    26) 650 (25
     338                            52
     156) 338 (2                    130
          312                       130
          26) 156 (6                ×
              156
              ×
```

HCF of 338 and 494 = 26   HCF of 338, 494 and 650 = 26

∴ Largest capacity of the container to measure the given petrol in 3 tanks is 26L.

Q -55 A room is 5 m 46 cm long and 3 m 92 cm broad. It is required to pave the floor with minimum square slabs. Find the number of slabs required.

Sol: To pave the floor with minimum square slabs, then that square slab size must be larger. This problem is an application of HCF.

HCF of 546 cm and 392 cm ∵ 1 m = 100 cm

```
        392) 546 (1
             392
        154) 392 (2          ∴ Number of slabs required = 14.
             308
         84) 154 (1
             84
         70) 84 (1
             70
         14) 70 (5
             70
              ×
```

Q -56 Find the maximum number of children who has 1044 apples and 1479 bananas is to be distributed in such a way that each children will get same number of apples and bananas.

Sol: Maximum number of children means we have to calculate the HCF of 1044 and 1479 to get same number of apples and bananas for each child.

```
        1044) 1479 (1
              1044
         435) 1044 (2      ∴ Required maximum number of children = 87.
              870
         174) 435 (2
              348
          87) 174 (2
              174
               ×
```

Q -57 Find the least multiple of 11, which when divided by 10, 15, 21 and 27 leaves the remainder 6 in each case.

Sol: LCM of 10, 15, 21, 27 = 1890

Required least number = 1890K + 6 which is a multiple of 11.

By trail and error method, take the values for K till the number is multiple of 11.
By taking K = 3, we get 11 multiple.

∴ Required least number = 1890 × 3 + 6 = 5676.

Q-58 **27 apple trees, 63 mango trees and 72 guava trees have to be planted in rows such that each row contains same number of trees of one kind only. Find the minimum number of rows in which the above trees are planted.**

Sol: In order to get the minimum number of rows, then no. of trees in each row must be more. This is an application of HCF.

HCF of 27, 63, 72 = 9

No. of rows of apple trees $= \dfrac{27}{9} = 3$, No. of rows of mango trees $= \dfrac{63}{9} = 7$

No. of rows of guava trees $= \dfrac{72}{9} = 8$

$\therefore$ Minimum number of rows $= 3 + 7 + 8 = 18$.

Q-59 **Three pieces of wood 32 m, 34.6 m and 42.8 m long have to be divided for making planks of same length. Find the largest possible length of each plank.**

Sol: Largest possible length of planks means it is an application of HCF.

HCF of 3200 cm, 3460 cm, 4280 cm $\because$ 1 m = 100 cm

```
3200) 3460 (1                    20) 4280 (214
      3200                           4280
       260) 3200 (12                    ×
            3120
             80) 260 (3
                 240
                  20) 80 (4
                      80
                       ×
```

HCF of 3200, 3460, 4280 = 20 cm

$\therefore$ Largest possible length of plank = 20 cm = 0.2 m.

Q-60 **If in each vessel 6 (or) 7 liters of milk is poured, then 4L were left. So, bigger vessels were taken to pour 9 (or) 10L of milk but still 4L of milk remained. What was the least number of liters of milk to be poured into vessel?**

Sol: By observing the given problem, it is an application of LCM.

Required number = LCM (6, 7, 9, 10) + 4 = 630 + 4 = 634

$\therefore$ Least number of liters of milk to be poured into vessel is 634L.

ASSESSMENT TEST

1. Find the LCM of 1.4, 0.25 and 3.2.

2. Find the HCF of 2.54 and 16.

3. Find the LCM of $\frac{3}{8}$, $\frac{9}{12}$ and $\frac{15}{24}$.

4. Find the HCF of $\frac{21}{4}$, $\frac{49}{6}$, $\frac{7}{10}$.

5. Find the greatest number which divides 966, 1173 and 1472 exactly.

6. Find the LCM of $x^2 + 2x - 8$, $x^2 + 8x + 16$ and $x^2 - 8x + 12$.

7. Find the LCM of $(2^3 \times 5^2 \times 7)$, $(2^2 \times 3^4 \times 5^3)$ and $(3^3 \times 7^2 \times 11^2 \times 13)$.

8. What is the HCF of $(2^5 \times 3^3 \times 7^2 \times 5)$, $(2^4 \times 3^2 \times 5^3)$ and $(2^3 \times 3^4 \times 5 \times 11^2)$?

9. What is the product of LCM and HCF of 48 and 120?

10. The product of two co-prime numbers is 221. Then, find the LCM of those numbers.

11. The HCF of two numbers is 15 and their LCM is 570. If one of the numbers is 75, then find the second number.

12. The LCM of two numbers is 255 and their HCF is 17. If the sum of the numbers is 136, then find their difference.

13. What is the greatest number less than 2200, which is divisible by both 24 and 28?

14. The ratio of two numbers is 4 : 5 and their HCF is 6. What is the LCM of those two numbers?

15. The LCM and HCF of two numbers are 189 and 9 respectively. If the ratio of two numbers is 3 : 7, then find the greatest number.

16. The LCM and HCF of two numbers are 504 and 12 respectively. If the first number is divided by 4, the quotient is 18. Then, find the second number.

17. Three numbers which are co-prime to each other are such that the product of first two numbers is 357 and that of last two is 777, then find the sum of three numbers.

18. LCM of two prime numbers x and y ($x < y$) is 899. What is the value of $5y - 2x$?

19. What is the least number which when tripled will be exactly divisible by 12, 18, 45 and 54?

20. There are four numbers. HCF of each possible pair is 7 and LCM of all the four numbers is 3240. Find the product of all the four numbers.

21. The difference of two numbers is $\frac{1}{7}$ of their sum. The sum of two numbers is 42. Find their LCM.

22. The ratio of two numbers is 6 : 7. If their LCM is 756, then what is their HCF?

23. The LCM of two numbers is 168 and their HCF is 28. If sum of the two numbers is 140, then find the sum of their reciprocals.

24. The sum of LCM and HCF of two numbers is 756 and their LCM is 35 times of their HCF. If one of the numbers is 147, then find the second number.

25. Find the least number which is to be added 3908, so that the number is exactly divisible by 14, 18 and 24.

26. Find the least number which when diminished by 12, is divisible by each one of the numbers 18, 24, 32, 48.

27. What is the least number which when increased by 11, is divisible by each one of the numbers 16, 22, 36 and 42?

28. The LCM and HCF of two numbers are 5168 and 17 respectively. If one of the numbers lies between 200 and 300, then find the two numbers.

29. Five bells toll at regular intervals of 4, 6, 8, 12 and 15 seconds respectively. If they toll together at 9 AM, then at what time will they toll together for the first time after 9 AM?

30. Five bells ring at regular intervals of 4, 5, 6, 8 and 10 seconds respectively. They starts ringing simultaneously. How many times will they ring together in 40 minutes?

31. Find the least number which when divided by 18, 24, 32, 36 leaves the same remainder 15 in each case.

32. Find the least number which when divided by15, 27, 48 and 63 leaves the remainders 8, 20, 41 and 56 respectively.

33. Find the smallest 6-digit number which is exactly divisible by 24, 36, 48 and 54.

34. Find the smallest 5-digit number which when divided by 5, 10, 15, 20 and 25 leaves the same remainder 4 in each case.

35. Find the least 4-digit number which when divided by 4, 8 and 9 leaves the remainders 2, 6 and 7 respectively.

36. Find the greatest number which will divide 463 and 559, so as to leave the same remainder 7 in each case.

37. Find the greatest number which on dividing 51, 119 and 204 leaves the same remainder.

38. Find the greatest number that will divide 1142, 1388, 1517 leaves the remainders 8, 11 and 5 respectively.

39. Find the greatest number of 4 digits which when divided by 5, 6, 8, 10 leaves no remainder.

40. Find the greatest 4–digit number which when divided by 8, 12, 16 and 24 leaves the same remainder 7 in each case.

41. Find the largest 5-digit number which when divided by 8, 9, 10 and 12 leaves the remainders 3, 4, 5 and 7 respectively.

42. Find the least multiple of 5, which when divided by 4, 6, 9 and 12 leaves the remainder 2 in each case.

43. Find the maximum number of families who has 1302 tables and 1581 chairs is to be distributed in such a way that each family will get same number of tables and chairs.

44. 32 banana trees, 48 orange trees and 56 mango trees have to be planted in rows such that each row contains same number of trees of one kind only. Find the minimum number of rows in which the above trees are planted.

45. Three pieces of wood 18.7m, 24.5m and 39.4m long have to be divided for making planks of same length. Find the largest possible length of each plank.

46. A room is 4m 73cm long and 3m 19cm broad. It is required to pave the floor with minimum square slabs. Find the number of slabs required.

47. The sum of two numbers is 532 and their HCF is 28. Find the number of such pairs which satisfy the above condition.

48. The product of two numbers is 6174 and their HCF is 21. Find the number of such pairs which satisfies the given condition.

49. In a go down, there are 252 kg, 322 kg and 392 kg of sugar is available in three different vessels. What will be the maximum capacity of the container which can measure the above mentioned sugar in three vessels?

50. If in each box 4 (or) 5 dozen mangoes is kept, then 3 dozen were left. So, bigger boxes were taken to kept 7 (or) 8 dozen mangoes but still 3 dozen mangoes remained. What was the least number of dozens of mangoes to be kept into the box?

KEY

1. 112
2. 0.02
3. $11\frac{1}{4}$
4. $2\frac{9}{20}$
5. 23
6. $(x-2)(x-6)(x+4)^2$
7. $2^3 \times 3^4 \times 5^3 \times 7^2 \times 11^2 \times 13^1$
8. $2^3 \times 3^2 \times 5^1$
9. 5760
10. 221
11. 114
12. 34
13. 2184
14. 120
15. 63
16. 84
17. 75
18. 97
19. 180
20. 1111320
21. 72
22. 18
23. $\frac{5}{168}$
24. 105
25. 124
26. 300
27. 11077
28. 272 and 323
29. 9:02 AM
30. 21 times
31. 303

32. 15113

33. 100800

34. 10204

35. 1006

36. 24

37. 17

38. 33

39. 9960

40. 9991

41. 99715

42. 290

43. 93

44. 17

45. 0.1 meter

46. 11

47. 9 pairs

48. 2 pairs

49. 14 kg

50. 283 dozen

PERCENTAGES

- ✓ Percent means as the name itself indicating that we are calculating the values for every 100. It is denoted by the symbol '%'.

- ✓ In percentages topic, if we don't know any value, consider that value as 100.

- ✓ To convert any fraction (or) decimal into percentage, then multiply that fraction (or) decimal with 100 and put the percentage symbol (%) at the end.

$$\textbf{Example: } \frac{1}{2} = \frac{1}{2} \times 100\% = 50\%$$

$$0.42 = 0.42 \times 100\% = 42\%$$

- ✓ To convert any percentage into fraction (or) decimal, then divide that value with 100 and remove the percentage symbol (%) at the end.

$$\textbf{Example: } 30\% = \frac{30}{100} = \frac{3}{10} \text{ (or) } 0.3$$

$$45\% = \frac{45}{100} = \frac{9}{20} \text{ (or) } 0.45$$

Fractions and their respective percentages:

Fractions	Percentages
1	100 %
$\frac{1}{2}$	50 %
$\frac{1}{3}$	$33\frac{1}{3}$ %
$\frac{1}{4}$	25 %
$\frac{1}{5}$	20 %
$\frac{1}{6}$	$16\frac{2}{3}$ %
$\frac{1}{7}$	$14\frac{2}{7}$ %
$\frac{1}{8}$	$12\frac{1}{2}$ %

Fractions	Percentages
$\frac{1}{9}$	$11\frac{1}{9}$ %
$\frac{1}{10}$	10 %
$\frac{1}{11}$	$9\frac{1}{11}$ %
$\frac{1}{12}$	$8\frac{1}{3}$ %
$\frac{1}{13}$	$7\frac{9}{13}$ %
$\frac{1}{14}$	$7\frac{1}{7}$ %
$\frac{1}{15}$	$6\frac{2}{3}$ %

❖ Basically, there are three models of problems can be seen in percentages.

Model - 1

➢ What percent of x is y?

➢ What percent is y of x?

➢ y is what percent of x?

✓ Answer for above three questions is same because the meaning of all three statements are same but the way of asking the question is different.

$$\text{Required percentage} = \frac{y}{x} \times 100\%$$

Shortcut:

Simplest way to remember is

$$\text{Required percentage} = \frac{\text{is}}{\text{of}} \times 100\%$$

Note:

Denominator – Number which comes after 'of'.

Numerator – Number which belongs to 'is'.

Example: What percent of 80 is 16?

Sol: Required percentage $= \dfrac{\text{is}}{\text{of}} \times 100\% = \dfrac{16}{80} \times 100\% = 20\%$.

Model - 2

✓ X is what percentage more than Y?

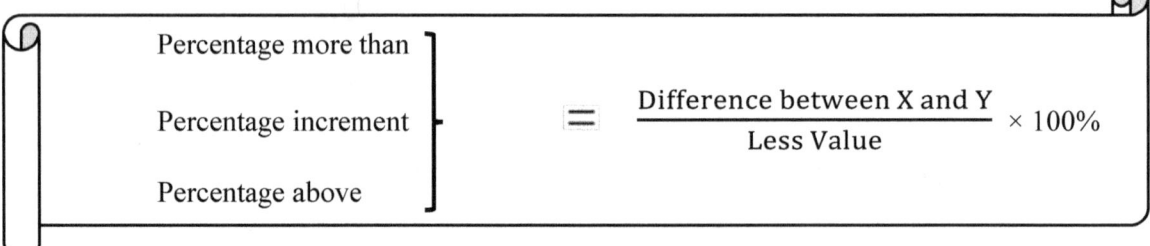

$$\left. \begin{array}{l} \text{Percentage more than} \\[4pt] \text{Percentage increment} \\[4pt] \text{Percentage above} \end{array} \right\} = \frac{\text{Difference between X and Y}}{\text{Less Value}} \times 100\%$$

Example: A is 20% less than B, then by what percent B is more than that of A?

Sol: Percentage more than $= \dfrac{\text{Difference between X and Y}}{\text{Less Value}} \times 100\%$

    A   B
    80   100   ∵ A – 20% less than B

Percentage more than $= \dfrac{100 - 80}{80} \times 100\% = 25\%$.

Model - 3

✓ X is what percentage less than Y?

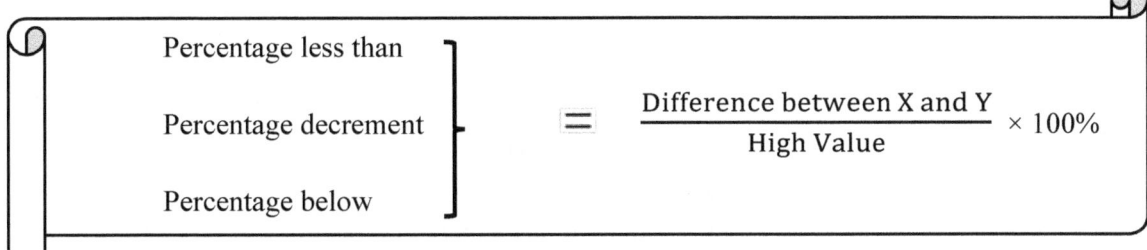

Percentage less than
Percentage decrement $=$ $\dfrac{\text{Difference between X and Y}}{\text{High Value}} \times 100\%$
Percentage below

Example: A is 25% more than B, then by what percent B is less than that of A?

Sol: Percentage less than $= \dfrac{\text{Difference}}{\text{High Value}} \times 100\%$

	A	B	
	125	100	∵ A – 25% more than B

Percentage less than $= \dfrac{125 - 100}{125} \times 100\% = 20\%$.

Some important formulae to be remember:

✓ The value of any number (or) price of any commodity increases (or) decreases by $x_1\%$, $x_2\%$, $x_3\%$ and so on. Then,

New value $=$ Initial value $\times \dfrac{100 \pm x_1}{100} \times \dfrac{100 \pm x_2}{100} \times \dfrac{100 \pm x_3}{100} \times$ ....

'+' means increases & '–' means decreases

> **Note:** This method is applicable for all types of problems, if any value increases or decreases continuously.

✓ The value of any number (or) price of any product is increases (or) decreases by $x\%$ and $y\%$, then

Net effect $= \pm x \pm y + \dfrac{(\pm x)(\pm y)}{100}\%$

Use, Positive sign for increment & Negative sign for decrement.

> **Note:** This method is applicable for the problems with only two variables.

✓ The value of a number first increased by $x\%$ and later decreased by same $x\%$, then the new value is always decreases.

Decreased percentage $= \dfrac{x^2}{100}\%$

✓ The price of product increases by x%, then the percentage decrease in consumption so as to maintain the same expenditure as before. Then,

$$\text{Required percentage} = \frac{x}{100 + x} \times 100\%$$

✓ The price of product decreases by x%, then the percentage increase in consumption so as to maintain the same expenditure as before. Then,

$$\text{Required percentage} = \frac{x}{100 - x} \times 100\%$$

SOLVED EXAMPLES

Q-1 Express the following percentages into fraction.

a) 35%   b) 56%   c) $6\frac{2}{3}\%$ d) $8\frac{1}{3}\%$

Sol: a) $35\% = \dfrac{35}{100} = \dfrac{7}{20}$

b) $56\% = \dfrac{56}{100} = \dfrac{14}{25}$

c) $6\frac{2}{3}\% = \dfrac{20}{3}\% = \dfrac{20}{3} \times \dfrac{1}{100} = \dfrac{1}{15}$

d) $8\frac{1}{3}\% = \dfrac{25}{3}\% = \dfrac{25}{3} \times \dfrac{1}{100} = \dfrac{1}{12}$.

Q-2 What is the value of 15% of $\dfrac{23}{138}$ of 6400 + 28% of $\dfrac{17}{119}$ of 5200?

Sol: $\dfrac{15}{100} \times \dfrac{23}{138} \times 6400 + \dfrac{28}{100} \times \dfrac{17}{119} \times 5200 = 160 + 208 = 368.$

Q-3 The height of A is 50% more than B, then what percent of A is B?

Sol: Let us consider B = 100, then

$A = \dfrac{150}{100} \times B = \dfrac{150}{100} \times 100 = 150$ $\because$ A – 50% more than B

$$\begin{array}{cc} \mathbf{A} & \mathbf{B} \\ 150 & 100 \end{array}$$

Required percentage $= \dfrac{\text{is}}{\text{of}} \times 100\%$

$\therefore$ Required percentage $= \dfrac{100}{150} \times 100\% = \dfrac{2}{3} \times 100\% = 66\frac{2}{3}\%$.

Q-4 There are three natural numbers. The first and second numbers are less than the third number by 30% and 20% respectively. What percent of second number is first number?

Sol: Here, 1st and 2nd numbers are comparing with 3rd number.

Therefore, consider third number T = 100

$$\begin{array}{ccc} \mathbf{F} & \mathbf{S} & \mathbf{T} \\ 70 & 80 & 100 \end{array}$$

$\because$ 1st – 30% less than 3rd

2nd – 20% less than 3rd

Required Percentage $= \dfrac{\text{is}}{\text{of}} \times 100\%$

$\therefore$ Required percentage $= \dfrac{F}{S} \times 100\% = \dfrac{70}{80} \times 100\% = 87.5\%$.

Q-5 A man purchased two products X and Y. The cost of product X is 50% more than Y, then by what percentage is the cost of product Y less than that of product X?

Sol: Consider, cost of product Y = 100

X	**Y**	∵ X – 50% more than Y
150	100	

$$\boxed{\text{Percentage Less than} = \frac{\text{Difference}}{\text{High Value}} \times 100\%}$$

∴ Required percentage less than $= \frac{150 - 100}{150} \times 100\% = 33\frac{1}{3}\%$.

Q-6 There are three persons A, B and C. A's salary is 25% less than B's salary. C's salary is $4\frac{1}{6}\%$ more than B's salary. By what percent A's salary is less than that of C's salary?

Sol: Here, A's salary and C's salary comparing with B's salary.

Consider, B's salary = 100

A	**B**	**C**	∵ A – 25% less than B & C – $4\frac{1}{6}\%$ More than B
75	100	$104\frac{1}{6} = \frac{625}{6}$	

$$\boxed{\text{Percentage Less than} = \frac{\text{Difference}}{\text{High Value}} \times 100\%}$$

Difference

∴ Required percentage less than $= \frac{\frac{625}{6} - 75}{\frac{625}{6}} \times 100\% = 28\%$.

Q-7 The income of Ramesh is $16\frac{2}{3}\%$ less than that of Rahul. By what percentage is the income of Rahul more than that of Ramesh?

Sol: Consider, Rahul income = 100

Ramesh	**Rahul**	
$83\frac{1}{3} = \frac{250}{3}$	100	∵ Ramesh – $16\frac{2}{3}\%$ less than Rahul

$$\boxed{\text{Percentage More than} = \frac{\text{Difference}}{\text{Less Value}} \times 100\%}$$

∴ Required percentage more than $= \frac{100 - \frac{250}{3}}{\frac{250}{3}} \times 100\% = 20\%$.

Q-8 | The expenditures of Prem and Sneha are 52% and 36% less than that of Anvesh respectively. By what percent is the expenditure of Sneha more than that of Prem?

Sol: In this problem expenditures of Prem and Sneha comparing with Anvesh.

Consider, expenditure of Anvesh = 100.

Prem	**Sneha**	**Rahul**	$\because$ Prem – 52% Less than Anvesh
48	64	100	Sneha – 36% Less than Anvesh

Difference

$$\text{Percentage More than} = \frac{\text{Difference}}{\text{Less Value}} \times 100\%$$

$\therefore$ Required percentage more than = $\frac{64-48}{48} \times 100\% = 33\frac{1}{3}\%$.

Q-9 | The price of non-stick pan is Rs. 200 more than that of a dinner set. If the cost of 8 dinner sets and 8 non-stick pans together is Rs. 11200. By what percentage is the price of dinner set less than that of non-stick pan?

Sol: Consider, Price of Non-stick pan = N, Price of dinner set = D

Given that, 8D + 8N = 11200

D + N = 1400 $\Rightarrow$ D + (D + 200) = 1400 $\because$ N = D + 200

$\Rightarrow$ D = 600, N = 800

$$\text{Percentage Less than} = \frac{\text{Difference}}{\text{High Value}} \times 100\%$$

$\therefore$ Required percentage less than = $\frac{800-600}{800} \times 100\% = \frac{200}{800} \times 100\% = 25\%$.

Q-10 | 676 sweets were distributed equally among children in such a way that the number of sweets received by each child is 25% of the total number of children. How many sweets did each child receive?

Sol:

$$\boxed{\text{Total sweets} = \text{Number of children} \times \text{Number of sweets per children}}$$

$676 = n \times (\frac{25}{100} \times n)$ $\Rightarrow$ $\frac{n^2}{4} = 676$ $\Rightarrow$ n = 52

$\therefore$ Number of sweets per children = $\frac{n}{4} = \frac{52}{4} = 13$ sweets.

Q - 11   A student is multiplied a number by $\frac{2}{3}$ instead of $\frac{3}{2}$. Approximately what percentage is the error in the calculation?

Sol: Consider a number $= x$

Wrong answer $= \frac{2}{3}x$, Correct answer $= \frac{3}{2}x$

$$\text{Percentage error} = \frac{\text{Difference}}{\text{correct answer}} \times 100\%$$

∴ Approximate percentage error $= \dfrac{\frac{3}{2}x - \frac{2}{3}x}{\frac{3}{2}x} \times 100\% \cong 55\%.$

Q - 12   If the numerator of a fraction is increased by 20% and its denominator is diminished by 10%, the new fraction is $\frac{28}{27}$. Find the original fraction.

Sol: Consider, Original fraction $= \frac{x}{y}$ ∵ Numerator – increased by 20%

New fraction $= \dfrac{x \times \frac{120}{100}}{y \times \frac{90}{100}} = \dfrac{28}{27}$ ∵ Denominator – decreased by 10%

∴ Original fraction, $\dfrac{x}{y} = \dfrac{7}{9}.$

Q - 13   When the shopkeeper increases the price of an article by 40%, then sales are decreased by 20%. What is the percentage change in Revenue?

Sol: Revenue = Price × Sales

Consider, Initial price = 100, Initial sales = 100

Initial Revenue = 100 × 100 = 10000

New price = 140   ∵ Price – increased by 40%

New Sales = 80   ∵ Sales – decreased by 20%

New Revenue = 140 × 80 = 11200

∴ New Revenue is increases as comparing with initial revenue.

$$\text{Percentage increment} = \frac{\text{Difference}}{\text{Less Value}} \times 100\%$$

∴ New Revenue increment percentage $= \dfrac{11200 - 10000}{10000} \times 100\% = 12\%.$

Shortcut – 1:

$$\text{New Value} = \text{Original value} \times \frac{100 \pm x_1}{100} \times \frac{100 \pm x_2}{100} \times \frac{100 \pm x_3}{100} \times \dots\dots$$

Consider, Original Revenue = 100

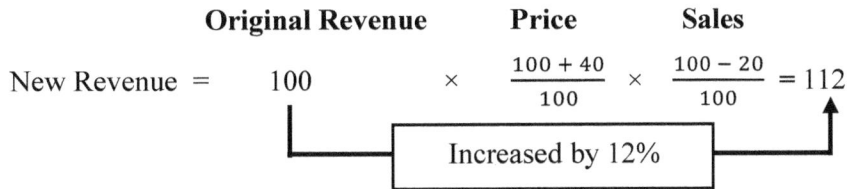

	Original Revenue		**Price**		**Sales**	
New Revenue =	100	×	$\frac{100+40}{100}$	×	$\frac{100-20}{100}$	= 112

Increased by 12%

∴ New Revenue is increased by 12%.

Shortcut – 2:

$$\text{Net effect in percentage} = \pm x \pm y + \frac{(\pm x)\,(\pm y)}{100}\,\%$$

Net effect in revenue $= +40 - 20 + \frac{(+40)(-20)}{100}\% = +12\%$.

Here the answer is positive sign, which indicates revenue is increased by 12%.

Q - 14 | **The base of a triangle is increased by 30% and its height is decreased by 15%. Find the percentage change in the area of triangle.**

Sol: Area of triangle $= \frac{1}{2} \times \text{base} \times \text{height}$

Consider, Initial base = 100, Initial height = 100

Initial Area $= \frac{1}{2} \times 100 \times 100 = 5000$

New base = 130 ∵ Base – increased by 30%

New height = 85 ∵ Height – decreased by 15%

New Area $= \frac{1}{2} \times 130 \times 85 = 5525$.

New area is increases as comparing with initial area

$$\text{Percentage increment} = \frac{\text{Difference}}{\text{Less Value}} \times 100\%$$

∴ Required percentage increment $= \frac{5525 - 5000}{5000} \times 100\% = 10.5\%$.

Shortcut – 1:

Consider original area = 100

	Original Area		**Base**		**Height**	
New Area	=	100	×	$\dfrac{100+30}{100}$	×	$\dfrac{100-15}{100}$ = 110.5

Increased by 10.5%

∴ New Area is increased by 10.5%.

Shortcut – 2:

$$\text{Net effect in Percentage} = \pm x \pm y + \frac{(\pm x)\,(\pm y)}{100}\,\%$$

Net effect in revenue $= +30 - 15 + \dfrac{(+30)(-15)}{100}\,\% = +10.5\%$.

Here the answer is positive sign, which indicates area is increased by 10.5%.

Q - 15 | **A positive integer first decreased by 30% and then increased by 30%. Find the overall change in it?**

Sol: Consider Initial Value = 100

	Initial value		**Decrease**		**Increase**	
New Value	=	100	×	$\dfrac{100-30}{100}$	×	$\dfrac{100+30}{100}$ = 91

Decreased by 9%

∴ New value is decreased by 9%.

Shortcut – 1:

$$\text{Net effect in Percentage} = \pm x \pm y + \frac{(\pm x)\,(\pm y)}{100}\,\%$$

Net effect in positive integer $= -30 + 30 + \dfrac{(-30)(+30)}{100}\,\% = -9\%$.

Here the answer is negative sign, which indicates area is decreased by 9%.

Shortcut – 2:

If increased and decreased percentages are same, then the value is always decreases.

$$\text{Decreased percentage} = \frac{x^2}{100}\,\%$$

Here $x = 30$

∴ Required decreased percentage $= \dfrac{30^2}{100}\,\% = 9\%$.

Q - 16 | The length and breadth of a cuboid are decreased by 10% each and height is increased by 20% then find the percentage change in the volume of cuboid?

Sol: Volume of cuboid = length × breadth × height

Consider initial length = 100, breadth = 100 and height = 100

Initial Volume = 100 × 100 × 100 =1000000

∵ Length – decreased by 10%, Breadth – decreased by 10% & Height – increased by 20%

New Volume = 90 × 90 × 120 = 972000

New volume is decreases comparing with initial volume

$$\text{Percentage decrement} = \frac{\text{Difference}}{\text{High Value}} \times 100\%$$

∴ Percentage decrease in volume = $\frac{1000000 - 972000}{1000000} \times 100\% = 2.8\%$.

Shortcut:

Consider initial volume = 100

	Initial Volume		Length		Breadth		Height	
New Volume =	100	×	$\frac{100 - 10}{100}$	×	$\frac{100 - 10}{100}$	×	$\frac{100 + 20}{100}$	= 97.2%

Decreased by 2.8%

∴ New volume is decreased by 2.8%.

Q - 17 | The price of sugar is increased by 20%. How much percent should a person reduce his consumption in order to maintain the same expenditure as before?

Sol:

Expenditure = Price × Consumption

Consider initial price = 100, Consumption = 100

As per the question, Initial expenditure = New expenditure

Initial Price × Initial Consumption = New Price × New Consumption

$100 \times 100 = (100 + 20)(100 - x)$ ∵ New Price – increased by 20%

$x = 16\frac{2}{3}\%$ New Consumption – decreased by x%

Therefore, consumption is decreased by $16\frac{2}{3}\%$.

Shortcut – 1:

Consider, initial expenditure = 100

New Expenditure		**Initial Expenditure**		**Price**		**Consumption**
100	=	100	×	$\dfrac{100 + 20}{100}$	×	$\dfrac{100 - x}{100}$

$$x = 16\dfrac{2}{3}\%$$

∴ Consumption is reduced by $16\dfrac{2}{3}\%$.

Shortcut – 2:

$$\text{Percentage decrease} = \dfrac{x}{100 + x} \times 100\%$$

Here, $x = 20$

∴ Percentage decrease in consumption $= \dfrac{20}{100 + 20} \times 100\% = 16\dfrac{2}{3}\%$.

Q - 18 **If the length of a rectangle is decreased by 10%, then by what percent should its breadth be increased so as to maintain the same area?**

Sol: Area of rectangle = length × breadth

Consider, initial length = 100 & initial breath = 100

As per the question, Initial Area = New Area

Initial Length × Initial Breadth = New Length × New Breadth

$$100 \times 100 = (100 - 10)(100 + x) \qquad ∵ \text{ New Length – decreased by 10\%}$$

$$x = 11\dfrac{1}{9}\% \qquad\qquad\qquad \text{New Breadth – increased by } x\%$$

Therefore, breadth is increased by $11\dfrac{1}{9}\%$.

Shortcut – 1:

Consider, initial area = 100

New Area		**Initial Area**		**Length**		**Breadth**
100	=	100	×	$\dfrac{100 - 10}{100}$	×	$\dfrac{100 + x}{100}$

$$x = 11\dfrac{1}{9}\%$$

∴ Breadth is increased by $11\dfrac{1}{9}\%$.

Shortcut – 2:

$$\boxed{\text{Percentage increase} = \frac{x}{100 - x} \times 100\%}$$

Here, $x = 10$

∴ Percentage increase in breadth = $\frac{10}{100 - 10} \times 100\% = 11\frac{1}{9}\%$.

Q - 19   **Mahi spends 20% of his income on house rent, 15% on medical bills, 25% on children education and 10% on miscellaneous expenses. Finally Mahi left with Rs. 15000, then find his income.**

Sol:

$$\boxed{\text{Income} = \text{Expenditure} + \text{Savings}}$$

Consider, Mahi total income = 100%

Total expenditures = 20% + 15% + 25% + 10% = 70%

Remaining income left = 100% – 70% = 30%

$$30\% \longrightarrow \text{Rs. } 15000$$

∴ Mahi total income = 100% $\longrightarrow$? $= \frac{15000 \times 100}{30} = $ Rs. 50000.

Q - 20   **Deepu spent 19% of amount with her on food items, 16% on clothes and 15% on transport. 20% of the remaining amount she spent on books and the remaining Rs. 24000 was kept aside for savings. What was the total amount with Deepu?**

Sol: Consider, Deepu amount = 100%

Initial expenditures = 19% + 16% + 15% = 50%

Remaining amount = 100% – 50% = 50%

Expenditures on books = 20% of remaining amount = $\frac{20}{100} \times 50\% = 10\%$

Total expenditures = 50% + 10 % = 60%

Remaining amount left = 100% – 60% = 40%

$$40\% \longrightarrow \text{Rs. } 24000$$

∴ Total amount of Deepu = 100% $\longrightarrow$? $= \frac{24000 \times 100}{40} = $ Rs. 60000.

Q - 21   Varun spent 15%of amount with him on groceries, 25% of the balance he spent on medical bills and 35% of the remaining he spent on children education and finally Varun left with Rs. 23205. How much amount does Varun has initially?

Sol:

> **Note:** If the question consists of the words like rest, balance (or) remaining, then we can use increase (or) decrease method.

Consider, Varun initial amount = Rs. x

Remaining value = Initial value $\times \dfrac{100 \pm x_1}{100} \times \dfrac{100 \pm x_2}{100} \times \dfrac{100 \pm x_3}{100} \times \ldots\ldots$

$23205 = x \times \dfrac{100 - 15}{100} \times \dfrac{100 - 25}{100} \times \dfrac{100 - 35}{100}$

$\therefore$ Varun initial amount x = Rs. 56000.

Q - 22   The value of machine is increases by 10% per annum. If its present value is Rs. 260150, then what was the value of machine two years ago?

Sol: Consider, value of machine two years ago = Rs. x

Present value = 2 years ago value $\times \dfrac{100 + 10}{100} \times \dfrac{100 + 10}{100}$

$260150 = x \times \dfrac{110}{100} \times \dfrac{110}{100}$ $\because$ Value – increases by 10%

$\therefore$ Value of machine two years ago x = Rs. 215000.

Q - 23   The value of machine is depreciates by 10% per annum. If the present value of machine is Rs. 400000, then what will be the value of machine after three years?

Sol: Consider, Value of machine after 3 years = Rs. x

After 3 years value x = Present value $\times \dfrac{100-10}{100} \times \dfrac{100-10}{100} \times \dfrac{100-10}{100}$

$x = 400000 \times \dfrac{90}{100} \times \dfrac{90}{100} \times \dfrac{90}{100} = 291600$ $\because$ Value – decreased by 10%

$\therefore$ Value of machine after three years = Rs. 291600.

Q - 24   The population of a town is increased by 15% per annum. If the present population is 240000, then what will be the population after 2 years?

Sol: Consider, population of a town after 2 years = x

Population after 2 years = Present population $\times \dfrac{100 + 15}{100} \times \dfrac{100 + 15}{100}$

$x = 240000 \times \dfrac{115}{100} \times \dfrac{115}{100} = 317400$ $\because$ Population – increased by 15%

$\therefore$ Population of a town after two years is 317400.

Q - 25 The population of a town is decreased by 12% during first year and increased by 25% during second year. If the present population is 501600, then what was the population 2 years ago?

Sol: Consider, population of a town 2 years ago = x

Present population = 2 years ago population $\times \frac{100-12}{100} \times \frac{100+25}{100}$

$501600 = x \times \frac{88}{100} \times \frac{125}{100}$ ∵ 1st year – decreased by 12% & 2nd year – increased by 25%

∴ Population of a town two years ago x is 456000.

Q - 26 The population of a city is decreased by 15% during first year, increased by 25% during second year and decreased by 10% during third year. What will be the population of city 3 years hence, if the present population is 576000?

Sol: Consider, population of a city 3 years hence = x

Population 3 years hence = Present population $\times \frac{100-15}{100} \times \frac{100+25}{100} \times \frac{100-10}{100}$

$x = 576000 \times \frac{85}{100} \times \frac{125}{100} \times \frac{90}{100} = 550800$

∵ 1st yr – decreased by 15%, 2nd yr – increased by 25% & 3rd yr – decreased by 10%

∴ Population of a city three years hence is 550800.

Q - 27 Sneha scored 64 marks in English, 90 marks in Mathematics, 83 marks in Telugu, 75 marks in Hindi, 69 marks in Science and 87 marks in Social Studies. The maximum marks in each subject is 100, then what percentage of marks scored by Sneha in all subjects together?

Sol: Percentage of marks = $\frac{\text{Total obtained marks}}{\text{Total maximum marks}} \times 100\%$

Percentage of marks = $\frac{64 + 90 + 83 + 75 + 69 + 87}{6 \times 100} \times 100\% = 78\%.$

Q - 28 In an exam, a student got 24% of marks and fails by 38 marks. Another student got 39% of marks and got 52 marks more than the pass mark. Find the maximum marks and pass marks in the exam.

Sol: Maximum marks in the exam = 100%

Pass marks of first student = Pass marks of second student

24% + 38 marks = 39% – 52 marks $\Rightarrow$ 15% = 90 marks $\Rightarrow$ 1% = 6 marks

∴ Maximum marks = 100% = 6 × 100 = 600 marks.

Pass marks = 24% + 38 marks (or) 39% – 52 marks

∴ Pass marks = 24 × 6 + 38 (or) 39 × 6 – 52 = 182 marks.

Shortcut:

We can solve this type of problems easily by representing given values diagrammatically.

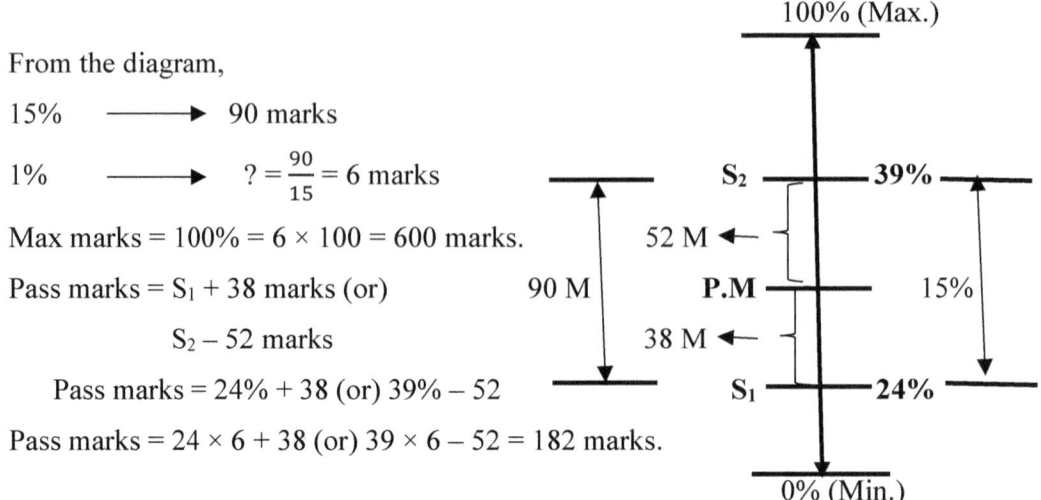

From the diagram,

15% $\longrightarrow$ 90 marks

1% $\longrightarrow$ $? = \dfrac{90}{15} = 6$ marks

Max marks = 100% = 6 × 100 = 600 marks.

Pass marks = S_1 + 38 marks (or)

S_2 − 52 marks

Pass marks = 24% + 38 (or) 39% − 52

Pass marks = 24 × 6 + 38 (or) 39 × 6 − 52 = 182 marks.

Q - 29 **In an exam, 40% and 35% students passed in physics and chemistry respectively. 19% students are passed in both the subjects, if the number of students who failed in the exam are 352. Find the total number of students who appeared in the exam.**

Sol:

Given that, pass percentage in physics n (p) = 40%

Pass percentage in chemistry n (c) = 35%

Pass percentage in both subjects n (p ∩ c) = 19%

Total pass percentage n (p ∪ c) = n (p) + n (c) − n (p ∩ c)

n (p ∪ c) = 40% + 35% − 19% = 56%

Consider appeared students = 100%

Fail percentage = 100% − 56% = 44%   ∵ Fail = Appeared − Pass

Number of students failed   = 44% $\longrightarrow$ 352

∴ Number of students appeared = 100% $\longrightarrow$ $? = \dfrac{352 \times 100}{44} = 800.$

Shortcut:

The simplest method to solve this type of problems by using Venn diagrams.

Pass percentage

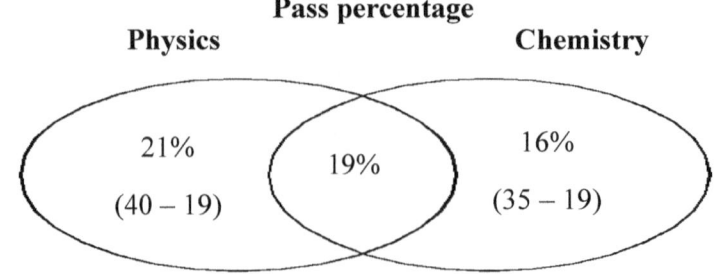

From the diagram,

$$\text{Total pass percentage} = 21\% + 19\% + 16\% = 56\%$$

$$\text{Total fail percentage} = 100\% - 56\% = 44\%$$

Failed students   =   44% $\longrightarrow$ 352

Appeared students   =   100% $\longrightarrow$ $? = \dfrac{352 \times 100}{44} = 800$

Therefore, total number of appeared students = 800.

Q - 30 | **In an exam, Sonu got 156 marks which was 16 marks more than the pass marks. Rekha got 64% of marks which was 116 marks more than the pass marks. What is the minimum pass percentage in the exam?**

Sol: Sonu marks = Pass marks + 16 marks = 156 marks

Pass marks = 156 – 16 = 140 marks

Rekha percentage of marks = Pass marks + 116 marks

64% = 140 + 116 = 256 marks

256 marks $\longrightarrow$ 64%

Pass marks = 140 marks $\longrightarrow$ $? = \dfrac{64 \times 140}{256} = 35\%$

Therefore, minimum pass percentage is 35%.

Q - 31 | **Manu gave 20% of amount available with him to Kiran, 25% of remaining to Srikanth, 35% of remaining to Raju and 50% of remaining to Abhi. Which person got the highest amount?**

Sol: Consider initial amount with Manu = Rs. 1000

Kiran = 20% of 1000 = $\dfrac{20}{100} \times 1000$ = Rs. 200,   Remaining amount = 1000 – 200 = Rs. 800

Srikanth = 25% of 800 = $\dfrac{25}{100} \times 800$ = Rs. 200,   Remaining amount = 800 – 200 = Rs. 600

Raju = 35% of 600 = $\dfrac{35}{100} \times 600$ = Rs. 210,   Remaining amount = 600 – 210 = Rs. 390

Abhi = 50% of 390 = $\dfrac{50}{100} \times 390$ = Rs. 195,   Remaining amount = 390 – 195 = Rs. 195

∴ Among all the persons "**Raju**" got the highest amount of Rs. 210.

Q - 32 Chandu secured 65% of votes in an election and was elected by a majority of 16920 votes. All votes polled were valid, then find the number of votes polled, if there are only two contestants Chandu and Madhu.

Sol:

$$\boxed{\text{Total number of votes} = \text{Valid votes} + \text{Invalid votes}}$$

In this problem, invalid votes = 0   ∵ All votes polled were valid

Valid votes (100%)   Given that, majority votes = 16920

Chandu   Madhu   Majority = 30% ⟶ 16920 votes

65%   35%   Total valid = 100% ⟶ ?

Majority 30%   Total valid votes $= \dfrac{16920 \times 100}{30} = 56400$

∴ Total number of votes = 56400.

Q - 33 In an election there are two contestants A and B. It was found that 9425 votes were invalid. B polled 26% of valid votes and lost the election by 42960 votes. Find total number of votes polled in an election.

Sol:

$$\boxed{\text{Total number of votes} = \text{Valid votes} + \text{Invalid votes}}$$

Valid votes (100%)   Given that, Invalid votes = 9425, B lost by 42960 votes

A   B   B lost by = 48% ⟶ 42960 votes

74%   26%   Valid votes = 100% ⟶ ?

Lost by 48%   Total valid votes $= \dfrac{42960 \times 100}{48} = 89500$

∴ Total number of votes = 89500 + 9425 = 98925 votes.

Q - 34 In an election between two candidates, 80% of the voters cast their votes, out of which 5% of votes were invalid. A candidate get 24225 votes which were 85% of total valid votes. Find the total number of votes enrolled in that election?

Sol:

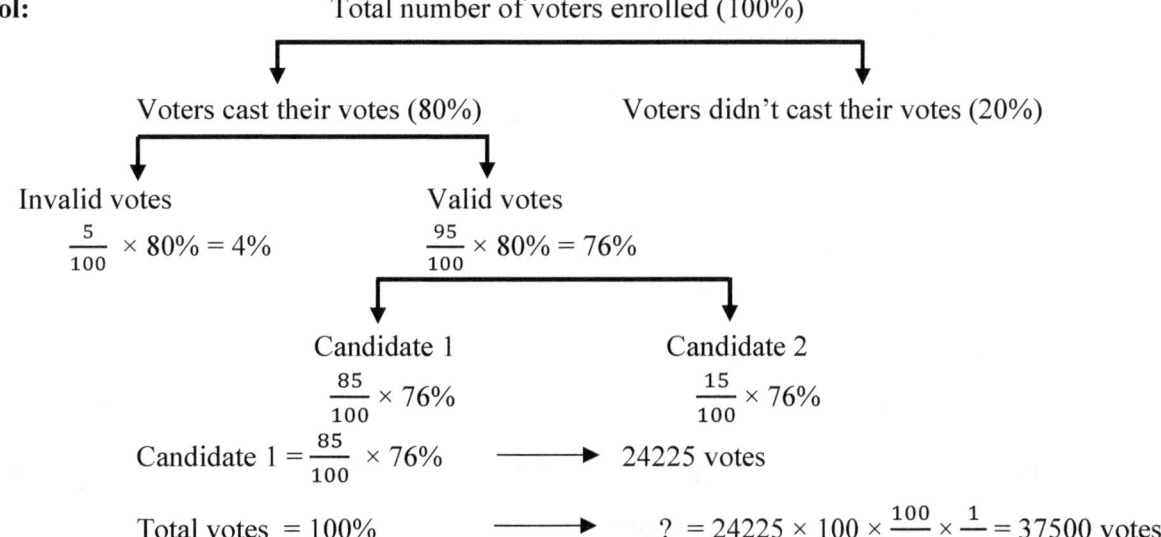

Total number of voters enrolled (100%)

Voters cast their votes (80%)   Voters didn't cast their votes (20%)

Invalid votes   Valid votes
$\frac{5}{100} \times 80\% = 4\%$ $\frac{95}{100} \times 80\% = 76\%$

Candidate 1   Candidate 2
$\frac{85}{100} \times 76\%$ $\frac{15}{100} \times 76\%$

Candidate $1 = \frac{85}{100} \times 76\%$ $\longrightarrow$ 24225 votes

Total votes $= 100\%$ $\longrightarrow$ $? = 24225 \times 100 \times \frac{100}{85} \times \frac{1}{76} = 37500$ votes.

$\therefore$ Total number of voters enrolled in an election are 37500.

Shortcut:

Consider, total number of voters $= x$

	Total	**Casting**	**Valid**	**Candidate 1**	
No. of votes for candidate 1 $=$	x $\times$	$\frac{80}{100}$ $\times$	$\frac{100-5}{100}$ $\times$	$\frac{85}{100}$	$=$ 24225

$x = 24225 \times \frac{100}{80} \times \frac{100}{95} \times \frac{100}{85} = 37500$

$\therefore$ Total number of voters enrolled in an election are 37500.

Q - 35 **There are two candidates participated in an election. It was found that 25% of votes were invalid. One who won the election got 70% of total valid votes, if the total number of votes are 84000. Find the number of valid votes that the other candidate got.**

Sol:

$$\boxed{\text{Total number of votes} = \text{Valid votes} + \text{Invalid votes}}$$

Invalid votes $= 25\%$ of total votes $= \frac{25}{100} \times 84000 = 21000$

Valid votes $=$ Total votes $-$ Invalid votes $= 84000 - 21000 = 63000$

One candidate got 70% of valid votes $= \frac{70}{100} \times 63000 = 44100$

$\therefore$ No. of valid votes for other candidate $= 63000 - 44100 = 18900$.

Shortcut:

Consider, total number of voters $= x$

	Total	**Valid**	**Candidate 2**	
No. of valid votes for 2nd candidate $=$	84000 $\times$	$\frac{75}{100}$ $\times$	$\frac{100-70}{100}$	$=$ 18900.

❖ We can also represent this problem diagrammatically.

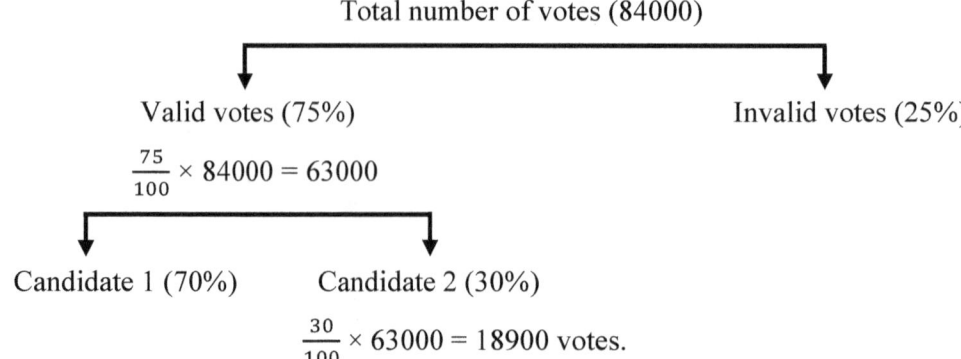

Total number of votes (84000)

Valid votes (75%)   Invalid votes (25%)

$\frac{75}{100} \times 84000 = 63000$

Candidate 1 (70%)   Candidate 2 (30%)

$\frac{30}{100} \times 63000 = 18900$ votes.

Q - 36 **The price of tea powder is increased by 25%. How many kilograms of tea powder can be bought now with the money, which was sufficient to buy 40kg of tea powder previously?**

Sol:

$$\boxed{\text{Total cost} = \text{Price per kg} \times \text{Quantity}}$$

As per the question, there is no change in cost before and after increment of price.

Initial cost = New cost

Initial price/kg × Initial Quantity = New price/kg × New Quantity

Consider, initial price/kg = 100

$100 \times 40 = (100 + 25) \times$ New Quantity   ∵ Price – increased by 25%

∴ New quantity after increment of price = 32 kg.

Q - 37 **A reduction of 20% in the price of sugar enables a shopkeeper to buy 15 kgs more for Rs. 3000. Find the reduced price of sugar per kg.**

Sol: Reduction in price = 20% of 3000 = $\frac{20}{100} \times 3000$ = Rs.600

Because of reduction of Rs. 600 Shopkeeper got 15 kgs more

Reduced price per 15 kg ⟶ Rs. 600

∴ Reduced price per kg ⟶ ? = $\frac{600}{15}$ = Rs. 40.

Q - 38 | The price of wheat is decreased by $16\frac{2}{3}$ % due to this reduction Madhuri can able to purchase 8 kg more for Rs. 2400. What is the original price of wheat per kg?

Sol: Reduction in price = $16\frac{2}{3}$ % of 2400 = $\frac{50}{3} \times \frac{1}{100} \times 2400$ = Rs. 400

Because of reduction of Rs. 400 Madhuri got 8 kgs more.

Reduced price of wheat per 8 kg $\longrightarrow$ Rs. 400

Reduced price of wheat per kg $\longrightarrow$? = $\frac{400}{8}$ = Rs. 50.

Original price	Reduced price
100%	$100\% - 16\frac{2}{3}\% = 83\frac{1}{3}\% = \frac{250}{3}\%$

Reduced price per kg = $\frac{250}{3}$ % $\longrightarrow$ Rs. 50

Original price per kg = 100% $\longrightarrow$? = $50 \times 100 \times \frac{3}{250}$ = Rs. 60

Therefore, Original price of wheat per kg = Rs. 60.

Q - 39 | Two – third of first number is equal to 40% of second number. The second number is equal to five – seventh of the third number. The value of third number is 6370. What is 35% of first number?

Sol: Given that, $\frac{2}{3} \times F = \frac{40}{100} \times S$ and $S = \frac{5}{7} \times T$ and T = 6370

$\because$ F – First number, S – Second number & T – Third number

$S = \frac{5}{7} \times 6370 = 4550$

$\frac{2}{3} \times F = \frac{40}{100} \times 4550 \qquad \Rightarrow \qquad F = 2730$

Hence, 35% of first number = $\frac{35}{100} \times 2730 = 955.5$.

Q - 40 | 18% of Amar's monthly salary is equal to 25% of Bhargav's monthly salary. Bhargav's monthly salary is three – fourth of Charan's monthly salary. If Charan's annual salary is 4.32 lakhs, then what is Amar's monthly salary?

Sol: Given that, 18% of A = 25% of B, $B = \frac{3}{4} \times C$ and C's annual salary = Rs. 4,32,000

C's monthly salary = $\frac{432000}{12}$ = Rs. 36000 $\qquad \because$ A – Amar, B – Bhargav & C – Charan

$B = \frac{3}{4} \times C = \frac{3}{4} \times 36000$ = Rs. 27000

$\frac{18}{100} \times A = \frac{25}{100} \times 27000 \qquad \Rightarrow \qquad A$ = Rs. 37500

Therefore, Amar's monthly salary = Rs. 37500.

Q - 41   **In a function, 60% of people who attended are men. If 50% of men and 58% of total number of persons had left the function, then what percent of women had left the function?**

Sol: Consider, total number of persons = 100

Total number of men = $\frac{60}{100} \times 100 = 60$,   Total number of women = $100 - 60 = 40$

Total number of persons left the functions = $\frac{58}{100} \times 100 = 58$

Number of men left = $\frac{50}{100} \times 60 = 30$, Number of women left = $58 - 30 = 28$

Percentage of women left the function = $\frac{\text{Women left the function}}{\text{Total no.of women}} \times 100\%$

$$= \frac{28}{40} \times 100\% = 70\%.$$

Q - 42   **The expenditure of Balu is 80% of his total income. If his income increases by 20% and expenditure increases by 30%, then by what percent does his savings increase or decrease?**

Sol: Consider, initial income = 100 | Savings = Income – Expenditure |

	Income	Expenditures	Savings
Initial	100	80	20
	↓(20% more)	↓ (30% more)	
New	$100 \times \frac{120}{100} = 120$	$80 \times \frac{130}{100} = 104$	$120 - 104 = 16$

New savings are decreases as comparing with initial savings

| Percentage decrement = $\frac{\text{Difference}}{\text{High Value}} \times 100\%$ |

Therefore, percentage decrease in savings = $\frac{(20-16)}{20} \times 100\% = 20\%$.

Q - 43   **A salesman's commission is 10% on all sales upto Rs. 20000 and 8% on all sales exceeding Rs. 20000. He remits Rs. 58940 to his parent company after deducting his commission. Find the total sales.**

Sol: Consider, total sales = Rs. x

Remaining amount = Total sales amount – Commission

$58940 = x - [\frac{10}{100} \times 20000 + \frac{8}{100} \times (x - 20000)]$

∴ Total sales x = Rs. 64500.

Q - 44 | Due to an increase of 25% in the price of bananas, 5 bananas less are available for Rs. 140. Find the present rate of bananas per dozen.

Sol: Increase in price = 25% of 140 = $\frac{25}{100} \times 140$ = Rs. 35

Because of increase of Rs. 35, 5 bananas less are available

Present rate of 5 bananas $\longrightarrow$ Rs. 35

Present rate of 12 bananas $\longrightarrow$? $= \frac{(35 \times 12)}{5}$ = Rs. 84

∴ Present rate of bananas per dozen = Rs. 84.

Q - 45 | Fresh fruit contains 75% water and dry fruit contains 48% water. How much dry fruit can be obtained from 52kg of fresh fruits?

Sol:

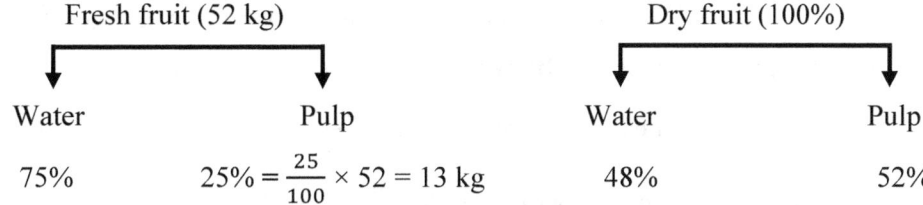

Fresh fruit (52 kg)		Dry fruit (100%)	
Water	Pulp	Water	Pulp
75%	25% = $\frac{25}{100} \times 52$ = 13 kg	48%	52%

Note: Pulp weight must be same in both fresh fruits and dry fruits.

Dry fruit pulp weight = 52% $\longrightarrow$ 13 kg

Total weight of dry fruit = 100% $\longrightarrow$? $= \frac{13 \times 100}{52}$ = 25 kg.

Q - 46 | A large watermelon weighs 24 kg with 90% of its weight being water. It is allowed to stand in the sun and some of the water evaporates, so that only 88% of its weight is water. What is its reduced weight?

Sol:

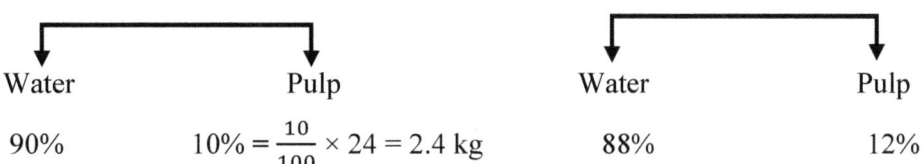

Before kept in sun watermelon (24 kg)		After kept in sun watermelon (100%)	
Water	Pulp	Water	Pulp
90%	10% = $\frac{10}{100} \times 24$ = 2.4 kg	88%	12%

Note: Pulp weight must be same in both the cases.

After kept in sun pulp weight = 12% $\longrightarrow$ 2.4 kg

Total weight = 100% $\longrightarrow$? $= \frac{2.4 \times 100}{12}$ = 20 kg

∴ Reduced weight of watermelon = 20 kg.

Q - 47 A certain number of oranges were bought from a fruit market. It is found that 20% of them are rotten and the remaining are distributed among A and B such that A gets 50% more oranges than B. What percentage of total number of oranges did B get?

Sol: Consider, total number of oranges bought = 100

No. of oranges rotten = 20% of total = $\frac{20}{100} \times 100 = 20$, Remaining oranges = $100 - 20 = 80$

Given that, A + B = 80, A = $\frac{150}{100} \times$ B   ∵ A – 50% more than B

$\frac{150}{100} \times$ B + B = 80   ⇒   B = 32

Therefore, B gets 32% of total number of oranges.

Q - 48 The price of car is Rs. 575000. It was insured 80% of its price. The car was met with an accident and it is completely damaged. The insurance company paid 85% of the insurance. What was the difference between the price of car and the amount received?

Sol: Insured price of car = 80% of 575000 = $\frac{80}{100} \times 575000$ = Rs. 460000

Amount received = 85% of 460000 = $\frac{85}{100} \times 460000$ = Rs. 391000

Difference between price of car and amount received = 575000 – 391000 = Rs. 184000.

Shortcut:

	Price of car	Insurance	Amount received		
Amount received =	575000	× $\frac{80}{100}$	× $\frac{85}{100}$	=	Rs. 391000

Difference between price of car and amount received = 575000 – 391000 = Rs. 184000.

Q - 49 Shyam went to stationary shop and bought some things worth Rs. 3000 out of which 48 rupees went on sales tax on taxable purchases. If the tax rate was 12%, then what was the cost of the tax free items?

Sol: Consider, taxable purchases = Rs. x

$x \times \frac{12}{100} = 48$ ⇒ x = Rs.400

∴ Cost of tax-free items = 3000 – (400 + 48) = Rs. 2552.

Q - 50 | In an exam there are 140 questions. Each correct answer gets 4 marks, for each wrong answer 2 marks are deducted and for each unattempted question 1 mark is deducted. A student answered 60% of total number of questions correctly and 15% of total number of questions are not attempted, then find the percentage of marks obtained by student in the exam?

Sol:

$$\text{Percentage of marks obtained} = \frac{\text{Marks obtained}}{\text{Total maximum marks}} \times 100\%$$

Total maximum marks = $140 \times 4 = 560$ marks

No. of questions correctly answered $= \frac{60}{100} \times 140 = 84$

No. of questions unattempted $= \frac{15}{100} \times 140 = 21$

No. of questions wrongly answered $= \frac{25}{100} \times 140 = 35$

Marks Obtained $= 84 \times 4 - 21 \times 1 - 35 \times 2 = 245$

$\therefore$ Percentage of marks obtained $= \frac{245}{560} \times 100\% = 43\frac{3}{4}\%$.

ASSESSMENT TEST

1. Express the following decimals into percentages.

    a) 0.4   b) 0.32   c) 1.6   d) 0.75

2. Find the value of 24 of 32% of $\frac{19}{96}$ of 4200.

3. Find the value of 29% of $\frac{11}{174}$ of 3600 + 14% of $\frac{16}{112}$ of 2650.

4. If 30% of A is added to 50% of B, the answer is 70% of B. What percentage of A is B?

5. Find the value of 48% of $\frac{7}{12}$ of 3800.

6. First number is 60% more than second number, then second number is what percent of first number?

7. The savings of Anu is $33\frac{1}{3}$% more than that of Manu, then by what percentage is the savings of Manu is less than that of Anu?

8. Three students Rahul, Krishna and Balu. Rahul's marks are 14% more than Balu's marks. Krishna's marks are 5% less than Balu's marks. By what percentage Krishna's marks are less than that of Rahul's marks?

9. The salary of Naveen is 20% less than that of Praveen. By what percentage is the salary of Praveen more than that of Naveen?

10. The weights of A and B are 40% and 34% less than that of weight of C respectively, then by what percentage is the weight of B more than that of weight of A?

11. If the numerator of a fraction is increased by 15% and its denominator decreased by 4%, the new fraction thus obtained is $\frac{23}{24}$. What is the original fraction?

12. The population of a city increased from 342500 to 513750 in a decade. Find the average percent increase of population per year.

13. The length of a rectangle is decreased by 40% and its breadth is increased by 20%. Find the percentage change in its area.

14. The radius of a cylinder is increased by 20% and height is decreased by 20%. Find the percentage change in the volume of cylinder.

15. There are three natural numbers. The first and second numbers are more than the third number by 20% and 50% respectively. What percent of second number is first number?

16. The price of wheat has increased by 40%. By what percentage must a consumer reduce the consumption of wheat, so as not to increase the expenditure?

17. The price of petrol is decreased by 25%. By what percentage should a person increase his consumption of petrol, so as to restore the same expenditure as before?

18. A batsman scored 180 runs which includes 13 boundaries and 8 sixes. What percent of his total score did he make by running between the wickets?

19. In a school, 30% of students are below 10 years. The number of students above 10 years is 3 times the number of students of 10 years age which is 140. What is the total number of students in the school?

20. Arun spent 30% of his income on furniture items, 20% of remaining on shopping and 10% of the rest he invests on share market. After all these expenditures finally, he left with Rs. 17640. What is the income of Arun?

21. Rohit spends 28% of his salary on electronic goods, 20% of the remaining he spends on house rent and 16% of the rest he gave to charity and finally he left with Rs. 24192. What is the salary of Rohit?

22. The value of machine is increases by 15% per annum. If its present value is Rs. 520000, then find the value of machine after two years.

23. The value of sports bike is decreases by 20% per annum. If the present value of bike is Rs. 166400, then what was the value of bike 3 years ago?

24. The population of a village is increases by 20% during first year and decreases by 10% during second year. What percent of population increase (or) decrease after two years?

25. The population of a town is increased by 8% per annum. Find population of a town two years ago, if the present population is 402408.

26. The population of a metro city is increases by 14% during first year, decreases by 25% during second year and increases by 12% during third year. If the present population is 900000, then what will be the population after 3 years?

27. If the present population of a small town is 186300. The population of town is decreased by 10% during first year, increased by 15% during second year and increased by 20% during third year. What was the population of town before 3 years?

28. Navya scored 85 marks in English, 91 marks in Hindi, 97 marks in Mathematics, 82 marks in Science, 92 marks in Social studies. The maximum marks in each subject is 100, then what percentage of marks scored by Navya in all subjects together?

29. In a test, 19% and 18% students failed in Geography and Biology respectively, while 12% students failed in both the subjects. If the number of students passing the exam is 675, then find the total number of students who appeared for the exam.

30. In an exam, a student got 32% of marks and fails by 52 marks. Another student got 56% of marks and got 68 marks more than the pass mark. Find the maximum marks and pass marks in the exam.

31. Teja got 174 marks and fails by 46 marks in a certain exam. Uma got 65% of marks which was 105 marks more than the pass mark. What is the minimum passing percentage in that exam?

32. Bharath and Ram are the only two persons participated in an election. Bharath polled 82% of valid votes and was elected by a majority of 59520 votes. It was found that 5864 votes were invalid, then find the total number of votes polled in an election.

33. In an election between two candidates, 15% of voters did not cast their votes. 10% of the votes polled were invalid. The successful candidate got 60% of the valid votes and won the election by a majority of 4437 votes. Find the number of voters enrolled.

34. In an election, 40% of voters voted for candidate A whereas 80% of the remaining voted for candidate B. The remaining voters did not vote. If the difference between who voted for A and those who did not vote is 18200. Find the number of voters were eligible for casting vote in that election.

35. In an examination, 10% of the candidates were found ineligible and 80% of the eligible candidates belongs to general category. If 10080 eligible candidates belongs to other categories, then how many candidates applied for the examination?

36. Karthik's monthly income is twice that of Sravan's monthly income. Sravan's monthly income is 15% less than that of Akshay's monthly income. Akshay's annual income is Rs. 312000, then find Karthik's monthly income.

37. An iron rod is cut into two pieces. The length of longer piece is 60% of the length of the rod. By what percentage is the shorter piece is less than that of longer piece?

38. A reduction of $33\frac{1}{3}$ % in the price of an apple enables a person to buy 9 apples more for Rs. 450. Find the original price of apples per dozen.

39. Fresh grapes contain 85% water, whereas dry grapes contain 35% water. If the weight of dry grapes is 600 kilograms, then what is the total weight of fresh grapes?

40. If the sales tax be reduced from $4\frac{3}{4}$ % to $4\frac{2}{3}$ %, then what difference does it make to a person who purchases an article with marked price of Rs. 10800?

41. 500 sarees were being transported 125 of them got damaged. What percentage did the damaged sarees form of those which were not damaged?

42. Mr. Sravan spends 15% of his income on rent, 20% of the remaining income on food, 10% of the remaining on education and 12% of the remaining on medical expenses. If he saves Rs. 84150, then what is his income?

43. A teacher distributes some sweets among his four students A, B, C and D. B gets 10% less than A. C gets 20% more than B. D gets 25% more than C. By what percent is the number of sweets received by D more than that of A?

44. A policeman rejects 0.04% of the meters as defective. How many will he examine to reject 3?

45. In an exam there are 120 questions. Each correct answer gets 3 marks. For each wrong answer 1 mark is deducted and no marks are given for questions left unanswered. A student answered 65% of total number of questions correctly and 10% of total number of questions are not attempted, then find the percentage of marks obtained by the student in the exam.

KEY

1. a) 40%   b) 32%   c) 160%   d) 75%
2. 6384
3. 119
4. 150%
5. 1064
6. 62.5%
7. 25%
8. $16\frac{2}{3}$ %
9. 25%
10. 10%
11. $\frac{4}{5}$
12. 5%
13. 28% decreases
14. 15.2% increases
15. 80%
16. $28\frac{4}{7}$ %
17. $33\frac{1}{3}$ %
18. $44\frac{4}{9}$ %
19. 800
20. Rs. 35000
21. Rs. 50000
22. Rs. 687700
23. Rs. 325000
24. 8% increases
25. 345000
26. 861840
27. 150000
28. 89.4%
29. 900
30. 500 marks, 212 marks

31. 44%

32. 98864

33. 29000

34. 65000

35. 56000

36. Rs. 44200

37. $33\frac{1}{3}\%$

38. Rs. 300

39. 2600 kgs

40. Rs. 9

41. $33\frac{1}{3}\%$

42. Rs. 156250

43. 35%

44. 7500 m

45. $56\frac{2}{3}\%$

RATIO, PROPORTION AND VARIATION

RATIO: Ratio is the comparison of two (or) more variables by division. It is denoted by "column" (:).

- ✓ Two numbers are in the ratio a : b, then first term 'a' is called as "antecedent" and second term 'b' is called as "consequent".

- ✓ If two numbers are in the ratio a : b, then

  First number = a × k, Second number = b × k   where 'k' is the common factor.

- ✓ If 'P' is divided in the ratio of a : b, then

  First term = $\dfrac{a}{a+b}$ × P, Second term = $\dfrac{b}{a+b}$ × P.

- ✓ If 'P' is divided in the ratio of a : b : c, then

  First term = $\dfrac{a}{a+b+c}$ × P, Second term = $\dfrac{b}{a+b+c}$ × P, Third term = $\dfrac{c}{a+b+c}$ × P.

> **Note:** The ratio does not change, if we multiply (or) divide with same number.
>
> **Example:** $\dfrac{2}{3} = \dfrac{2 \times 5}{3 \times 5} = \dfrac{2}{3}$ and $\dfrac{4}{7} = \dfrac{4 \div 5}{7 \div 5} = \dfrac{4}{7}$

Types of Ratios:

If three numbers are in the ratio $x : y : z$, then

1. Duplicate Ratio   ⟶ $x^2 : y^2 : z^2$

2. Sub – duplicate Ratio   ⟶ $\sqrt{x} : \sqrt{y} : \sqrt{z}$

3. Triplicate Ratio   ⟶ $x^3 : y^3 : z^3$

4. Sub – triplicate Ratio   ⟶ $\sqrt[3]{x} : \sqrt[3]{y} : \sqrt[3]{z}$

5. Inverse Ratio   ⟶ $\dfrac{1}{x} : \dfrac{1}{y} : \dfrac{1}{z}$

> **Note:** To convert fraction ratios into normal ratios, then multiply each fraction with LCM.

$$\frac{1}{x} : \frac{1}{y} : \frac{1}{z} = \frac{1}{x} \times xyz : \frac{1}{y} \times xyz : \frac{1}{z} \times xyz = yz : xz : xy$$

6. **Compound Ratio:** If two (or) more ratios are given, then compound ratio is product of given ratios.

   Example: The compound ratio of 2 : 3, 4 : 5 and 7 : 9 is

   $$\frac{2}{3} \times \frac{4}{5} \times \frac{7}{9} = \frac{56}{135}$$

PROPORTION: Equality of two ratios are called as proportion. It is denoted by double column (::).

✓   If a : b :: c : d, then we can say that a, b, c and d are in proportion.

   'd' is called as fourth proportional.

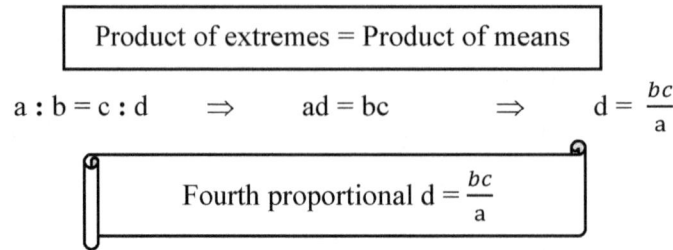

$$\boxed{\text{Product of extremes} = \text{Product of means}}$$

$$a : b = c : d \quad \Rightarrow \quad ad = bc \quad \Rightarrow \quad d = \frac{bc}{a}$$

$$\text{Fourth proportional } d = \frac{bc}{a}$$

✓   If a : b :: b : c, then 'c' is called as third proportional.

$$a : b = b : c \quad \Rightarrow \quad ac = b^2 \quad \Rightarrow \quad c = \frac{b^2}{a}$$

$$\text{Third proportional } c = \frac{b^2}{a}$$

✓   If a : b :: b : c, then 'b' is called as mean proportional.

$$a : b = b : c \quad \Rightarrow \quad ac = b^2 \quad \Rightarrow \quad b = \sqrt{ac}$$

$$\text{Mean proportional } b = \sqrt{ac}$$

Note: 1. To calculate fourth proportional we must require 3 values.

2. To calculate third proportional (or) mean proportional we must require 2 values.

VARIATION:

❖   Basically there are two types of variations.

   1. Direct variation   2. Inverse variation

1.   Direct Variation:

   Two numbers x and y are said to be direct variation as if x increases, y also increases and if x decreases, y also decreases.

$$x \, \alpha \, y \qquad \Rightarrow \qquad x = ky \qquad\qquad \text{Where 'k' is a constant.}$$

2. Inverse Variation:

Two numbers x and y are said to be inverse variation as if x increases, y decreases and if x decreases, y increases.

$$x \, \alpha \, \frac{1}{y} \qquad \Rightarrow \qquad x = \frac{k}{y} \qquad \qquad \text{Where 'k' is a constant.}$$

Joint Variation:

If x is directly varies with y and inversely varies with z, then it is called as joint variation.

$$x \, \alpha \, y \, , x \, \alpha \, \frac{1}{z} \quad \Rightarrow \quad x \, \alpha \, \frac{y}{z} \quad \Rightarrow \quad x = k \times \frac{y}{z} \qquad \text{Where 'k' is a constant.}$$

Some important points to be remember:

✓ If $\dfrac{a}{b} = \dfrac{c}{d}$, then

     1. Invertendo $\longrightarrow$ $\dfrac{b}{a} = \dfrac{d}{c}$

     2. Alternendo $\longrightarrow$ $\dfrac{a}{c} = \dfrac{b}{d}$

     3. Componendo $\longrightarrow$ $\dfrac{a+b}{b} = \dfrac{c+d}{d}$

     4. Dividendo $\longrightarrow$ $\dfrac{a-b}{b} = \dfrac{c-d}{d}$

     5. Componendo and Dividendo $\longrightarrow$ $\dfrac{a+b}{a-b} = \dfrac{c+d}{c-d}$

✓ If $\dfrac{a}{b} = \dfrac{c}{d} = \dfrac{e}{f} = \ldots\ldots = k$, then $\dfrac{a+c+e+\ldots}{b+d+f+\ldots} = k$

✓ If the given ratios are a : b and c : d, then

     i.   a : b > c : d, if ad > bc

     ii.   a : b < c : d, if ad < bc

     iii. a : b = c : d, if ad = bc

✓ If two numbers are in the ratio a : b and 'k' is added to both the numbers, then the ratio changes to c : d. Then,

$$\text{First number} = \frac{ka(c-d)}{ad-bc} \qquad \text{Second number} = \frac{kb(c-d)}{ad-bc}$$

✓ If two numbers are in the ratio a : b and 'k' is subtracted from both the numbers, then the ratio changes to c : d. Then,

$$\text{First number} = \frac{ka(d-c)}{ad-bc} \qquad \text{Second number} = \frac{kb(d-c)}{ad-bc}$$

✓ Two persons A and B, their incomes are in the ratio a : b and their expenditures are in the ratio c : d. If each of them saves the same amount Rs. S, then

$$\text{Income of A} = \frac{aS(d-c)}{ad-bc} \qquad \text{Income of B} = \frac{bS(d-c)}{ad-bc}$$

$$\text{Expenditure of A} = \frac{cS(b-a)}{ad-bc} \qquad \text{Expenditure of B} = \frac{dS(b-a)}{ad-bc}$$

SOLVED EXAMPLES.

Q - 1 a) **What is the duplicate ratio of 4 : 7?**

b) **What is the sub – duplicate ratio of 169 : 289?**

c) **What is the triplicate ratio of 5 : 9?**

d) **What is the sub – triplicate ratio of 1331 : 1728?**

Sol: a) $4 : 7 \longrightarrow$ Duplicate ratio $= 4^2 : 7^2 = 16 : 49$

b) $169 : 289 \longrightarrow$ Sub – duplicate ratio $= \sqrt{169} : \sqrt{289} = 13 : 17$

c) $5 : 9 \longrightarrow$ Triplicate ratio $= 5^3 : 9^3 = 125 : 729$

d) $1331 : 1728 \longrightarrow$ Sub – triplicate ratio $= \sqrt[3]{1331} : \sqrt[3]{1728} = 11 : 12$

Q - 2 **Find the compound ratio of 3 : 5, 7 : 12 and 10 : 13.**

Sol: Compound ratio means product of given ratios.

$\therefore$ Required compound ratio $= \dfrac{3}{5} \times \dfrac{7}{12} \times \dfrac{10}{13} = \dfrac{7}{26}$ $\Rightarrow$ $7 : 26.$

Q - 3 **If a : b = 5 : 7, b : c = 3 : 2, then find a : b : c.**

Sol: Since, 'b' is the common variable, then its value must be equal in both ratios.

$a : b = (5 : 7) \times 3 = 15 : 21$

$b : c = (3 : 2) \times 7 = 21 : 14$

$\therefore a : b : c = 15 : 21 : 14.$

Shortcut:

$$\begin{array}{cccc} & \mathbf{a} & \mathbf{b} & \mathbf{c} \\ a : b = 5 : 7 = & 5 & 7 & \square \\ b : c = 3 : 2 = & \square & 3 & 2 \end{array}$$

$$\begin{array}{cccc} a : b & = & 5 & 7 & \boxed{7} \\ & & \downarrow & \downarrow & \downarrow \\ b : c & = & \boxed{3} & 3 & 2 \end{array}$$

Note:
- ✓ Right side boxes are filled with previous numbers.
- ✓ Left side boxes are filled with next numbers.

$\therefore a : b : c = 5 \times 3 : 7 \times 3 : 7 \times 2 = 15 : 21 : 14.$

Q-4 If A : B = 3 : 4, B : C = 8 : 9 and C : D = 5 : 2, then what is the value of A : B : C : D?

Sol:

	A	B	C	D		A	B	C	D
A : B = 3 : 4 =	3	4	☐	☐	⇒	3	4	4	4
B : C = 8 : 9 =	☐	8	9	☐	⇒	8	8	9	9
C : D = 5 : 2 =	☐	☐	5	2	⇒	5	5	5	2

∴ A : B : C : D = $3 \times 8 \times 5 : 4 \times 8 \times 5 : 4 \times 9 \times 5 : 4 \times 9 \times 2 = 30 : 40 : 45 : 18$.

Q-5 If X : Y = $\frac{2}{9} : \frac{5}{6}$, Y : Z = $\frac{3}{2} : \frac{4}{3}$ and Z : W = $\frac{3}{4} : \frac{5}{8}$, then what is the value of X : W?

Sol:

> **Note:** To convert fraction ratios into normal ratios multiply every fraction with LCM.

$X : Y = \frac{2}{9} : \frac{5}{6} = \frac{2}{9} \times 18 : \frac{5}{6} \times 18 = 4 : 15$ ∵ LCM of 9, 6 = 18

$Y : Z = \frac{3}{2} : \frac{4}{3} = \frac{3}{2} \times 6 : \frac{4}{3} \times 6 = 9 : 8$ ∵ LCM of 2, 3 = 6

$Z : W = \frac{3}{4} : \frac{5}{8} = \frac{3}{4} \times 8 : \frac{5}{8} \times 8 = 6 : 5$ ∵ LCM of 4, 8 = 8

		X	Y	Z	W
X : Y = 4 : 15	⇒	4	15	15	15
Y : Z = 9 : 8	⇒	9	9	8	8
Z : W = 6 : 5	⇒	6	6	6	5

∴ X : W = $4 \times 9 \times 6 : 15 \times 8 \times 5 = 9 : 25$.

Q-6 If $\frac{A}{5} = \frac{B}{7} = \frac{C}{8}$, then find A : B : C.

Sol: Let us consider, $\frac{A}{5} = \frac{B}{7} = \frac{C}{8} = k$ ⇒   A = 5k, B = 7k, C = 8k

A : B : C = 5 : 7 : 8.

Q-7 If 5x = 7y = 10z, then what is the value of x : y : z?

Sol: Let us consider, $5x = 7y = 10z = k$ ⇒ $x = \frac{K}{5}, y = \frac{K}{7}, z = \frac{K}{10}$

$x : y : z = \frac{1}{5} : \frac{1}{7} : \frac{1}{10} = \frac{1}{5} \times 70 : \frac{1}{7} \times 70 : \frac{1}{10} \times 70$ ∵ LCM of 5, 7, 10 = 70

$x : y : z = 14 : 10 : 7$.

Q - 8   If 4A = 7B and 5B = 3C, then find A : C.

Sol:

		A	B	C

$4A = 7B \quad \Rightarrow \quad A : B = 7 : 4 \quad \Rightarrow \quad$ 7 , 4 , $\boxed{4}$

$5B = 3C \quad \Rightarrow \quad B : C = 3 : 5 \quad \Rightarrow \quad$ $\boxed{3}$, 3 , 5

$A : C = 7 \times 3 : 4 \times 5 = 21 : 20.$

Q - 9   **Find the fourth proportional to the numbers**

a) 48, 18, 32   **b) 0.06, 1.2, 0.08**

Sol:

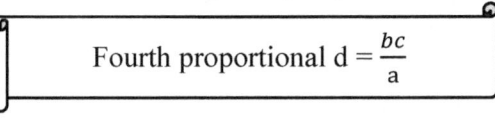
$$\text{Fourth proportional } d = \frac{bc}{a}$$

a) Fourth Proportional $d = \dfrac{18 \times 32}{48} = 12.$ $\because a = 48, b = 18, c = 32$

b) Fourth Proportional $d = \dfrac{1.2 \times 0.08}{0.06} = 1.6.$ $\because a = 0.06, b = 1.2, c = 0.08$

Q - 10   **Find the third proportional to the numbers**

a) 16, 24   **b) 0.6, 0.9**

Sol:

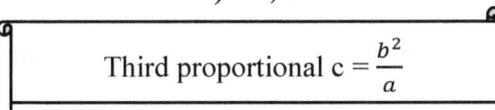
$$\text{Third proportional } c = \frac{b^2}{a}$$

a) Third Proportional $c = \dfrac{24 \times 24}{16} = 36.$ $\because a = 16, b = 24$

b) Third Proportional $c = \dfrac{0.9 \times 0.9}{0.6} = 1.35.$ $\because a = 0.6, b = 0.9$

Q - 11   **Find the mean proportional between**

a) 121, 289   **b) 2.56, 0.81**

Sol:

$$\text{Mean proportional } b = \sqrt{ac}$$

a) Mean Proportional $b = \sqrt{121 \times 289} = 11 \times 17 = 187.$ $\because a = 121, c = 289$

b) Mean Proportional $b = \sqrt{2.56 \times 0.81} = 1.6 \times 0.9 = 1.44.$ $\because a = 2.56, c = 0.81$

Q - 12   **If $x : y = 7 : 8$, then find the value of $(9x - 4y) : (5x + 3y)$.**

Sol: $x : y = 7 : 8 \quad \Rightarrow \quad x = 7k, y = 8k$

$\dfrac{9x - 4y}{5x + 3y} = \dfrac{9 \times 7k - 4 \times 8k}{5 \times 7k + 3 \times 8k} = \dfrac{31k}{59k} \quad \Rightarrow \quad 31 : 59.$

Shortcut:

> **Note:**
>
> ✓ For these kinds of problems, if the degrees of numerator and denominator are equal we can directly substitute their respective values.
>
> ✓ If the degrees of numerator and denominator are not equal, then the ratio is not possible to calculate. In that case answer **"can't be determined."**

$$\frac{9x - 4y}{5x + 3y} = \frac{9 \times 7 - 4 \times 8}{5 \times 7 + 3 \times 8} = 31 : 59.$$

Q - 13 If a : b = 3 : 5, then find

a) $\dfrac{2a^2 + b^2}{5a^2 - b^2}$

b) $\dfrac{3a^2 + 4b^2}{5a - 2b}$

Sol: a) a : b = 3 : 5

$$\frac{2a^2 + b^2}{5a^2 - b^2} = \frac{2 \times 3^2 + 5^2}{5 \times 3^2 - 5^2} = \frac{18 + 25}{45 - 25} \quad \Rightarrow \quad 43 : 20.$$

b) $\dfrac{3a^2 + 4b^2}{5a - 2b}$

Here, degrees of numerator and denominator are not equal.

Hence the answer is **can't be determined**.

Q - 14 **Three-fourth of a number is equal to 42% of the second number. Find the ratio of first number to the second number.**

Sol: Let, First number = F, Second number = S

$$\frac{3}{4} \times F = \frac{42}{100} \times S \quad \Rightarrow \quad F : S = 14 : 25.$$

Q - 15 **If 24% of (5A – 2B) = 36% of (A + 3B), then what is the ratio of A : B?**

Sol: $\dfrac{24}{100} \times (5A - 2B) = \dfrac{36}{100} \times (A + 3B) \quad \Rightarrow \quad 2(5A - 2B) = 3(A + 3B)$

$10A - 4B = 3A + 9B \quad \Rightarrow \quad 7A = 13B \quad \Rightarrow \quad A : B = 13 : 7.$

Q - 16 If $(9x^2 - 5y^2) : (3x^2 + 8y^2) = 3 : 5$, then find $x : y$.

Sol: $\dfrac{9x^2 - 5y^2}{3x^2 + 8y^2} = \dfrac{3}{5} \quad \Rightarrow \quad 45x^2 - 25y^2 = 9x^2 + 24y^2$

$36x^2 = 49y^2 \quad \Rightarrow \quad x^2 : y^2 = 49 : 36$

$\therefore x : y = 7 : 6.$

Q - 17 | If $(a + b) : (b + c) : (c + a) = 4 : 10 : 7$ and $(a + b + c) = 18$. Find the value of b.

Sol: Let, $a + b = 4k$, $b + c = 10k$, $c + a = 7k$

$4k + 10k + 7k = 2(a + b + c)$

$21k = 2 \times 18 \quad \Rightarrow \quad k = \dfrac{12}{7}$ $\because a + b + c = 18$

$\therefore b = (a + b + c) - (c + a) = 18 - 7 \times \dfrac{12}{7} = 6.$

Q - 18 | If $\dfrac{2x}{3y} = \dfrac{7}{8}$, then find the values of $\dfrac{2x - y}{4x + y} + \dfrac{6}{25}$.

Sol: $\dfrac{2x}{3y} = \dfrac{7}{8} \quad \Rightarrow \quad x : y = 21 : 16$

Since degrees of both numerator and denominator are equal, then we can directly substitute the values of x and y.

$\therefore \dfrac{2x - y}{4x + y} + \dfrac{6}{25} = \dfrac{2 \times 21 - 16}{4 \times 21 + 16} + \dfrac{6}{25} = \dfrac{26}{100} + \dfrac{6}{25} = \dfrac{1}{2}.$

Q - 19 | Two numbers are in the ratio 5 : 8. The sum of those two numbers is 728. Find the two numbers.

Sol: Consider, two no's, First number $F = 5k$, Second number $S = 8k$ $\because F : S = 5 : 8$

Sum of two numbers $F + S = 5k + 8k = 728 \quad \Rightarrow \quad k = 56$

$\therefore$ First number $F = 5 \times 56 = 280$ & Second number $S = 8 \times 56 = 448.$

<u>Shortcut:</u>

Given that, two numbers ratio $F : S = 5 : 8$

Total sum = 13 parts $\longrightarrow$ 728

1 part $\longrightarrow$ $? = \dfrac{728}{13} = 56$

$\therefore$ First number $F = 5 \times 56 = 280$ & Second number $S = 8 \times 56 = 448.$

Q - 20 | Two numbers are in the ratio 9 : 13. If the larger number is 72 more than the smaller number. Find the two numbers.

Sol: Consider, two no's, First number $F = 9k$, Second number $S = 13k$ $\because F : S = 9 : 13$

Difference of two numbers $= 13k - 9k = 72 \quad \Rightarrow \quad k = 18$

$\therefore$ First number $F = 9 \times 18 = 162$ & Second number $S = 13 \times 18 = 234.$

Shortcut:

Given that, two numbers ratio F : S = 9 :13

Difference of two no's = 4 parts $\longrightarrow$ 72

           1 part $\longrightarrow$ $? = \frac{72}{4} = 18$

$\therefore$ First number F = 9 × 18 = 162 & Second number S = 13 × 18 = 234.

Q - 21 | **Three natural numbers are in the ratio 2 : 3 : 5. If the sum of their squares is 1862, then find the product of first and second numbers.**

Sol: Consider, three no's, First number F = 2k, Second number S = 3k & Third number T = 5k

Sum of their squares = $(2k)^2 + (3k)^2 + (5k)^2 = 1862$ $\because$ F : S : T = 2 : 3 : 5

         $38k^2 = 1862$ $\Rightarrow$ k = 7

First number F = 2 × 7 = 14 and Second number S = 3 × 7 = 21

$\therefore$ Product of 1st and 2nd numbers = 14 × 21 = 294.

Shortcut:

Given that, three numbers ratio F : S : T = 2 : 3 : 5

Squares ratio = $2^2 : 3^2 : 5^2 = 4 : 9 : 25$

Sum of their squares = 38 sq. parts $\longrightarrow$ 1862

          1 sq. part $\longrightarrow$ $? = \frac{1862}{38} = 49$

          1 part $\longrightarrow$ $\sqrt{49} = 7$

First number F = 2 × 7 = 14 and Second number S = 3 × 7 = 21

$\therefore$ Product of 1st and 2nd numbers = 14 × 21 = 294.

Q - 22 | **Two numbers are in the ratio 4 : 7. If 3 is added to both the numbers, their ratio changes to 5 : 8. Find the sum of two numbers.**

Sol: Let us consider, First number F = 4k & Second number S = 7k $\because$ F : S = 4 : 7

According to question,

     $\frac{4k + 3}{7k + 3} = \frac{5}{8}$ $\Rightarrow$ 32k + 24 = 35k + 15 $\Rightarrow$ k = 3

$\therefore$ Sum of two numbers F + S = 4k + 7k = 11k = 11 × 3 = 33.

Shortcut – 1:

> **Note:**
>
> ✓ If the same value is added (or) subtracted from both the numbers, then the difference between those numbers must be equal before and after the addition (or) subtraction.
>
> ✓ If the difference between the numbers are not equal, we have to make it as equal by multiplying first ratio with second difference and vice versa.

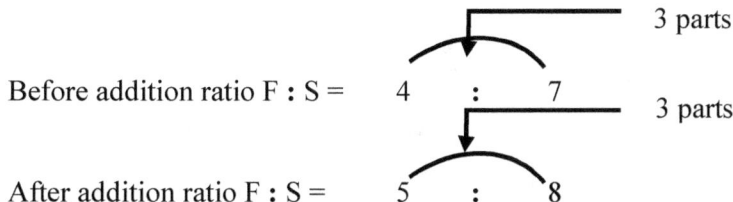

Before addition ratio F : S =   4   :   7   3 parts

3 parts

After addition ratio F : S =   5   :   8

Here the difference between two numbers are equal before and after addition.

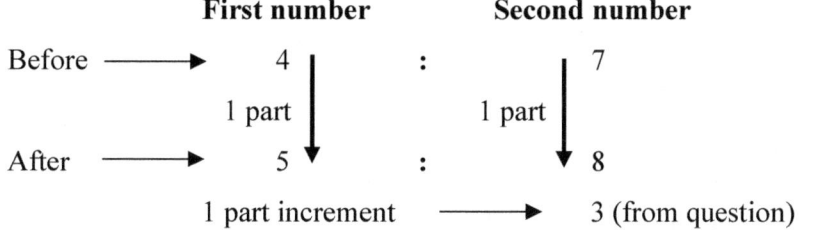

	First number		**Second number**
Before ⟶	4	:	7
	1 part		1 part
After ⟶	5	:	8

1 part increment ⟶ 3 (from question)

First Number F = 4 parts = 4 × 3 = 12 & Second Number S = 7 parts = 7 × 3 = 21

∴ Sum of two numbers = 12 + 21 = 33.

Shortcut – 2:

> **Note:** If the two numbers are in the ratio a : b and 'k' is added to both the numbers, then the ratio changes to c : d. Then,
>
> First number $= \dfrac{ka(c-d)}{ad-bc}$ Second number $= \dfrac{kb(c-d)}{ad-bc}$

Here a = 4, b = 7, c = 5, d = 8, k = 3

First number $= \dfrac{3 \times 4\,(5-8)}{4 \times 8 - 7 \times 5} = 12$ Second number $= \dfrac{3 \times 7\,(5-8)}{4 \times 8 - 7 \times 5} = 21$

∴ Sum of two numbers = 12 + 21 = 33.

Q - 23 | Two numbers are in the ratio 9 : 5. If 4 is subtracted from each number, their ratio changes to 7 : 3. Find the greatest number.

Sol: Let us consider, First number F = 9k & Second number S = 5k $\because$ F : S = 9 : 5

According to question,

$$\frac{9k-4}{5k-4} = \frac{7}{3} \qquad \Rightarrow \qquad 27k-12 = 35k-28 \qquad \Rightarrow \qquad k = 2$$

$\therefore$ Greatest number F = 9 × 2 = 18.

Shortcut – 1:

Before subtraction ratio = 9 : 5   4 parts

                             2 parts

After subtraction ratio = 7 : 3   4 parts

Here the difference between two numbers are equal before and after subtraction.

2 parts decrement $\longrightarrow$ 4 (from question) $\Rightarrow$ 1 part = 2

$\therefore$ Greatest number F = 9 Parts = 9 × 2 = 18.

Shortcut – 2:

> **Note:** If the two numbers are in the ratio a : b and 'k' is subtracted from both the numbers, then the ratio changes to c : d. Then,
>
> First number $= \dfrac{ka(d-c)}{ad-bc}$ Second number $= \dfrac{kb(d-c)}{ad-bc}$

Here a = 9, b = 5, c = 7, d = 3, k = 4

$\therefore$ Greatest number F $= \dfrac{4 \times 9\,(3-7)}{9 \times 3 - 5 \times 7} = 18.$

Q - 24 | From each of the two given numbers, one – third of smallest number is subtracted. After such subtraction, the larger number is twice the smaller number. Find the ratio of smaller to larger.

Sol: Consider, Smaller number = S, Larger number = L.

According to question,

$$\left(L - \frac{S}{3}\right) = 2\left(S - \frac{S}{3}\right) \qquad \Rightarrow \qquad L - \frac{S}{3} = \frac{4}{3}S \qquad \Rightarrow \qquad L = \frac{5}{3}S$$

$\therefore$ Required ratio S : L = 3 : 5.

Q - 25 The salaries of Kunal and Amit are in the ratio 3 : 7. If the salary of each person is increased by Rs. 5400, the new ratio obtained is 21 : 43. What is Amit's present salary?

Sol: Consider, Kunal salary = K, Amit salary = A

Ratio K : A = 3 : 7 $\Rightarrow$ K = 3x, A = 7x

From the question,

$\frac{3x + 5400}{7x + 5400} = \frac{21}{43}$ $\Rightarrow$ $129x + 232200 = 147x + 113400$

$18x = 118800$ $\Rightarrow$ $x = 6600$

$\therefore$ Amit's present salary = 7x + 5400 = 7 × 6600 + 5400 = Rs. 51600.

Shortcut:

		Kunal		Amit	4 parts
Before increment	$\longrightarrow$	3	:	7	
					22 parts
After increment	$\longrightarrow$	21	:	43	

$\because$ Differences are not equal, so multiply 1st ratio with 2nd difference and vice versa.

			Kunal		Amit	
Before	$\longrightarrow$	(3 : 7) × 22 =	66	:	154	
						18 parts
After	$\longrightarrow$	(21 : 43) × 4 =	84	:	172	

18 parts $\longrightarrow$ Rs. 5400 $\Rightarrow$ 1 part $= \frac{5400}{18}$ = Rs. 300

$\therefore$ Amit's present salary = 172 parts = 172 × 300 = Rs. 51600.

Q - 26 Find the least whole number which when subtracted from both the terms of the ratio 5 : 4 gives a ratio less than 14 : 19.

Sol: Consider 'x' is subtracted from both the terms, then

$\frac{5-x}{4-x} < \frac{14}{19}$ $\Rightarrow$ $95 - 19x < 56 - 14x$

$5x > 39$ $\Rightarrow$ $x > 7.8$

$\therefore$ Least whole number to be subtracted is 8.

Q-27 | Three numbers are in the ratio 4 : 7 : 11. If first number is increased by 40%, second number is decreased by 10% and the third number is decreased by 30%. Find the new ratio.

Sol: Consider, 3 numbers A, B and C are in the ratio A : B : C = 4 : 7 : 11

New ratio $= 4 \times \frac{(100 + 40)}{100} : 7 \times \frac{(100 - 10)}{100} : 11 \times \frac{(100 - 30)}{100}$

∵ A – increased by 40%, B – decreased by 10%, C – decreased by 30%

$= 4 \times 140 : 7 \times 90 : 11 \times 70$

∴ New ratio = 8 : 9 : 11.

Q-28 | What is the ratio of circumference and area of circle, if the radius of circle is 4 cm?

Sol: Circumference of circle : Area of circle $= 2\pi r : \pi r^2 = 2 : r$

∴ Required ratio = 2 : 4 = 1 : 2.   ∵ radius r = 4 cm

Q-29 | The ratio of number of boys and girls in a class is 8:7. If 90% of the boys and 85% of girls are passed in the test, then what percentage of students are failed in the test?

Sol: Fail percentage $= \frac{No.\,of\ students\ failed}{Total\ students} \times 100\%$

Required ratio $= \frac{8 \times \frac{10}{100} + 7 \times \frac{15}{100}}{8 + 7} \times 100\%$ ∵ Failed boys – 10%, Failed girls – 15%

∴ Fail Percentage $= 12\frac{1}{3}\%$.

Q-30 | If a, b, c, d and e are in continued proportion, then find a : e (in terms of a and b).

Sol: If a, b, c, d and e are in continued proportion, then $\frac{a}{b} = \frac{b}{c} = \frac{c}{d} = \frac{d}{e}$

$a : e = \frac{a}{e} = \frac{a}{b} \times \frac{b}{c} \times \frac{c}{d} \times \frac{d}{e}$ $\Rightarrow$ $\frac{a}{e} = \frac{a}{b} \times \frac{a}{b} \times \frac{a}{b} \times \frac{a}{b} = \frac{a^4}{b^4}$ $\Rightarrow$ $a : e = a^4 : b^4$.

Q-31 | The ratio of males and females in a town is 5 : 7 respectively and the percentage of children among males and females is 10% and 25% respectively. If the number of adult females in the town is 168000, then what is the total population?

Sol: Ratio of males and females = 5 : 7

No. of adult females = 75%   ∵ Female children – 25%

No. of adult females $= 7 \times \frac{75}{100}$ parts $\longrightarrow$ 168000

Total population = 12 parts $\longrightarrow$ $? = \frac{168000 \times 12 \times 100}{7 \times 75} = 384000$

∴ Total population = 384000.

Q - 32 **Jai , Krish and Varma shares a certain amount in the ratio 4 : 7 : 8. If the difference between Jai and Krish is Rs. 720, then what is the total amount?**

Sol: Ratio of shares = 4 : 7 : 8

Consider, Jai = 4k, Krish = 7k, Varma = 8k

Difference between Jai and Krish = 7k – 4k = Rs. 720 $\Rightarrow$ 3k = 720 $\Rightarrow$ k = 240

∴ Total amount = 4k + 7k + 8k = 19k = 9 × 240 = Rs. 4560.

Shortcut:

Jai : Krish : Varma = 　4　:　7　:　8

3 parts

Difference = 3 parts ⟶ Rs. 720

∴ Total amount = 19 parts ⟶ $? = \dfrac{720 \times 19}{3} = $ Rs. 4560.

Q - 33 **The number of students in three sections A, B and C are in the ratio of 5 : 6 : 3. If 7 students are shifted from B to A, the ratio of students in their two sections are interchanged. Find the number of students in section B initially.**

Sol: A : B : C = 5 : 6 : 3 $\Rightarrow$ A = 5x, B = 6x, C = 3x where 'x' is the common factor

$\dfrac{5x + 7}{6x - 7} = \dfrac{6}{5}$ ∵ 7 students are shifted from B to A then, ratio is interchanged i.e; 6 : 5

25x + 35 = 36x – 42 $\Rightarrow$ 11x = 77 $\Rightarrow$ x = 7

∴ Initial no. of students in section B = 6x = 6 × 7 = 42.

Shortcut:

　　　　　　　　　　　　　　　A　　**B**

Before shifting ⟶ 5 : 6　　　　1 part ⟶ 7 students

　　　　　　　1 part　　1 part

After shifting ⟶ 6 : 5

∴ No. of students in section B initially = 6 parts = 6 × 7 = 42.

Q - 34 **The ratio of number of boys and girls in a school was 5 : 3. Some new boys and girls were admitted to the school, in the ratio 5 : 7. At this total number of students in the school became 1200 and the ratio of boys and girls changed to 7 : 5. Find the number of students in the school before new admission.**

Sol: Boys : Girls = 5 : 3 $\Rightarrow$ Boys = 5x, Girls = 3x

No. of boys admitted = 5y, No. of girls admitted = 7y $\because$ New admission ratio = 5 : 7

Now, total strength = 5x + 3x + 5y + 7y = 1200

8x + 12y = 1200 $\Rightarrow$ 2x + 3y = 300 $\rightarrow$ (1)

Also, new ratio of boys and girls = 7 : 5

$\dfrac{Total\ boys}{Total\ girls} = \dfrac{5x + 5y}{3x + 7y} = \dfrac{7}{5}$ $\Rightarrow$ 25x + 25y = 21x + 49y $\Rightarrow$ x = 6y $\rightarrow$ (2)

Substitute (2) in (1), then

2 × 6y + 3y = 300 $\Rightarrow$ y = 20 & x = 6 × 20 = 120

$\therefore$ Total strength before new admission = 8x = 8 × 120 = 960.

Q - 35   **The incomes of A and B are in the ratio 4 : 3 and their expenditures are in the ratio 7 : 5, if each saves Rs. 3000. Find their incomes.**

Sol:

$$\boxed{\text{Savings = Income – Expenditure}}$$

		A		B

Incomes $\longrightarrow$ 4 : 3 $\Rightarrow$ $I_A = 4x$, $I_B = 3x$

Expenditures $\longrightarrow$ 7 : 5 $\Rightarrow$ $E_A = 7y$, $E_B = 5y$

Savings of A = Savings of B $\because$ Each saves Rs. 3000

4x – 7y = 3x – 5y $\Rightarrow$ x = 2y

$S_A = 4x – 7y$ = Rs. 3000 $\Rightarrow$ 4 × 2y – 7y = 3000 $\Rightarrow$ y = 3000

x = 2 × 3000 = 6000

$\therefore$ A's Income $I_A = 4x = 4 × 6000 =$ Rs. 24000.

B's Income $I_B = 3x = 3 × 6000 =$ Rs. 18000.

Shortcut:

Note: Two persons A and B, their incomes are in the ratio a : b and their expenditures are in the ratio c : d. If each of them saves the same amount Rs. S, then

Income of A = $\dfrac{aS(d - c)}{ad - bc}$ Income of B = $\dfrac{bS(d - c)}{ad - bc}$

Here a = 4, b = 3, c = 7, d = 5, S = Rs. 3000

$\therefore$ Income of A = $\dfrac{4 \times 3000\,(5 - 7)}{4 \times 5 - 3 \times 7}$ = Rs. 24000.

Income of B = $\dfrac{3 \times 3000\,(5 - 7)}{4 \times 5 - 3 \times 7}$ = Rs. 18000.

Q - 36   A customer bought some Sugar and Rice for Rs. 1340. The ratio of weight of Sugar and Rice is 8 : 5 and the price of equal amount of Sugar and Rice in the ratio 4 : 7. At what price the sugar was bought?

Sol:

$$\boxed{\text{Total Cost} = \text{Weight} \times \text{Price}}$$

	Sugar		**Rice**
Weight $\rightarrow$	8	:	5
Price $\rightarrow$	4	:	7

Total cost $\rightarrow$ 8×4 : $5 \times 7 = 32 : 35$

Total cost = 67 parts $\longrightarrow$ Rs. 1340

$\therefore$ Cost of sugar = 32 parts $\longrightarrow$ $? = \dfrac{1340 \times 32}{67} = \text{Rs. } 640.$

Q - 37   In a box there are three types of coins 25 paisa, 50 paisa and 1 rupee. The ratio of the number of coins are 8 : 6 : 9 respectively. If the total amount in the box is Rs. 210, then find the number of 25 paisa coins.

Sol:

$$\boxed{\text{Total amount} = \text{No. of coins} \times \text{Coin denomination}}$$

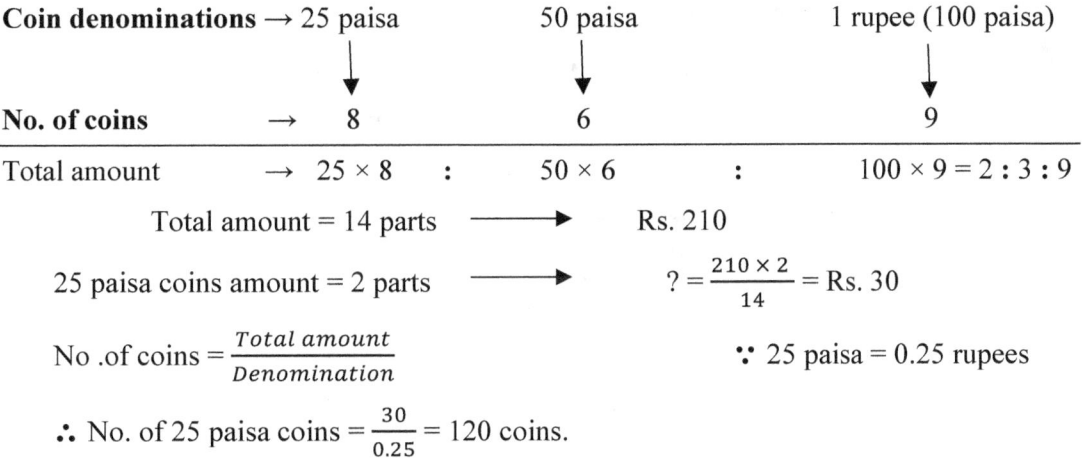

Coin denominations $\rightarrow$ 25 paisa   50 paisa   1 rupee (100 paisa)

No. of coins $\rightarrow$ 8   6   9

Total amount $\rightarrow$ 25×8 : 50×6 : $100 \times 9 = 2 : 3 : 9$

Total amount = 14 parts $\longrightarrow$ Rs. 210

25 paisa coins amount = 2 parts $\longrightarrow$ $? = \dfrac{210 \times 2}{14} = \text{Rs. } 30$

No .of coins $= \dfrac{Total\ amount}{Denomination}$ $\because$ 25 paisa = 0.25 rupees

$\therefore$ No. of 25 paisa coins $= \dfrac{30}{0.25} = 120$ coins.

> **Note:** Denominations must be either in terms of paisa (or) in terms of rupees.

Q - 38 In a bag for every 5 notes of 20 rupees, there are 3 notes of 50 rupees. For every 8 notes of 50 rupees, there are 7 notes of 100 rupees. If the value of 20 rupees notes is Rs. 3200, then find the total amount in the bag.

Sol:

Total amount = No. of notes × Note denomination

		20 rupees		**50 rupees**		**100 rupees**
No of notes	→	5		3		3
		8		8		7
No. of notes	→	5 × 8	:	3 × 8	:	3 × 7
No. of notes	→	40	:	24	:	21
Note Denomination	→	20		50		100
Total amount	→	40 × 20	:	24 × 50	:	21 × 100 = 8 : 12 : 21

20 rupees notes total amount = 8 parts ⟶ Rs. 3200

Total amount = 41 parts ⟶ $? = \dfrac{3200 \times 41}{8} = $ Rs. 16400

∴ Total amount in the bag is Rs. 16400.

Q - 39 The incomes of *x*, *y* and *z* are in the ratio 7 : 10 : 13 and their expenditures are in the ratio 3 : 5 : 6. If *x* saves one – fourth of his income, then find the ratio of their savings.

Sol:
Incomes = 7 : 10 : 13 ⟹ $I_x = 7a, I_y = 10a, I_z = 13a$

Expenditures = 3 : 5 : 6 ⟹ $E_x = 3b, E_y = 5b, E_z = 6b$

Where a, b are common factors in 2 ratios

Savings = Income – Expenditure

$S_x = \dfrac{1}{4} \times 7a$

$E_x = 3b = \dfrac{3}{4} \times 7a \quad \Rightarrow \quad b = \dfrac{7a}{4}$ ∵ *x* saves $\dfrac{1}{4}$ of income

$S_y = I_y - E_y = 10a - 5b = 10a - 5 \times \dfrac{7}{4}a = \dfrac{5a}{4}$

$S_z = I_z - E_z = 13a - 6b = 13a - 6 \times \dfrac{7}{4}a = \dfrac{10a}{4}$

$S_x : S_y : S_z = \dfrac{7a}{4} : \dfrac{5a}{4} : \dfrac{10a}{4} = 7 : 5 : 10.$

Shortcut:

Incomes = 7 : 10 : 13, Expenditures = 3 : 5 : 6

Consider their incomes are $I_x = 700$, $I_y = 1000$, $I_z = 1300$

$S_x = \frac{1}{4} \times 700 = 175$, $E_x = 700 - 175 = 525$

Expenditure of x = 3 parts $\longrightarrow$ 525

1 part $\longrightarrow$? $= \frac{525}{3} = 175$

E_y = 5 parts = 5 × 175 = 875, E_z = 6 parts = 6 × 175 = 1050

$S_y = I_y - E_y = 1000 - 875 = 125$, $S_z = I_z - E_z = 1300 - 1050 = 250$

$S_x : S_y : S_z = 175 : 125 : 250 = 7 : 5 : 10$.

Diagrammatical representation:

Incomes	$\rightarrow$ 7 : 10 : 13 =	700	1000	1300	$\because x$ saves $\frac{1}{4}$ of income
Expenditure	$\rightarrow$ 3 : 5 : 6 =	525	875	1050	
Savings	$\rightarrow$	175 :	125 :	250 = 7 : 5 : 10.	

Q - 40 An employer reduces the number of employees in the ratio 22 : 21 and increases their wages in the ratio 15 : 16. If the original wage bill was Rs. 26400, then find the ratio in which the wage bill is increased?

Sol:

> Total wage bill = No. of employees × Employee wages

Employee ratio = 22 : 21, Wage ratio = 15 : 16

Let, original employees = 22x, new employees = 21x

Original wages = 15y, new wages = 16y

Original wage bill = 22x × 15y = Rs. 26400 $\Rightarrow$ $xy = 80$

New wage bill = 21x × 16y = 336xy = 336 × 80 = Rs. 26880

∴ Ratio of wage bills = 26400 : 26880 = 55 : 56.

Shortcut:

		Original		New
Employees	$\rightarrow$	22	:	21
Wages	$\rightarrow$	15	:	16
Total wages	$\rightarrow$	22 × 15	:	21 × 16 = 55 : 56.

Q - 41 A father divides his total money to his family members in such a way that half of the money goes to his wife, $\frac{1}{3}^{rd}$ of remaining among his two daughters equally and the rest among his three sons equally. If each son gets Rs. 40000, how much money will each daughter gets?

Sol: Consider, total amount with father = Rs. x

To wife $= \frac{1}{2} \times x = \frac{x}{2}$ Remaining $= x - \frac{x}{2} = \frac{x}{2}$

To 2 daughters $= \frac{1}{3} \times \frac{x}{2} = \frac{x}{6}$ Remaining $= \frac{x}{2} - \frac{x}{6} = \frac{x}{3}$

Each daughter $= \frac{1}{2} \times \frac{x}{6} = \frac{x}{12}$

To 3 sons $= \frac{x}{3}$ $\Rightarrow$ Each son $= \frac{1}{3} \times \frac{x}{3} = \frac{x}{9}$

Given that, each son gets Rs. 40000 $\Rightarrow$ $\frac{x}{9} = 40000$ $\Rightarrow$ Rs. 360000

$\therefore$ Each daughter $= \frac{x}{12} = \frac{360000}{12} = $ Rs. 30000.

Q - 42 An amount of Rs. 24860 is divided among x, y, z such that x receives $\frac{1}{3}^{rd}$ as much as y and z together and y receives $\frac{5}{6}^{th}$ as much as x and z together. What is the amount received by z?

Sol: According to question, $x = \frac{1}{3}(y + z)$ $\Rightarrow$ $x : (y + z) = 1 : 3$

Consider $x = k$, $y + z = 3k$

$x + (y + z) = k + 3k = 4k = $ Rs. 24860 $\Rightarrow$ k = 6215

$x = $ Rs. 6215

$y = \frac{5}{6}(x + z)$ $\Rightarrow$ $y : (x + z) = 5 : 6$

$y = 5k$, $x + z = 6k$

$(x + z) + y = 11k = 24860$ $\Rightarrow$ k = 2260

$y = 5k = 5 \times 2260 = $ Rs. 11300

$\therefore z = (x + y + z) - (x + y) = 24860 - (6215 + 11300) = $ Rs. 7345.

Shortcut:

$$x = \frac{1}{3}(y + z) \quad \Rightarrow \quad x : (y + z) = 1 : 3$$

Total = $x + y + z = 4$ parts $\longrightarrow$ Rs. 24860

$x = 1$ part $\longrightarrow \quad ? = \dfrac{24860 \times 1}{4} = $ Rs. 6215

$$y = \frac{5}{6}(x + z) \quad \Rightarrow \quad y : (x + z) = 5 : 6$$

Total = $x + y + z = 11$ parts $\longrightarrow$ Rs. 24860

$y = 5$ parts $\longrightarrow \quad ? = \dfrac{24860 \times 5}{11} = $ Rs.11300

$\therefore z = (x + y + z) - (x + y) = 24860 - (6215 + 11300) = $ Rs. 7345.

Q - 43 If the positions of the digits of a two digit number are interchanged, then the ratio of newly formed number and original number is 7 : 4. Also, the difference between new and original number is 27. Find the original number.

Sol: Consider, new number = $7x$, original number = $4x$ $\quad \because$ Ratio = 7 : 4

Difference: $7x - 4x = 27 \quad \Rightarrow \quad 3x = 27 \quad \Rightarrow \quad x = 9$

$\therefore$ Original number = $4x = 4 \times 9 = 36$.

Q - 44 Three persons A, B and C divides an amount of Rs. 520 in such a way that B gets 40 more than A and C gets 80 more than B, then find the ratio of their shares.

Sol: According to question,

$B = A + 40$, $C = B + 80 = (A + 40) + 80 = A + 120$

$A + B + C = 520 \quad \Rightarrow \quad A + (A + 40) + (A + 120) = 520$

$3A = 360 \quad \Rightarrow \quad A = 120$, $B = 160$, $C = 240$

$\therefore A : B : C = 120 : 160 : 240 = 3 : 4 : 6$.

Q - 45 Rocky have 2845 gold coins and divides to three persons A, B and C. Out of the total coins received by each of them, all three members lost 15 coins each. Now the ratio of gold coins with them is 3 : 5 : 6. How many coins did B received from Rocky?

Sol: Total coins = 2845

Total no. of coins lost = 15 + 15 + 15 = 45 $\qquad$ Remaining coins = 2845 – 45 = 2800

Remaining ratio of A, B, C = 3 : 5 : 6

Remaining total = 14 parts $\longrightarrow$ 2800

B = 5 parts $\longrightarrow \quad ? = \dfrac{2800 \times 5}{14} = 1000$

$\therefore$ Total coins received by B from Rocky = 1000 + 15 = 1015.

Q - 46 An amount of Rs. 5760 is divided among P, Q, R. If their shares are reduced by Rs. 15, Rs. 25 and Rs. 20 respectively, the remaining amount will be in the ratio of 4 : 6 : 9. Find the share of R.

Sol: Total amount = Rs. 5760

Total reduced amount = 15 + 25 + 20 = Rs. 60

Remaining amount = 5760 – 60 = Rs. 5700

Given that, Remaining ratio = 4 : 6 : 9

Remaining total = 19 parts $\longrightarrow$ Rs. 5700

R's share = 9 parts $\longrightarrow$ $? = \dfrac{5700 \times 9}{19} = $ Rs. 2700

∴ Total amount of R = 2700 + 20 = Rs. 2720.

Q - 47 The fares of first, second and third class between two stations were 11 : 9 : 6 and the number of first, second, third class passengers were in the ratio 2 : 3 : 5. If the total amount received was Rs. 51350, then find the amount received from third class passengers.

Sol:

$$\boxed{\text{Total amount = No. of passengers} \times \text{Fare}}$$

Ratio of fares = 11 : 9 : 6 $\Rightarrow$ 1st class = 11x, 2nd class = 9x, 3rd class = 6x

No. of passengers = 2 : 3 : 5 $\Rightarrow$ 1st class = 2y, 2nd class = 3y, 3rd class = 5y

Amount of 1st class = 11$x \times$ 2y = 22xy Amount of 2nd class = 9$x \times$ 3y = 27xy,

Amount of 3rd class = 6$x \times$ 5y = 30xy

Total amount = 22xy + 27xy + 30xy = Rs. 51350

$\Rightarrow$ 79xy = Rs. 51350 $\Rightarrow$ xy = Rs. 650

∴ Amount received from 3rd class passengers = 30xy = 30 × 650 = Rs. 19500.

Shortcut:

		1st class		2nd class		3rd class
Fares	$\rightarrow$	11	:	9	:	6
No. of passengers	$\rightarrow$	2	:	3	:	5
Total amount	$\rightarrow$	11 × 2	:	9 × 3	:	6 × 5 = 22 : 27 : 30.

Total amount   = 79 parts $\longrightarrow$ Rs. 51350

3rd class amount = 30 parts $\longrightarrow$ $? = \dfrac{30 \times 51350}{79} = $ Rs. 19500.

Q - 48 In a library, there are three different types of books A, B and C in the ratio 15 : 9 : 11 and its prices are in the ratio 3 : 7 : 12. If number of books of type B is 18 and prices of each book of type is Rs. 3600, then find the total price of all the books of type A.

Sol:

$$\boxed{\text{Total price of all books} = \text{No. of books} \times \text{Price per book}}$$

No. of books A : B : C = 15 : 9 : 11 $\Rightarrow$ A = $15x$, B = $9x$, C = $11x$

Prices   A : B : C = 3 : 7 : 12 $\Rightarrow$ A = $3y$, B = $7y$, C = $12y$

Given that, no. of books of type B = $9x$ = 18 $\Rightarrow$ $x = 2$

Type A no. of books = $15x = 15 \times 2 = 30$

Also, type C price per book = $12y$ = Rs. 3600 $\Rightarrow$ $y = 300$

Type A price per book = $3y = 3 \times 300 = $ Rs.900

$\therefore$ Total price of all books of type A = $30 \times 900 = $ Rs. 27000.

Q - 49 X varies directly with square of Y and inversely with Z. If Y = 6, Z = 9, then X = 12. Find the value of X when Y = 8, Z = 12.

Sol:

$X \alpha Y^2, X \alpha \dfrac{1}{Z} \Rightarrow X \alpha \dfrac{Y^2}{Z} \Rightarrow X = k\dfrac{Y^2}{Z}$ Where 'k' is a constant

$Y = 6, Z = 9, X = 12$

$12 = k \times \dfrac{6 \times 6}{9} \Rightarrow k = 3$

$\therefore X = k\dfrac{Y^2}{Z} = 3 \times \dfrac{8 \times 8}{12} = 16.$ $\because Y = 8, Z = 12$

Q - 50 The monthly income of Sushma is Rs. 24000 and her expenditure is Rs. 16000. Her expenses are varies directly with square of her monthly income. What is the income of Sushma, if her expenses are equal to half of the income?

Sol:

$E \alpha I^2$ $\Rightarrow$ $E = kI^2$ $I = $ Rs. 24000, $E = $ Rs. 16000

$16000 = k (24000)^2$ $\Rightarrow$ $k = \dfrac{1}{36000}$

$E = kI^2$ $\Rightarrow$ $\dfrac{I}{2} = kI^2$ $\because E = \dfrac{I}{2}$

Therefore, Income $I = \dfrac{1}{2k} = \dfrac{36000}{2} = $ Rs. 18000.

Q - 51 | **What number must be subtracted from each of the numbers 72, 46, 52, 34 so that the remainders are in proportion?**

Sol: Consider, number to be subtracted is '*x*'

$$\frac{72 - x}{46 - x} = \frac{52 - x}{34 - x} \qquad \because \text{Proportion means equality of 2 ratios}$$

$$\Rightarrow \qquad 2448 - 34x - 72x + x^2 = 2392 - 46x - 52x + x^2$$

$$\Rightarrow \qquad 8x = 56 \qquad \Rightarrow \qquad x = 7.$$

> **Note:** This problem can be solved easily through options.

Q - 52 | **There are two types of expenses for booking rooms in a hotel. One is fixed and other is depends on number of guests. If there are 15 guests, total expenses are Rs. 9000 and if there are 35 guests average expenses per guest is Rs. 400. What is the total expenses in the hotel when there are 50 guests?**

Sol: Consider, fixed price = k, price per guest = *x*

For 15 guests, total expenses = $k + 15x = 9000 \quad \rightarrow (1)$

For 35 guests, total expenses = $k + 35x = 14000 \rightarrow (2)$

$(2) - (1) \qquad \Rightarrow \qquad 20x = 5000 \qquad\qquad \because 35 \times 400 = 14000$

Price per guest *x* = Rs. 250 ...... Substitute in (1)

$k + 15 \times 250 = 9000 \qquad \Rightarrow \qquad$ Fixed price k = Rs. 5250

$\therefore$ For 50 guests, total expenses = $k + 50x = 5250 + 50 \times 250 = $ Rs. 17750.

Q - 53 | **The time period (T) of a pendulum is directly proportional to the square root of length of the string. If the time period of such a pendulum is 60 seconds and length of the string is 25 cm. Find the length of the string, if the time period is 72 seconds.**

Sol: $T \alpha \sqrt{l} \qquad \Rightarrow \qquad T = k\sqrt{l} \qquad\qquad$ where 'k' is a constant

$60 = k\sqrt{25} \qquad \Rightarrow \qquad k = 12 \qquad\qquad \because T = 60 \text{ seconds}, l = 25 \text{ cm}$

T = 72 seconds, *l* = ?

$T = k\sqrt{l} \qquad \Rightarrow \qquad 72 = 12\sqrt{l} \qquad \Rightarrow \qquad \sqrt{l} = 6 \qquad \Rightarrow \qquad l = 36 \text{ cm}$

$\therefore$ Length of the string is 36 cm.

Q - 54 | The value of diamond is directly proportional to the square of its weight. It breaks into three pieces and their weights are in the ratio 2 : 4 : 7. So that there is a loss of 25 lakhs. Find the actual value of diamond.

Sol: Given that, value α (weight)^2

Weight ratio = 2 : 4 : 7 = 2x, 4x, 7x

Without breakage total weight = $2x + 4x + 7x = 13x$

Actual value of diamond $(13x)^2 = 169x^2$

With breakage total value = $(2x)^2 + (4x)^2 + (7x)^2 = 69x^2$

Loss = $169x^2 - 69x^2 = 100x^2$ $\because$ Loss = 25 lakhs

$100x^2 = 25$ lakhs $\Rightarrow$ $x^2 = 25000$

$\therefore$ Actual value of diamond = $169x^2$ = 169 × 25000 = Rs. 42.25 lakhs.

Q - 55 | Time period of a pendulum is directly proportional to the square root of length of the string and inversely proportional to square root of gravitational constant. When the gravitational is 16 m/sec^2 and length of string is 49 m and then time period is 14 seconds. What is the time period, if the length of string is 225 m and gravitational constant is 144 m/sec^2?

Sol: $T \alpha \sqrt{l}$, $T \alpha \dfrac{1}{\sqrt{g}}$ $\Rightarrow$ $T \alpha \dfrac{\sqrt{l}}{\sqrt{g}}$ $\Rightarrow$ $T = k\sqrt{\dfrac{l}{g}}$ Where 'k' is a constant

Given that, g = 16 m/sec^2, l = 49 m, T = 14 seconds

$14 = k\sqrt{\dfrac{49}{16}}$ $\Rightarrow$ $k = 8$

g = 144 m/sec^2, l = 225 m, T = ?

$T = k\sqrt{\dfrac{l}{g}}$ $\Rightarrow$ $T = 8\sqrt{\dfrac{225}{144}} = 10$ seconds

$\therefore$ Time period = 10 seconds.

ASSESSMENT TEST

1. In a ratio 5 : 7, if antecedent is 30, then find the value of consequent.

2. If p : q = 2 : 5, q : r = 4 : 3 and r : s = 6 : 7, then find p : q : r : s.

3. If $a : b = \frac{1}{2} : \frac{3}{5}$, $b : c = \frac{4}{7} : \frac{9}{14}$ and $c : d = \frac{3}{4} : \frac{7}{8}$, then find the value of a : d.

4. If 3A = 4B = 6C, then what is the value of A : B : C?

5. If 8A = 17B and 13B = 12C, then find A : C.

6. Find the fourth proportional to the numbers

   a) 72, 64, 27   b) 0.056, 0.63, 0.512

7. Find the third proportional to the numbers

   a) 54, 81   b) 0.8, 0.16

8. Find the mean proportional between

   a) 529, 676   b) 1.089, 32.4

9. If 15% of (3A + B) = 25% of (4A − B), then find A : B.

10. If p : q = 6 : 11, then find (3q − 4p) : (q + 2p).

11. If $(8x^2 − 4y^2) : (4y^2 − 5x^2) = 7 : 5$, then find the value of $x : y$.

12. What number must be added to each of the numbers 57, 68, 93, 110, so that the remainders are in proportion?

13. Two – fifth of a number is equal to 54% of the second number. What is the ratio of first number to the second number?

14. Two numbers are in the ratio of 7 : 11. If 8 is added to both the numbers their ratio changes to 9 : 14. What is the smallest number?

15. Three numbers are in the ratio 3 : 5 : 7. If the difference between the squares of second and third numbers is 1536, then find the three numbers.

16. The ratio of boys and girls in a school is 3 : 4. If 50 boys increases the ratio becomes 42 : 51, then find the number of boys after increment.

17. Find the least whole number which when subtracted from both the terms of the ratio 8 : 9 gives a ratio less than 12 : 17.

18. If the ratio of length and perimeter of a rectangle is 7 : 22, then what is the ratio of length and breadth of a rectangle?

19. The incomes of A, B and C are in the ratio 5 : 6 : 8. If the income of A is increased by 20%, B is increased by 25% and C is decreased by 10%. Find the new ratio of their incomes.

20. There are two numbers such that their difference, their sum and their product are in the ratio 5 : 9 : 28. Find the product of two numbers.

21. An amount of Rs. 1600 is distributed among x, y and z in the ratio 6 : 5 : 14. What is the difference between the shares of y and z?

22. A, B and C play cricket. The ratio of A's runs to B's runs and B's runs to C's runs is 4 : 5 each. Altogether they score 549 runs, then how many runs does C score?

23. An amount of Rs. 34870 is divided among P, Q and R such that P receives two – third of Q and R together and R receives four – seventh of P and Q together. What is the amount received by Q?

24. If the positions of the digits of a two – digit number are interchanged, then the ratio of original number and newly formed number is 2 : 9. Also, the difference between original number and new number is 63. Find the newly formed number.

25. Three persons P, Q and R divides an amount of Rs. 530 in such a way that Q gets 20 less than P and R gets 60 more than Q. Find the ratio of their shares.

26. A shopkeeper distributes some pens among four sons w, x, y, z in the ratio $\frac{1}{4} : \frac{1}{6} : \frac{1}{7} : \frac{1}{9}$. What is the minimum number of pens that the shopkeeper should have?

27. Out of 150 applications for a post 90 are males and 100 members have driving license. What is the ratio between the minimum and maximum number of males having driving license?

28. The incomes of p, q and r are in the ratio 10 : 13 : 15 and their expenditures are in the ratio 8 : 9 : 12. If p saves one – fifth of his income, then find the ratio of their savings.

29. Ramu distributes Rs. 5250 to his two sons, three daughters and four nephews. If each daughter receives 3 times that of each nephew and each son receives 4 times that of each nephew. How much does each son receives?

30. The incomes of Anil and Sunil are in the ratio 7 : 6 and their expenditures are in the ratio 10 : 7. If each saves Rs. 6600, then find their expenditures.

31. The sum of monthly incomes of A, B and C is Rs. 11300. For every rupee that A earns, B earns 70 paisa and for every rupee that B earns, C earns 80 paisa. Find the monthly income of B.

32. The number of students in three sections A, B and C are in the ratio 5 : 7 : 9. If 8 students are shifted from B to A, the ratio of students in their two sections are interchanged. Find the number of students in section A initially.

33. The ratio of males and females in a city is 8 : 11 respectively and the percentage of children among males and females are 15% and 20% respectively. If the number of adult males in the city is 234600, then what is the population of the city?

34. The monthly salary of Mr. Nithish is Rs. 40000 and his expenses are Rs. 25000. His expenses are varies directly with square of his monthly income. What is the salary of Mr. Nithish, if his expenses are equal to one – fourth of the income?

35. In a bag for every three notes of Rs. 10 there are two notes of Rs. 20. For every four notes of Rs. 20 there are three notes of Rs. 50. If the total amount in the bag was Rs. 7250, then find number of Rs. 50 notes.

36. x varies directly with square root of y and inversely with z, if $y = 16$, $z = 7$ then $x = 24$. Find the value of x, when $y = 225$ and $z = 21$.

37. An amount of Rs. 3640 is divided among A, B, C. If their shares are decreased by Rs. 32, Rs. 45 and Rs. 53 respectively, the remaining amount will be in the ratio of 6 : 11 : 9. How much amount did C gets initially?

38. The electricity bill of a certain establishment is partly fixed and partly varies as the number of units of electricity consumed. When in a certain month 760 units are consumed, the bill is Rs. 2400. In another month 850 units are consumed and the bill is Rs. 2670. In yet another month 600 units are consumed. What is the bill for that month?

39. Mr. Sriram have 4180 gold coins and divides them among his three sons Ramesh, Mahesh and Suresh. Out of total coins received by each of them. Ramesh donates 62 coins, Mahesh donates 37 coins and Suresh donates 21 coins. Now the ratio of gold coins with them are in the ratio 13 : 7 : 8. How many coins did Ramesh receive from Mr. Sriram?

40. There are 3 types of boxes x, y and z are in the ratio 7 : 9 : 13 and each box carries some balls in the ratio 6 : 11 : 4. If the number of boxes of type y is 27 and the number of balls from each box of type x is 2400, then what is the total number of balls carried by type z box?

41. An amount of Rs. 2450 is divided among 80 members consisting of boys, girls and children. Total amounts given to them in the ratio 24 : 21 : 25 respectively and amount received by each boy, girl and child are in the ratio 6 : 3 : 5. Find the amount received by each boy.

42. The fares of aeroplane, train and bus between two places were 12 : 9 : 7 and the ratio of number of passengers from aeroplane, train and bus were 2 : 5 : 9. If the total amount received from all modes of transportation was Rs. 32340, then what is the amount received from the passengers, who preferred aeroplane?

43. Time period (T) of pendulum is directly proportional to square root of length of the string. If the time period of such a pendulum is 98 seconds and length of the string is 49 m. Find the length of the string, if the time period is 112 seconds.

44. The value of diamond is directly proportional to the square of its weight. It breaks into three pieces and their weights are in the ratio 3 : 5 : 6, so that there is a loss of 43.8 lakhs. Find the actual value of the diamond.

45. Time period of pendulum is directly proportional to square root of length of the string and inversely proportional to square root of gravitational constant. When the gravitational constant is 64 m/sec^2 and length of string is 25 m and then time period is 15 seconds. What is the time period of pendulum, if the length of string is 441 m and gravitational constant is 324 m/sec^2?

KEY

1. 32
2. 16 : 40 : 30 : 35
3. 40 : 63
4. 4 : 3 : 2
5. 51 : 26
6. a) 24   b) 5.76
7. a) 121.5   b) 0.032
8. a) 598   b) 5.94
9. 8 : 11
10. 9 : 23
11. 4 : 5
12. 9
13. 27 : 20
14. 280
15. 24, 40, 56
16. 560
17. 6
18. 7 : 4
19. 20 : 25 : 24
20. 56
21. Rs. 576
22. 225
23. Rs. 8242
24. 81
25. 17 : 15 : 21
26. 169
27. 4 : 9
28. 2 : 4 : 3
29. Rs. 1000
30. Rs. 6000, Rs. 4200
31. Rs. 3500
32. 20

33. 655500
34. Rs. 16000
35. 75
36. 30
37. Rs. 1268
38. Rs. 1920
39. 1947
40. 62400
41. Rs. 42
42. Rs. 5880
43. 64 m
44. 57.8 lakhs
45. 28 seconds

AVERAGES

AVERAGE: In general average is calculated by representing a group of values with a single value and it is defined as

$$\text{Average} = \frac{Sum\ of\ all\ observations}{No.of\ observations}$$

Note: Average of any number of quantities will always be greater than the lowest value and lesser than the highest value.

Average Speed:

❖ Basically average speed is depends on two conditions.

    1) Distances are equal   2) Distances are not equal

1) Distances are equal:

  ✓ If there are 2 equal distances,

        $S_1 \rightarrow$ Speed during first half of the distance

        $S_2 \rightarrow$ Speed during second half of the distance

        $T_1 \rightarrow$ Time taken to cover first half of the distance

        $T_2 \rightarrow$ Time taken to cover second half of the distance

$$\text{Average Speed} = \frac{Total\ distance}{Total\ time} = \frac{D+D}{T_1+T_2}$$

$$\text{Where} \quad T_1 = \frac{D}{S_1}, \ T_2 = \frac{D}{S_2} \qquad \because S = \frac{D}{T} \qquad \Rightarrow \qquad T = \frac{D}{S}$$

$$\text{Average Speed} = \frac{2D}{\frac{D}{S_1} + \frac{D}{S_2}} = \frac{2D}{D[\frac{1}{S_1} + \frac{1}{S_2}]} = \frac{2S_1S_2}{S_1+S_2}$$

$$\text{Average speed} = \frac{2S_1S_2}{S_1+S_2}$$

  ✓ Similarly, if there are 3 equal distances.

$$\text{Average Speed} = \frac{Total\ distance}{Total\ time}$$

$$\text{Average Speed} = \frac{D+D+D}{T_1+T_2+T_3} = \frac{3D}{\frac{D}{S_1} + \frac{D}{S_2} + \frac{D}{S_3}} = \frac{3D}{D[\frac{1}{S_1} + \frac{1}{S_2} + \frac{1}{S_3}]}$$

$$\text{Average speed} = \frac{3S_1S_2S_3}{S_1S_2 + S_2S_3 + S_3S_1}$$

2) **Distances are not equal:**

✓ If distances are not equal, then

Average Speed = $\dfrac{Total\ distance}{Total\ time}$

$$\text{Average speed} = \dfrac{D_1 + D_2 + D_3}{T_1 + T_2 + T_3}$$

D_1	D_2	D_3
S_1	S_2	S_3
T_1	T_2	T_3

Some important formulae to be remember:

✓ Sum of 'n' natural numbers $= \dfrac{n(n+1)}{2}$.

✓ Sum of squares of 'n' natural numbers $= \dfrac{n(n+1)(2n+1)}{6}$.

✓ Sum of cubes of 'n' natural numbers $= \dfrac{n^2(n+1)^2}{4}$ or $\left[\dfrac{n(n+1)}{2}\right]^2$.

✓ a) The average of first 'n' consecutive odd numbers is '**n**'.

   b) The average of first 'n' consecutive even numbers is '**n+1**'.

✓ a) Average of odd numbers from 1 to n is

   Average $= \dfrac{last\ odd\ number + 1}{2}$.

   b) Average of even numbers from 1 to n is

   Average $= \dfrac{last\ even\ number + 2}{2}$.

✓ If the difference between any two consecutive numbers is same, then

   Average $= \dfrac{First\ number + last\ number}{2}$.

✓ The average of 'n' numbers is A.

   a) if each number is multiplied by 'k', then new average $= A \times k$

   b) if each number is added by 'k', then new average $= A + k$

   c) if each number is subtracted by 'k', then new average $= A - k$

   d) if each number is divided by 'k', then new average $= A \div k$

SOLVED EXAMPLES

Q - 1 **What is the average of all numbers 47, 59, 69, 57, 79, 73?**

Sol: $\text{Average} = \dfrac{Sum\ of\ observations}{No.of\ observations} = \dfrac{47 + 59 + 69 + 57 + 79 + 73}{6} = 64.$

Alternate method:

✓ By observing all the given numbers, take one approximate average round figure number which is in between lowest and highest numbers.

✓ Consider 60 is the approximate average round figure number.

✓ Now, compare every number with approximate average. If the number is less take –ve sign, if the number is more take +ve sign.

$\therefore \text{Average} = 60 + \dfrac{(-13 - 1 + 9 - 3 + 19 + 13)}{6} = 60 + \dfrac{24}{6} = 64.$

Q - 2 **Find the average of all prime numbers between 40 and 70.**

Sol: Prime numbers between 40 and 70 = 41, 43, 47, 53, 59, 61, 67

$\text{Average} = \dfrac{Sum\ of\ observations}{No.of\ observations} = \dfrac{41 + 43 + 47 + 53 + 59 + 61 + 67}{7} = 53.$

Alternate method:

Consider, approximate average = 50, then compare every number with 50

$\therefore \text{Average} = 50 + \dfrac{(-9 - 7 - 3 + 3 + 9 + 11 + 17)}{7} = 50 + \dfrac{21}{7} = 53.$

Q - 3 **Find the average of 73 natural numbers.**

Sol: $\text{Average} = \dfrac{Sum\ of\ 73\ natural\ numbers}{No.of\ observations}$ ∵ Sum of 'n' natural no's $= \dfrac{n(n + 1)}{2}$

Sum of 73 natural numbers $= \dfrac{73(73 + 1)}{2} = \dfrac{73 \times 74}{2}$

$\text{Average} = \dfrac{\frac{73 \times 74}{2}}{73} = 37.$

Shortcut:

Average of 1, 2, 3, …..73

For the above numbers difference between any two consecutive numbers is same, then

$$\text{Average} = \dfrac{First\ number + Last\ number}{2}$$

$\therefore \text{Average} = \dfrac{1 + 73}{2} = 37.$

Q - 4   **Find the average of squares of natural numbers from 1 to 35.**

Sol: Sum of squares of 'n' natural numbers $= \dfrac{n(n+1)(2n+1)}{6}$

Sum of squares from $1 - 35 = \dfrac{35 \times 36 \times 71}{6}$ $\because n = 35$

$\therefore$ Average $= \dfrac{\frac{35 \times 36 \times 71}{6}}{35} = \dfrac{36 \times 71}{6} = 426.$

Q - 5   **What is the average of cubes of natural numbers from 1 to 31?**

Sol: Sum of cubes of 'n' natural numbers $= \dfrac{n^2(n+1)^2}{4}$

Sum of cubes from $1 - 31 = \dfrac{31^2 \times 32^2}{4}$

$\therefore$ Average $= \dfrac{Sum\ of\ observations}{No.of\ observations} = \dfrac{\frac{31^2 \times 32^2}{4}}{31} = \dfrac{31 \times 32^2}{4} = 7936.$

Q - 6   **Find the average of first 80 consecutive odd numbers.**

Sol: Average of first 'n' consecutive odd numbers = n

Here n = 80, therefore Average = 80.

Q - 7   **Find the average of first 53 consecutive even numbers.**

Sol: Average of first 'n' consecutive even numbers = n + 1

Here n = 53, therefore Average = 53 + 1 = 54.

Q - 8   **What is the average of odd numbers from**
a) 1 to 65   b) 1 to 74

Sol: Average of odd numbers from 1 to n $= \dfrac{last\ odd\ number + 1}{2}$

a) Average of odd numbers from 1 to 65 $= \dfrac{65+1}{2} = 33.$

b) Average of odd numbers from 1 to 74 $= \dfrac{73+1}{2} = 37.$

$\because$ Last odd number from 1 to 74 is 73

Q - 9   **What is the average of even numbers from**
a) 1 to 84   b) 1 to 57

Sol: Average of even numbers from 1 to n $= \dfrac{Last\ even\ number + 2}{2}$

a) Average of even numbers from 1 to 84 $= \dfrac{84+2}{2} = 43.$

b) Average of even numbers from 1 to 57 $= \dfrac{56+2}{2} = 29.$

$\because$ Last even number from 1 to 57 is 56

Q - 10 **Find the average of first 16 multiples of 8.**

Sol: $\text{Average} = \dfrac{Sum\ of\ observations}{No.of\ observations} = \dfrac{8[1 + 2 + ... + 16]}{16}$

Sum of 16 natural numbers $= \dfrac{16 \times 17}{2}$ $\qquad\quad$ $\because$ Sum of 'n' natural numbers $= \dfrac{n(n + 1)}{2}$

$\therefore \text{Average} = \dfrac{8[\frac{16 \times 17}{2}]}{16} = 68.$

Shortcut:

First 16 multiples of 8 = 8, 16, 24, . . . ,120, 128

Since, the difference between any two consecutive numbers is same, then

$$\text{Average} = \dfrac{First\ number + Last\ number}{2}$$

$\therefore \text{Average} = \dfrac{8 + 128}{2} = 68.$

Q - 11 **Find the average of multiples of 16 in the natural number series from 150 to 350.**

Sol: First multiple of 16 after 150 = 160, Last multiple of 16 before 350 = 336

Average of numbers 160, 176, . . . , 336

$\because$ The difference between any two consecutive numbers is same,

$\therefore \text{Average} = \dfrac{First\ number + Last\ number}{2} = \dfrac{160 + 336}{2} = 248.$

Q - 12 **The average of 15 numbers is 9. What will be the new average, if each number is multiplied by 12?**

Sol:

Note: The average of 'n' numbers is A. If each number is multiplied by 'k', then

New average = A × k

Here, initial average A = 9 and k = 12

$\therefore$ New average = 9 × 12 = 108.

Q - 13 **There are 40 boys and 50 girls in a class. If the average weight of boys is 54 kg and the average weight of girls is 45 kg. What is the average weight of entire class?**

Sol: $\text{Average weight of entire class} = \dfrac{Sum\ of\ boys + Sum\ of\ girls}{Total\ strength}$

Sum of boys = Average of boys × No. of boys = 54 × 40 = 2160 kg

Sum of girls = Average of girls × No. of girls = 45 × 50 = 2250 kg

$\therefore \text{Average weight of entire class} = \dfrac{2160 + 2250}{40 + 50} = 49$ kg.

Q - 14 | **Naresh bought 15 pens, 6 books and 9 pencil boxes each at a rate of Rs. 12, Rs. 25 and Rs. 10 respectively. Find the average cost of all the items.**

Sol:

Average cost of all items $= \dfrac{Total\ cost\ of\ all\ items\ together}{No.of\ items}$

$\therefore$ Average $= \dfrac{15 \times 12 + 6 \times 25 + 9 \times 10}{15 + 6 + 9} = $ Rs. 14.

Q - 15 | **If the average of 8 consecutive odd numbers is 32. Find the largest of these numbers.**

Sol:

Consider, 8 consecutive odd numbers are $x, x + 2, x + 4, x + 6, x + 8, x + 10, x + 12, x + 14$

$\because$ Gap between any two consecutive odd numbers is 2

Average $= \dfrac{Sum\ of\ bservations}{No.of\ observations} = \dfrac{x + x + 2 + x + 4 + x + 6 + x + 8 + x + 10 + x + 12 + x + 14}{8}$

$32 = \dfrac{8x + 56}{8} \qquad \Rightarrow \qquad 8x = 32 \times 8 - 56 = 200 \qquad \Rightarrow \qquad x = 25$

$\therefore$ Largest number $= x + 14 = 25 + 14 = 39.$

Alternate method:

Consider, 8 consecutive odd numbers are $x - 7, x - 5, x - 3, x - 1, x + 1, x + 3, x + 5, x + 7$

Average $= \dfrac{x - 7 + x - 5 + x - 3 + x - 1 + x + 1 + x + 3 + x + 5 + x + 7}{8}$

$32 = \dfrac{8x}{8} \qquad \Rightarrow \qquad x = 32$

$\therefore$ Largest number $= x + 7 = 32 + 7 = 39.$

Shortcut:

✓ Average means middle number. Since 32 is an even number therefore 32 is not one of the 8 consecutive odd numbers.

25	27	29	31	33	35	37	39

Left side 4 consecutive odd no's ← | 32 Avg. | → Right side 4 consecutive odd no's

$\therefore$ Largest number $= 39.$

Q - 16 | **A boy is going to school from his house at a speed of 30 kmph and return home at a speed of 45 kmph. Find the average speed.**

Sol:

In the given problem, distances are equal in both the cases.

Average speed $= \dfrac{2S_1 S_2}{S_1 + S_2}$

$\therefore$ Average speed $= \dfrac{2 \times 30 \times 45}{30 + 45} = 36$ kmph.

Q - 17 A man covers three equal distances at a speeds of 15 kmph, 20 kmph and 30 kmph respectively, then find the average speed during the entire journey.

Sol: According to question, man covers 3 equal distances.

$$\text{Then, Average speed} = \frac{3S_1 S_2 S_3}{S_1 S_2 + S_2 S_3 + S_3 S_1}$$

$$\therefore \text{Average speed} = \frac{3 \times 15 \times 20 \times 30}{15 \times 20 + 20 \times 30 + 30 \times 15} = 20 \text{ kmph.}$$

Q - 18 An aeroplane travels distances 1200 km, 1500 km and 2000 km with a speeds of 600 kmph, 500 kmph and 400 kmph respectively. What is the average speed of entire journey?

Sol: In the given problem, distances are not equal.

$$\text{Then, Average Speed} = \frac{Total\ distance}{Total\ time} = \frac{D_1 + D_2 + D_3}{T_1 + T_2 + T_3}$$

$$T_1 = \frac{D_1}{S_1} = \frac{1200}{600} = 2 \text{ hours, } T_2 = \frac{D_2}{S_2} = \frac{1500}{500} = 3 \text{ hours, } T_3 = \frac{D_3}{S_3} = \frac{2000}{400} = 5 \text{ hours}$$

$$\therefore \text{Average Speed} = \frac{1200 + 1500 + 2000}{2 + 3 + 5} = 470 \text{ kmph.}$$

Q - 19 The average of 11 consecutive even numbers is 80, then what is the difference between smallest and largest numbers?

Sol: Consider, 11 consecutive even numbers are

$x - 10, x - 8, x - 6, x - 4, x - 2, x, x + 2, x + 4, x + 6, x + 8, x + 10$

$$\text{Average} = \frac{Sum\ of\ observations}{No. of\ observations} \qquad \Rightarrow \qquad 80 = \frac{11x}{11} \qquad \Rightarrow \qquad x = 80$$

Smallest number $= x - 10 = 80 - 10 = 70$ & Largest number $= x + 10 = 80 + 10 = 90$

$\therefore$ Difference between smallest and largest numbers $= 90 - 70 = 20$.

Shortcut – 1:

✓ Average means middle number. Since 80 is an even number therefore 80 is one of the 11 consecutive even numbers which is middle number.

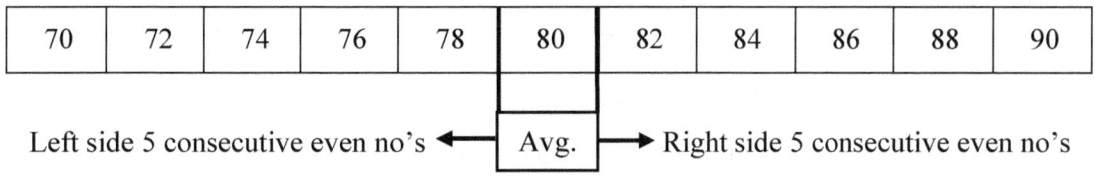

70	72	74	76	78	80	82	84	86	88	90

Left side 5 consecutive even no's ← Avg. → Right side 5 consecutive even no's

Smallest number = 70, Largest number = 90

$\therefore$ Difference = 90 – 70 = 20.

Shortcut – 2:

> **Note:**
>
> ✓ If there are n consecutive numbers, then $(n - 1)$ gaps are formed between smallest and largest numbers.
>
> ✓ The gap between any two consecutive even numbers (or) odd numbers is always '2'.

❖ There are 11 consecutive even numbers, therefore 10 gaps are formed between smallest and largest numbers.

∴ Difference = 10 gaps = $10 \times 2 = 20$.   ∵ Each gap = 2

Q - 20 **The average height of A and B is 155 cm, that of B and C is 162 cm and the average height of C and A is 170 cm. Find the height of A.**

Sol: Sum of observations = Average × No. of observations

$A + B = 155 \times 2 = 310$ cm   →   (1)

$B + C = 162 \times 2 = 324$ cm   →   (2)

$C + A = 170 \times 2 = 340$ cm   →   (3)

$(1) + (2) + (3) \Rightarrow 2(A + B + C) = 310 + 324 + 340 = 974$ cm

$A + B + C = 487$ cm

∴ Height of A = $(A + B + C) - (B + C) = 487 - 324 = 163$ cm.

Q - 21 **The average of 5 consecutive odd numbers A, B, C, D and E is 53. What is the product of B and E?**

Sol: Consider, 5 consecutive odd numbers are

$x - 4$	$x - 2$	x	$x + 2$	$x + 4$
A	B	C	D	E

Average $= \dfrac{x - 4 + x - 2 + x + x + 2 + x + 4}{5} = 53 \Rightarrow x = 53$

$B = x - 2 = 53 - 2 = 51$ & $E = x + 4 = 53 + 4 = 57$

∴ Product of B and E = $51 \times 57 = 2907$.

Shortcut:

✓ Average means middle number. Since 53 is an odd number, therefore 53 is one of the 5 consecutive odd numbers which is middle number.

A	B	C	D	E
49	51	53	55	57

∴ Product of B and E = $51 \times 57 = 2907$.

Q - 22 The average salary of entire staff in an office is Rs. 600 per day. The average salary of officers is Rs. 1500 and that of non – officers is Rs. 400. Find the number of non – officers, if the number of officers are 22.

Sol: Consider, No. of non – officers $= x$

Average of entire staff $= \dfrac{Sum\ of\ officers + Sum\ of\ non-officers}{Total\ staff}$

$600 = \dfrac{1500 \times 22 + 400 \times x}{22 + x}$ $\Rightarrow$ $600 = \dfrac{100(15 \times 22 + 4x)}{22 + x}$

$132 + 6x = 330 + 4x$ $\Rightarrow$ $2x = 198$ $\Rightarrow$ $x = 99$

$\therefore$ No. of non – officers $= 99$.

Q - 23 There are 96 students in a class. The average marks of failed students is 22 and that of passed students is 70. If the average marks in the exam for entire class is 52, then find the number of passed students.

Sol:
Total students = Passed students + Failed students

Consider, No. of passed students $= x$, then

Failed students $= 96 - x$ $\because$ Total no. of students $= 96$

Average of entire class $= \dfrac{Sum\ of\ passed\ students + Sum\ of\ failed\ students}{Total\ strength}$

$52 = \dfrac{70 \times x + 22(96 - x)}{96}$ $\Rightarrow$ $4992 = 70x + 2112 - 22x$

$\Rightarrow$ $48x = 2880$ $\Rightarrow$ $x = 60$

$\therefore$ No. of passed students $= 60$.

Q - 24 Out of four members, first is thrice the second, fourth is twice the second and third is half of first. If the average of all four numbers is 90, then find the third number.

Sol: Consider, second number is 'x'.

1st	2nd	3rd	4th
$3x$	X	$\dfrac{3x}{2}$	$2x$

$\because$ 1st – thrice of 2nd, 4th – twice of 2nd, 3rd – half of 1st

Average $= \dfrac{3x + x + \frac{3x}{2} + 2x}{4} = 90$ $\Rightarrow$ $\dfrac{15x}{2} = 90 \times 4$ $\Rightarrow$ $x = 48$

$\therefore$ Third number $= \dfrac{3x}{2} = \dfrac{3 \times 48}{2} = 72$.

Q - 25 **The average of 37 observations is 21. The average of first 18 of them is 19 and that of last 18 is 20. Find the 19th observation.**

Sol: Sum of 37 observations = (Sum of first 18 obs.) + 19th obs. + (Sum of last 18 obs.)

19th obs. = Sum of 37 obs. – (Sum of first 18 obs.) – (Sum of last 18 obs.)

∴ 19th observation = $37 \times 21 - (18 \times 19) - (18 \times 20) = 75$.

Shortcut:

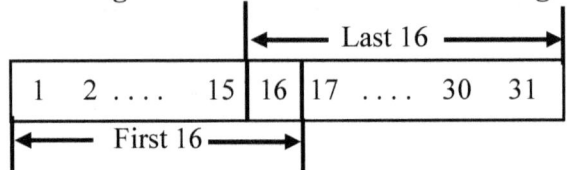

Note: If the middle observation is not the part of first and last half of the no's, then 'sign' of distributed value will be opposite.

For first 18 observations   →   Distributed value = +36

For last 18 observations   →   Distributed value = +18

∴ 19th observation = $21 + (+36 +18) = 75$.

Q - 26 **The average weight of 31 students in a class is 55 kg. If the average weight of first 16 students is 59 kg and that of last 16 student is 52 kg. What is the weight of middle student?**

Sol:

Here 16th value is repeated twice.

Weight of middle student = (Sum of first 16) + (Sum of last 16) – (Sum of 31 students)

∴ Weight of middle student = $16 \times 59 + 16 \times 52 - 31 \times 55 = 71$ kg.

Shortcut:

First 16 students	Last 16 students

1 2 16	16 17 31

55   55   55   55   55 ....   55 $\longrightarrow$ Overall average

$+4\downarrow$ $+4\downarrow$ $+4\downarrow$ $-3\downarrow$ $-3\downarrow$ $-3\downarrow$

59   59   59   52   52   52

$\overline{\qquad +4 \times 16 = +64 \qquad}$ $\overline{\qquad -3 \times 16 = -48 \qquad}$ $\longrightarrow$ Distributed value

Weight of middle student = Overall average + Distributed value

$\therefore$ Weight of middle student = $55 + (+64 - 48) = 71$ kg.

> **Note:** If the middle observation is one the part of first and last half of the numbers, then 'sign' of distributed value will not change.

Q - 27   **The average of 14 results is 42. It was found later that one result 94 was wrongly entered as 38. Find the new average.**

Sol: Sum of 14 results = $14 \times 42 = 588$ $\because$ Sum = Average × No. of obs.

New Average $= \dfrac{New\ Sum}{No.of\ obs.} = \dfrac{588 + 94 - 38}{14} = 46.$ $\because$ Correct no. = 94 & Wrong no. = 38

(or)

$\therefore$ New Average $= 42 + \dfrac{94 - 38}{14} = 46.$

> Use, +ve sign for correct result
> −ve sign for wrong result

Q - 28   **The average marks of 25 students in a class is increased by 1.8 when one of the students who got 40 marks is replaced by another student. How many marks secured by new student?**

Sol: Consider, average of 25 students = x

Sum of 25 students = $25x$ $\because$ Sum = Avg. × No. of observations

New Average $= \dfrac{New\ Sum}{No.of\ Obs.}$

$\dfrac{25x - 40 + S_n}{25} = x + 1.8$ $\because$ Average is increased by 1.8

S_n – Marks of new student

$25x - 40 + S_n = 25x + 25 \times 1.8$ $\Rightarrow$ $S_n = 40 + 45 = 85$

$\therefore$ New Student secured 85 marks.

Shortcut:

✓ Average of a class increased means new student marks definitely more than replaced student i.e., 40.

✓ Because of replacement of one student, all 25 students marks are increased by 1.8 each.

New student marks = 40 + (25 × 1.8) = 85.

Q - 29 | **The average wages of 15 workers are increased by Rs. 300 when two of them whose wages are Rs. 2600 and Rs. 3400 replaced by two new workers. Find the average wages of two new workers.**

Sol:　　　Consider, average of 15 workers = x, then

Sum of 15 workers = $15x$　　　　　　　　　　　∵ Sum = Avg. × No. of observations

New Average = $\dfrac{New\ Sum}{No.of\ Obs}$ 　　$\Rightarrow$　　$\dfrac{15x - 2600 - 3400 + W_1 + W_2}{15}$

Where W_1, W_2 → New workers wages　　　　∵ Average increased by 300

$15x - 6000 + W_1 + W_2 = 15x + 4500$　　$\Rightarrow$　　$W_1 + W_2 = 10500$

∴ Average = $\dfrac{W_1 + W_2}{2} = \dfrac{10500}{2}$ = Rs. 5250.

Shortcut:

✓ Average of workers increased means sum of new workers wages definitely more than the sum of replaced workers i.e., 2600 + 3400 = 6000.

✓ Because of replacement of 2 workers, all 15 workers wages are increased by Rs. 300 each.

Sum of new workers = 6000 + (15 × 300) = 10500

∴ Average = $\dfrac{Sum\ of\ new\ workers}{2} = \dfrac{10500}{2}$ = Rs. 5250.

Q - 30 | **The average weight of 8 persons sitting in a boat is 43 kg. If the weight of boat is also included, then the average is increased to 55 kg. What is the weight of the boat?**

Sol:　　　Sum of 8 persons = 43 × 8 = 344 kg　　　　∵ Sum = Avg. × No. of observations

Sum of 8 persons and boat = 55 × 9 = 495 kg

Weight of boat = Sum of 8 persons and boat – Sum of 8 persons

∴ Weight of boat = 495 – 344 = 151 kg.

Shortcut:

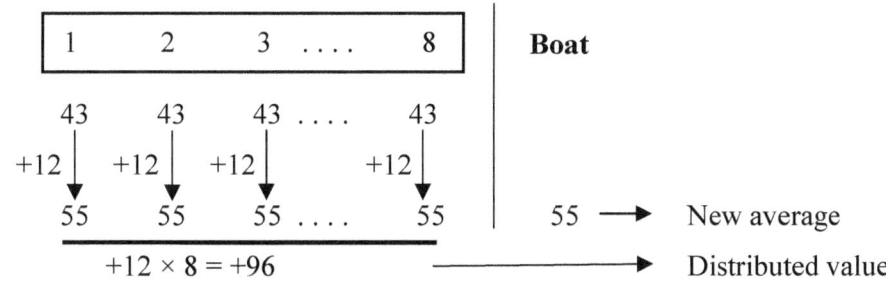

Boat weight = New average + Distributed value

∴ Boat weight = 55 + 96 = 151 kg.

Q - 31 **The average marks of a class are 70%. The average marks of 15 students who failed is 30%. If these students are excluded, the average marks of the class would increase by 15% points. Find the strength of the class**

Sol: Consider, Total strength of a class = x and also, given that Failed students = 15

Therefore, Passed students = $x - 15$ ∵ Pass = Total – Fail

Sum of entire class = $70 \times x$ ∵ Sum = Average × No. of observations

Sum of failed students = $30 \times 15 = 450$

Average of passed students $= \dfrac{Sum\ of\ passed\ students}{No.of\ passed\ students}$

$(70 + 15) = \dfrac{70x - 450}{x - 15}$ ⇒ $85x - 1275 = 70x - 450$ ⇒ $15x = 825$

∴ Total strength of class $x = 55$.

Q - 32 **A batsman makes a score of 128 runs in the 21ˢᵗ innings and thus increases his average by 4 runs. Find the average after 21ˢᵗ innings.**

Sol: Consider, average after 20 innings = x, after 21 innings = $x + 4$

Sum of 20 innings = $20 \times x = 20x$ ∵ Sum = Average × No. of observations

Average after 21ˢᵗ innings $= \dfrac{Sum\ of\ 20\ innings + 21st\ innings}{No.of\ innings}$

$x + 4 = \dfrac{20x + 128}{21}$ ⇒ $21x + 84 = 20x + 128$ ⇒ $x = 44$

∴ Average after 21 innings = $x + 4 = 44 + 4 = 48$.

Shortcut:

$$
\boxed{\begin{array}{ccc} 1 & 2 & \dots & 20 \end{array}} \quad \textbf{21}^{\textbf{st}} \textbf{ innings}
$$

Initial average $=$ x x . . . . x 128 runs

$+4\downarrow$ $+4\downarrow$ $+4\downarrow$

New average $= x + 4$ $x + 4$ . . . . $x + 4$ $x + 4$

$+4 \times 20 = +80 \quad \longrightarrow \quad$ Distributed value

$$\boxed{\text{Runs in } 21^{st} \text{ innings} = \text{New average} + \text{Distributed value}}$$

$128 = (x + 4) + 80 \qquad \Rightarrow \qquad x + 4 = 48$

$\therefore$ Average after 21st innings $= x + 4 = 48$.

Q - 33 | **The average runs of a batsman in 10 innings is 52. If the highest and lowest runs were excluded, then the average would decrease by 2. If the difference between these two innings is 74, then find the highest score.**

Sol: Sum of 10 innings $= 52 \times 10 = 520$ $\because$ Sum $=$ Average $\times$ No. of observations

Sum of 8 innings $= 520 - (H + L)$ $\because$ H – Highest runs, L – Lowest runs

Average of 8 innings $= \dfrac{520 - (H + L)}{8} = 52 - 2 = 50$ Average is decreased by 2

$520 - (H + L) = 8 \times 50 = 400 \quad \Rightarrow \quad H + L = 120 \quad \rightarrow \quad (1)$

Also, given that $H - L = 74 \quad \rightarrow \quad (2)$

$(1) + (2) \qquad \Rightarrow \qquad H + L = 120$

$ H - L = 74$

$ \overline{2H \quad = 194} \qquad \qquad \Rightarrow \qquad$ Highest runs $H = 97$.

Shortcut:

Distributed value $= -2 \times 8 = -16$ | H   L

$$\boxed{\begin{array}{ccc} 1 & 2 & \dots & 8 \end{array}} \qquad \boxed{\begin{array}{cc} 9 & 10 \end{array}}$$

52   52 . . . .   52 | 52   52

$-2\downarrow$ $-2\downarrow$ $-2\downarrow$

50   50 . . . .   50 | Excluded

> Here distributed value sign will change because of exclusion of H and L.

$H + L = (52 + 52) + 16 = 120 \quad \rightarrow \quad (1)$

Also, given that $H - L = 74 \quad \rightarrow \quad (2)$

$(1) + (2) \qquad \Rightarrow \qquad H + L = 120$

$ H - L = 74$

$ \overline{2H \quad = 194} \qquad \qquad \Rightarrow \qquad$ Highest runs $H = 97$.

Q - 34 The average runs scored by a cricket team in 30 matches is 290. What average should the team maintain in the next 10 matches, so that the new average becomes 300 per match?

Sol: Sum of 30 matches = 30 × 290 = 8700 ∵ Sum = Average × No. of obs.

Consider, average for next 10 matches = x

Sum of next 10 matches = $10x$

New average after 40 matches = $\dfrac{\text{Sum of 30 matches + Sum of next 10 matches}}{No.of\ matches}$

$300 = \dfrac{8700 + 10x}{40}$ ⇒ $8700 + 10x = 12000$ ⇒ $10x = 3300$

∴ Average for next 10 matches $x = 330$.

Shortcut:

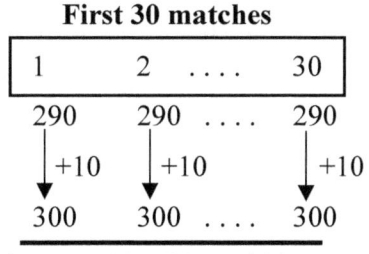

| First 30 matches | Next 10 matches |

Distributed value → +10 × 30 = +300   300 × 10 = 3000 → New sum

Sum of next 10 matches = New sum + Distributed value

∴ Average of next 10 matches = $\dfrac{Sum\ of\ next\ 10\ matches}{No.of\ matches} = \dfrac{3000 + 300}{10} = 330$.

Q - 35 A batsman had a certain average of runs for his 13 innings. In the 14th innings he made a score of 80 runs and thus his average of runs was decreased by 4. Find his average after 14th innings.

Sol: Consider, average after 13 innings = x, after 14 innings = $x - 4$

Sum of 13 innings = 13 × x = $13x$ ∵ Sum = Average × No. of obs.

Average after 14 innings = $\dfrac{\text{Sum of 13 innings + 14th innings}}{No.of\ innings}$

$x - 4 = \dfrac{13x + 80}{14}$ ⇒ $13x + 80 = 14x - 56$ ⇒ $x = 136$

∴ Average after 14 innings = $x - 4 = 136 - 4 = 132$.

Shortcut:

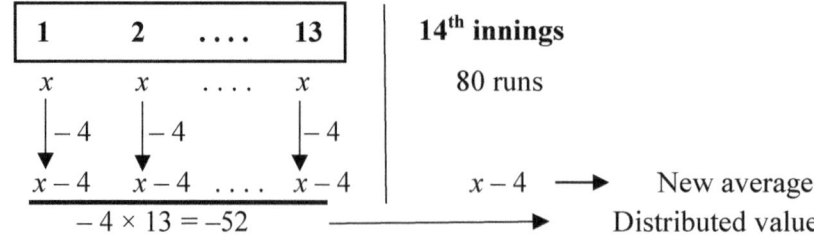

Runs in 14th innings = New average + Distributed value

$80 = -52 + (x - 4) \qquad \Rightarrow \qquad x - 4 = 132$

∴ Average after 14th innings = $x - 4 = 132$.

Q - 36

The average score of Ramesh, Mahesh and Suresh is 63. Ramesh's score is 15 less than Ajay and 10 more than Mahesh. If Ajay scored 30 marks more than the average score of Ramesh, Mahesh and Suresh. What is the sum of Mahesh's and Suresh's scores?

Sol: Sum of Ramesh, Mahesh, Suresh R + M + S = 63 × 3 = 189

Given that, R = A – 15, R = M + 10   ∵ R – Ramesh, M – Mahesh

A = 63 + 30 = 93, R = 93 – 15 = 78   S – Suresh, A – Ajay

78 = M + 10   ⇒   M = 68

S = (R + M + S) – (R + M) = 189 – (78 + 68) = 43

∴ Sum of Mahesh's and Suresh's scores, M + S = 68 + 43 = 111.

Q - 37

The average temperatures of the town from Monday to Thursday of a week is 53^{0}C and the average from Tuesday to Friday of a week is 56^{0}C. If the temperatures of the Monday and Friday are in the ratio 4 : 5, then what is the temperature on Friday?

Sol: Sum of temperatures from Monday to Thursday

M + T + W + Th = 53 × 4 = 212^{0}C   →   (1)

Sum of temperatures from Tuesday to Friday

T + W + Th + F = 56 × 4 = 224^{0}C   →   (2)

(2) – (1)   ⇒   F – M = 224 – 212 = 12^{0}C

Given that, Monday : Friday = 4 : 5

Difference = 1 part → 12^{0}C

∴ Friday = 5 parts = 5 × 12 = 60^{0}C.

Q - 38 The average price of 15 books is Rs. 24, while the average price of 12 of these books is Rs. 18. Of the remaining 3 books, if the price of first book is 40% more than second and price of third book is 20% more than second. What is the price of each of these 3 books?

Sol: Sum of 15 books = 24 × 15 = Rs. 360 ∵ Sum = Average × No. of observations

Sum of 12 books = 18 × 12 = Rs. 216

Sum of remaining 3 books = 360 – 216 = Rs. 144

Consider, second book is 100%

First : Second : Third = 140% : 100% : 120% ∵ 1st – 40% more than 2nd

F : S : T = 7 : 5 : 6 3rd – 20% more than 2nd

Total sum = 18 parts → 144 ⇒ 1 part → $\frac{144}{18} = 8$

∴ First book = 7 parts = 7 × 8 = Rs. 56, Second book = 5 parts = 5 × 8 = Rs. 40

Third book = 6 parts = 6 × 8 = Rs. 48.

Q - 39 25 members contribute a certain amount to orphanage on the occasion of Independence day. 23 of them contributes Rs. 250 each and the remaining two members contributes Rs. 23 more than the average contribution of all 25. What is the total money contributed by them?

Sol: Consider, average of all 25 members = x

Sum of 23 members = 250 × 23 = Rs. 5750 ∵ Sum = Avg. × No. of obs.

Average of 25 members = $\dfrac{\text{Sum of 23 members + sum of remaining 2 members}}{No.\ of\ persons}$

$x = \dfrac{5750 + (x + 23) + (x + 23)}{25}$ ∵ Remaining 2 members contribute 23 more than avg.

25x = 5796 + 2x ⇒ 23x = 5796 ⇒ x = Rs. 252

∴ Total money of 25 members = 25 × x = 25 × 252 = Rs. 6300.

Q - 40 The average cost of 13 apples is Rs. 20 per apple. 8 more apples with an average cost of Rs. 24 per apple are purchased and 4 apples with an average cost of Rs. 11 per apple are returned. What is the average cost per apple of all the apples?

Sol: Sum of 13 apples = 20 × 13 = Rs. 260   ∵ Sum = Avg. × No. of obs.

Sum of additional 8 apples = 24 × 8 = Rs. 192

Sum of returned 4 apples = 11 × 4 = Rs. 44

$$\text{Average} = \frac{\text{Sum of 13 apples} + \text{sum of 8 apples} - \text{Sum of 4 apples}}{Total\ no.\ of\ apples}$$

∴ Average $= \dfrac{260 + 192 - 44}{13 + 8 - 4} = \dfrac{408}{17} = $ Rs. 24.

Q - 41   **The average of 'n' results is 45. If one more result of value 151 is added, then the average is increased by 2. Find the value of 'n'.**

Sol: Sum of 'n' results = 45 × n = 45n   ∵ Sum = Average × No. of observations

$$\text{Average of (n + 1) results} = \frac{\text{Sum of 'n' results} + \text{(n+1)th result}}{No.\ of\ results}$$

$(45 + 2) = \dfrac{45n + 151}{n + 1}$ ∵ Average is increased by 2

$47n + 47 = 45n + 151$ ⇒ $2n = 104$ ⇒ $n = 52.$

Shortcut:

1	2	....	n	$(n + 1)^{th}$ **result**
45	45	....	45	151

+2↓ +2↓   +2↓

47   47   47   47 ⟶ New average

+2 × n = +2n   ⟶ Distributed value

$(n + 1)^{th}$ result = New average + Distributed value

$151 = 47 + 2n$ ⇒ $2n = 104$ ⇒ $n = 52.$

Q - 42   **There are 56 students who go to a mess. If 8 new students are admitted, the average expenditure of mess is increased by Rs. 72 per day and the average expenditure per head is decreased by 2 rupees. What is the original expenditure of the mess?**

Sol: Consider, original expenditure per head = x

Sum of 56 students expenses = 56 × x = 56x ∵ Sum = Avg. × No. of obs.

$$\text{New average} = \frac{\text{Sum of 56 students expenses} + \text{Additional expenses}}{Total\ strength}$$

$x - 2 = \dfrac{56x + 72}{56 + 8}$ ∵ Average is decreased by 2

$64x - 128 = 56x + 72$ ⇒ $x = 25$

∴ Original expenditure of mess = 56x = 56 × 25 = Rs. 1400.

Q - 43

The average number of persons who are visiting central library on MON, TUE and WED is 400 and the average number of visitors on THUR, FRI is 300. If the average number of visitors per day in the week is 450, then find the average number of persons who are visiting central library in weekdays (SAT and SUN).

Sol: Total no. of visitors on MON, TUE, WED is M + T + W = 400 × 3 = 1200

Total no. of visitors on THUR, FRI is Th + F = 300 × 2 = 600

Total no. of visitors on entire week = 450 × 7 = 3150

Average no. of visitors on weekends $= \dfrac{Total\ no.\ of\ visitors\ on\ weekends}{No.of\ days}$

∴ Average on SAT & SUN $= \dfrac{(Total\ week) - (M+T+W) - (Th+F)}{2} = \dfrac{3150 - 1200 - 600}{2} = 675.$

Q - 44

The average weight of P, Q, R and S is 52 kg. A new person T is joined in the group, then the average weight of the group is increased by 2 kg. Again a new person U replaces P, then the average of 5 persons is 55 kg. What is the average weight of Q, R, S and U?

Sol: Sum of 4 persons P, Q, R, S = 52 × 4 = 208 kg → (1)

Sum of 5 persons P, Q, R, S, T = 54 × 5 = 270 kg → (2) ∵ Average is increased by 2

Sum of 5 persons Q, R, S, T, U = 55 × 5 = 275 kg → (3) ∵ U replaces with P

(2) – (1) ⇒ T = 270 – 208 = 62 kg ……. substitute in (3)

Q + R + S + 62 + U = 275 ⇒ Q + R + S + U = 275 – 62 = 213 kg

∴ Average of Q, R, S, U $= \dfrac{Sum\ of\ Q,R,S,U}{No.of\ persons} = \dfrac{213}{4} = 53.25$ kg.

Q - 45

In the first 20 overs of a cricket match, the run rate is only 4.8. What should be the run rate in the next 30 overs to reach the target of 345 runs?

Sol:

Runs scored = No. of overs × Run rate

Runs scored in first 20 overs = 20 × 4.8 = 96

Required runs in next 30 overs = Target – Runs in first 20 overs = 345 – 96 = 249 runs

∴ Run rate for next 30 overs $= \dfrac{Required\ runs}{No.of\ overs} = \dfrac{249}{30} = 8.3.$

Q - 46 Bumrah's bowling average is 17.5. He took 5 wickets for 25 runs and thereby his average decreases by 0.5. How many wickets taken by him initially?

Sol:

$$\text{Bowling Average} = \frac{Total\ runs\ given}{No.of\ wickets}$$

Consider, initial no. of wickets = x

Total runs given = $17.5 \times x = 17.5x$

New bowling average = $\frac{17.5x + 25}{x + 5} = 17.5 - 0.5 = 17$ ∵ Average is decreased by 0.5

$17.5x + 25 = 17x + 85 \quad \Rightarrow \quad 0.5x = 60 \quad \Rightarrow \quad x = 120$

∴ Initial number of wickets = 120.

Q - 47 The average wickets per test match of a bowler was 6. If 4 matches, where he took no wickets are excluded and he takes 13 wickets per match in a 5 match series, his average increased to 7 wickets per match. How many matches did he play before the latest series?

Sol:

$$\text{Average wickets per match} = \frac{Total\ no.of\ wickets}{No.of\ matches}$$

Consider, initial no. of matches = x

Initial total no. of wickets = $6 \times x = 6x$

After exclusion of 4 matches and inclusion of 5 match series,

New average = $\frac{6x - (0 \times 4) + (13 \times 5)}{x - 4 + 5} = 7$ ∵ In 5 match series he took 13 wickets/match

$6x + 65 = 7x + 7 \quad \Rightarrow \quad x = 58$

∴ Initial no. of matches $x = 58$.

Q - 48 The average marks of 42 students are 51 but after checking there are two mistakes found. After adjustment, if a student got 28 marks more and other student got 70 marks less, then what will be the adjusted average?

Sol: Sum of marks of 42 students initially = $51 \times 42 = 2142$ ∵ Sum = Avg. × No. of obs.

After adjustment average = $\frac{2142 - 28 + 70}{42} = 52$

(or)   ∵ By mistake 1st student got 28 more, 2nd got 70 less marks

∴ Adjusted average = $51 + \frac{(-28 + 70)}{42} = 52$.

Q - 49 The average marks of the students in 4 sections A, B, C, D is 60. The average of students of A, B, C, D individually are 45, 50, 72 and 80 respectively. If the average marks of students of section A and B together is 48 and that of B and C together is 60, then what is the ratio of number of students in sections A and D?

Sol: Given that, average of students in A, B, C, D = 60

Also, average of students in B and C = 60

"To maintain overall average is 60, then A and D average also must be 60".

Consider, no. of students in A = x, no. of students in D = y

Average of section A = 45, Average of section D = 80

Average of A and D = $\dfrac{Sum\ of\ A + Sum\ of\ D}{No.of\ students\ in\ A\ and\ D}$

$60 = \dfrac{45x + 80y}{x + y}$ $\Rightarrow$ $60x + 60y = 45x + 80y$ ∵ Sum = Avg. × No. of obs.

$15x = 20y$ $\Rightarrow$ $x : y = 20 : 15 = 4 : 3$

∴ Ratio of students in A and D = 4 : 3.

Q - 50 The average score of Virat after 58 innings is 58 and in the 59th innings he scores 117 runs. In the 60th innings, find the minimum number of runs required to increase his average score by 1 than it was before 60th innings.

Sol: Sum of runs after 58 innings = 58 × 58 = 3364

Average after 59th innings = $\dfrac{Sum\ of\ 58\ innings + 59th\ innings}{No.of\ innings} = \dfrac{3364 + 117}{59} = 59$

Average after 60th innings = $\dfrac{Sum\ of\ 58\ innings + 59th\ innings + 60th\ innings}{No.of\ innings}$

$59 + 1 = \dfrac{3364 + 117 + x}{60}$ $\Rightarrow$ $x = 119$ ∵ Average is increased by 1 (59 + 1)

∴ Required minimum no. of runs in 60th innings $x = 119$.

ASSESSMENT TEST

1. Find the average of 120 natural numbers.

2. Find the average of squares of natural numbers from 1 to 53.

3. What is the average of cubes of natural numbers from 1 to 40?

4. If $38a + 38b = 4902$, what is the average of a and b?

5. a) What is the average of odd numbers from 1 to 95?

   b) What is the average of odd numbers from 1 to 126?

6. a) What is the average of even numbers from 1 to 108?

   b) What is the average of even numbers from 1 to 135?

7. Find the average of first 75 consecutive odd numbers.

8. Find the average of first 64 consecutive even numbers.

9. Find the average of first 21 multiples of 9.

10. Find the average of multiples of 13 in the natural number series from 200 to 500.

11. The average of 20 numbers is 17. What will be the new average, if each number is multiplied by 7?

12. The average of 9 consecutive even numbers is 44. What is the smallest number?

13. The average of 14 consecutive odd numbers is 64, then find the difference between second smallest and largest numbers.

14. A number 'p' is equal to 60% of the average of 5, 7, 9, 11 and a number 'q'. If the average of p and q is 40, then find the value of q.

15. The average of 8 consecutive even numbers A, B, C, D, E, F, G and H is 37. What is the product of C and G?

16. The average of 'x' numbers is y^2 and the average of 'y' numbers is x^2. Find the average of all the numbers.

17. The average cost of 'x' articles is Rs. 40 and that of 'y' articles is Rs. 62. The average cost of all articles combinedly is Rs. 54. What is $x : y$?

18. There are 80 boys and 70 girls in a class. If the average contribution of amount for boys is Rs. 68 and that of girls is Rs. 83. What is the average contribution of entire class?

19. Gopal rides a bike from his house at a speed of 60 kmph while going to office and return home at a speed of 40 kmph, then find the average speed during the entire journey.

20. A person covers three equal distances at a speeds of 18 kmph, 24 kmph and 36 kmph. What is the average speed during the entire journey?

21. A man starts from his house and covers 30 km at a speed of 6 kmph, next 24 km at a speed of 8 kmph and last 36 km at a speed of 9 kmph, then find the average speed of a man for the entire journey.

22. A person covers first $\frac{1}{3}$ rd of distance at a speed of 10 kmph, next $\frac{1}{6}$ th of distance at a speed of 15 kmph and the remaining distance at a speed of 20 kmph. Find the average speed during the whole journey.

23. Out of three numbers, second is twice the first and is also four times of third. If the average of three numbers is 91, then find the largest number.

24. The average weight of L and M is 54 kg, that of M and N is 58 kg and the average weight of N and L is 63 kg. Find the weight of N.

25. The average temperature from Sunday to Thursday is 44^0C and from Monday to Friday is 47^0C. If the temperature on Friday is 53^0C, then what is the temperature on Sunday?

26. The average weight of girls in a class is 45 kg and that of boys is 54 kg. If the ratio of number of boys and girls is 5 : 4, then find the average weight of students in the class.

27. The average of 'n' observations is 73. If one more observation of value 139 is added, then the average is increased by 1. Find the value of 'n'.

28. A company produces on an average 5900 articles per month for the first 5 months. How many articles it must produce on an average per month over the next 7 months, to maintain the overall average of 5375 articles per month?

29. Total strength of the class is 180. The average marks of boys is 39 and that of girls is 75. If the average marks of entire class is 60, then find the number of girls in the class.

30. The average salary of entire staff in a college is Rs. 650 per day. The average salary of teaching staff is Rs. 800 and that of non-teaching staff is Rs. 230. Find the number of teaching staff, if the number of non–teaching staff is 60.

31. Twelve friends went to shopping for buying clothes. Eleven of them spent Rs. 2500 on their clothes and the twelfth one spent Rs. 550 more than the average expenditure of all twelve. What is the money spent by twelfth person?

32. The average weight of 41 children is 30 kg. The average of first 20 of them is 31 kg and that of last 20 is 28 kg, then find the weight of 21st child.

33. The average of 23 results is 70. If the average of first 12 results is 67 and that of last 12 results is 71. What is twelfth result?

34. The average of 30 observations is 26. Later it was found that an observation 20 was wrongly entered as 80, then find the new average.

35. The average expenditure on 12 articles is decreased by 0.5 when the cost of one of the article Rs. 50 is replaced by another. Find the cost of new article.

36. The average price of 10 pairs of clothes is Rs. 1950. If the price of trolley bag is also included, then the average is increased by Rs. 50. What is the price of trolley bag?

37. A batsman makes a score of 94 runs in the 15th innings and thus his average is increases by 3 runs. What is his average after 15th innings?

38. The average score of a batsman in 16 innings is 43. If the highest and lowest scores were excluded, then the average of remaining innings would decrease by 3. If the difference between these two innings is 62, then find the lowest score.

39. There are 63 students who are going to mess. If 9 new students are admitted, the average expenditure of the mess is increased by Rs. 108 per day and the average expenditure per head is decreased by 1 rupee. What is the original expenditure of the mess?

40. The average weight of 35 students in a class is 48 kg. 5 of them whose average weight is 45 kg leave the class and other 5 students whose average weight is 50 kg join the class. What is the new average weight of the class?

41. In an exam, a student average marks were 74 per paper. If he had obtained 25 marks more in Maths paper and 17 marks more for his Science paper, his average per paper would have been 77. How many papers were there in the exam?

42. The average height of 3 men P, Q and R is 172 cm. Another man S joins the group and the average now becomes 170 cm. If another man T, whose height is 6 cm more than that of S, replaces P. Now the average height of Q, R, S and T becomes 169 cm. Find the height of P.

43. The average marks of 35 students is 47. But after checking, there are 2 mistakes found. After adjustment, if a student got 60 marks more and other student got 25 marks less. What will be the adjusted average?

44. Murali's bowling average is 18.2. He took 6 wickets for 30 runs and there by his average decreases by 0.2. How many wickets taken by him initially?

45. The average wickets per test match of a bowler was 4. If two matches, where he took no wickets are excluded and he takes 10 wickets per match in a 4 match series, his average increases to 5 wickets per match. How many matches did he play before the latest series?

KEY

1. 60.5
2. 963
3. 16810
4. 64.5
5. a) 48   b) 63
6. a) 55   b) 68
7. 75
8. 65
9. 99
10. 351
11. 119
12. 36
13. 24
14. 68
15. 1428
16. xy
17. 4 : 7
18. Rs. 75
19. 48 kmph
20. 24 kmph
21. 7.5 kmph
22. 14.4 kmph
23. 52
24. 67 kg
25. 38^0C
26. 50 kg
27. 65
28. 5000
29. 105
30. 168
31. Rs. 3100
32. 50 kg
33. 46

34. 24

35. Rs. 44

36. Rs. 2500

37. 52

38. 33

39. Rs. 1260

40. $48\frac{5}{7}$ kg

41. 14

42. 174 cm

43. 46

44. 390

45. 30

PROBLEMS ON AGES

✓ Problems on ages is completely depends on few topics i.e. ratios and averages.

✓ In ages topic, the difference between the ages of any two persons at any point of time must be equal.

✓ If the difference between the ages are not equal, then we have to make it as equal by multiplying first ratio with second difference and second ratio with first difference.

Some important formulae to be remember:

✓ If the ratio of the ages of A and B at present is a : b. After 'T' years, the ratio will be c : d. Then,

   i. Present age of A = a $\times \dfrac{T\,(c-d)}{ad-bc}$

   ii. Present age of B = b $\times \dfrac{T(c-d)}{ad-bc}$

✓ If the ratio of the ages of A and B at present is a : b. Before 'T' years, the ratio was c : d. Then,

   i. Present age of A = a $\times \dfrac{T\,(c-d)}{bc-ad}$

   ii. Present age of B = b $\times \dfrac{T(c-d)}{bc-ad}$

✓ The product of the ages of A and B is 'T' years and the ratio of their present ages is a : b. Then,

   i. Present age of A = a $\times \sqrt{\dfrac{T}{ab}}$

   ii. Present age of B = b $\times \sqrt{\dfrac{T}{ab}}$

✓ If the man's age is $x\%$ of what he was 't_1' years ago and $y\%$ of what he will be after 't_2' years. Then, the present age is $\dfrac{xt_1 + yt_2}{x-y}$ years $(x > y)$.

SOLVED EXAMPLES

Q - 1 **The ratio of ages of father and son is 9 : 4. If the age of son is 28 years, then find the age of father.**

Sol: Given that, F : S = 9 : 4

Let, F = 9x, S = 4x (x – common factor)   **Present Ages**

S = 4x = 28 years $\Rightarrow$ x = 7   Father – F, Son – S

$\therefore$ Father's age F = 9x = 9 × 7 = 63 years.

Shortcut:

F : S = 9 : 4

Son, S = 4 Parts $\longrightarrow$ 28 years

Father, F = 9 Parts $\longrightarrow$? $= \dfrac{9 \times 28}{4} = 63$ years.

Q - 2 **The present ages of Rehan and Aman are in the ratio 5 : 3. What is the sum of their ages, if the age of Aman is 36 years old?**

Sol: Given that, R : A = 5 : 3

Let, R = 5x, A = 3x (x – common factor)   **Present Ages**

A = 3x = 36 years $\Rightarrow$ x = 12   Rehan – R, Aman – A

$\therefore$ Sum of their ages = 5x + 3x = 8x = 8 × 12 = 96 years.

Shortcut:

R : A = 5 : 3

Aman, A = 3 parts $\longrightarrow$ 36 years

Sum of their ages = 8 parts $\longrightarrow$? $= \dfrac{36 \times 8}{3} = 96$ years.

Q - 3 **6 years ago, the ratio of the ages of mother and daughter is 11:5. If the present age of daughter is 21years, then find the age of mother after 6 years.**

Sol: 6 years ago, M – 6 : D – 6 = 11 : 5

Let, M – 6 = 11x, D – 6 = 5x **Present Ages**

Given that, D = 21 years $\Rightarrow$ 21 – 6 = 5x $\Rightarrow$ x = 3   Mother – M, Daughter – D

M – 6 = 11x = 11 × 3 = 33 years

M = 33 + 6 = 39 years

$\therefore$ Mother's age after 6 years is M + 6 = 39 + 6 = 45 years.

Shortcut:

Given that, M – 6 : D – 6 = 11 : 5 and D = 21 years

D – 6 = 21 – 6 = 15 years

D – 6 = 5 parts $\longrightarrow$ 15 years

M – 6 = 11 parts $\longrightarrow$? $= \dfrac{15 \times 11}{5} = 33$ years

M – 6 = 33 years $\Rightarrow$ M = 39 years

∴ Mother's age after 6 years M + 6 = 39 + 6 = 45 years.

Q - 4 **The present ages of a family of father, mother and their only son are in the ratio 5 : 4 : 2 respectively. The age of mother after 15 years will be 47 years, then what is the age of son before 5 years?**

Sol: Given that, F : M : S = 5 : 4 : 2   **Present Ages**

Let, F = 5*x*, M = 4*x*, S = 2*x*   Father – F, Mother – M, Son – S

After 15 years mother's age is M + 15 = 47 years

Present age of mother M = 47 – 15 = 32 years

4*x* = 32 $\Rightarrow$ *x* = 8

∴ Son's age before 5 years = S – 5 = 2*x* – 5 = 2 × 8 – 5 = 11 years.

Shortcut:

F : M : S = 5 : 4 : 2

After 15 years Mother's age M + 15 = 47 years $\Rightarrow$ M = 47 – 15 = 32 years

M = 4 parts $\longrightarrow$ 32 years

S = 2 parts $\longrightarrow$? $= \dfrac{32 \times 2}{4} = 16$ years.

∴ Son's age before 5 years = S – 5 = 16 – 5 = 11 years.

Q - 5 **At present the ratio between wife and husband is 5 : 6. 15 years back, the ratio was 10 : 13. What will be the age of wife 6 years hence?**

Sol: Given that, W : H = 5 : 6   **Present Ages**

Let, W = 5*x*, H = 6*x*   (*x* – common factor)   Wife – W, Husband – H

According to question,

$\dfrac{W - 15}{H - 15} = \dfrac{10}{13}$ $\Rightarrow$ $\dfrac{5x - 15}{6x - 15} = \dfrac{10}{13}$

65*x* – 195 = 60*x* – 150 $\Rightarrow$ 5*x* = 45 $\Rightarrow$ *x* = 9

∴ Wife's age 6 years hence = W + 6 = 5*x* + 6 = 5 × 9 + 6 = 51 years.

Shortcut – 1:

> **Note:**
> ✓ The difference between the ages of any 2 persons at any point of time must be equal.
> ✓ If difference between the ages of any 2 persons are not equal, then we have to make it as equal by multiplying the 1st ratio with 2nd difference and 2nd ratio with 1st difference.

$$
\begin{array}{ccc}
 & \text{W} & \text{H} \\
\end{array}
$$

Present → 5 : 6 ⟶ 1 part

15 years ago → 10 : 13 ⟶ 3 parts

Here, difference between their ages are not equal. Therefore, we have to make it as equal.

	W H		W H
Present →	(5 : 6) × 3	=	15 : 18
15 years ago →	(10 : 13) × 1	=	10 : 13

5 parts

5 parts ⟶ 15 years

Wife W = 15 parts ⟶ $? = \dfrac{15 \times 15}{5} = 45$ years

∴ Wife's age 6 years hence = 45 + 6 = 51 years.

Shortcut – 2:

> **Note:**
> ✓ If the ratio of the present ages of A and B is a : b. Before 'T' years, the ratio was c : d. Then,
> i) Present age of A = $a \times \dfrac{T(c-d)}{bc-ad}$ ii) Present age of B = $b \times \dfrac{T(c-d)}{bc-ad}$

	W	H
Present →	5 :	6
15 years ago →	10 :	13

Here a = 5, b = 6, c = 10, d = 13, T = 15

Present age of wife = $a \times \dfrac{T(c-d)}{bc-ad} = 5 \times \dfrac{15(10-13)}{6 \times 10 - 5 \times 13} = 45$ years

∴ Wife's age 6 years hence = 45 + 6 = 51 years.

Q - 6 The present ages of A, B and C are in the ratio 5 : 9 : 6. 4 years ago, the sum of their ages was 108. Find their present ages.

Sol: Given that, Present A : B : C = 5 : 9 : 6

Let, A = 5x, B = 9x, C = 6x   (x – common factor)

4 years back sum = (A – 4) + (B – 4) + (C – 4) = 108

A + B + C = 108 + 12 = 120 years

$5x + 9x + 6x = 120$ $\Rightarrow$ $20x = 120$ $\Rightarrow$ $x = 6$

$\therefore$ Present ages: A = 5x = 5 × 6 = 30 years.

B = 9x = 9 × 6 = 54 years.

C = 6x = 6 × 6 = 36 years.

Q - 7 The sum of the ages of Ram, Gopal and Varma is 85 years. 7 years back, the ratio of their ages was 3 : 5 : 8. What is the present age of Gopal?

Sol: Given that, Present sum R + G + V = 85 years   **Present Ages**

7 years back ratio = 3 : 5 : 8   Ram – R, Gopal – G, Varma – V

7 years back sum = 85 – 7 – 7 – 7 = 64 years $\because$ Each person reduced by 7 years

7 years back sum = 16 parts $\longrightarrow$ 64 years

7 years back Gopal = G – 7 = 5 parts $\longrightarrow$? $= \dfrac{64 \times 5}{16} = 20$ years

$\therefore$ Present age of Gopal G = 20 + 7 = 27 years.

Q - 8 At present, Bhuvana's age is 2.5 times that of Sravani's age. After 22 years, the respective ratio between Sravani and Bhuvana age then will be 7 : 12. What is the average age of Bhuvana and Sravani at present?

Sol: According to question,   **Present Ages**

B = 2.5 S $\Rightarrow$ B : S = 2.5 : 1 $\Rightarrow$ B : S = 5 : 2   Bhuvana – B, Sravani – S

Let, B = 5x, S = 2x   (x – common factor)

$\dfrac{S + 22}{B + 22} = \dfrac{7}{12}$ $\Rightarrow$ $\dfrac{2x + 22}{5x + 22} = \dfrac{7}{12}$

$24x + 264 = 35x + 154$ $\Rightarrow$ $11x = 110$ $\Rightarrow$ $x = 10$

B = 5x = 5 × 10 = 50 years   &   S = 2x = 2 × 10 = 20 years

$\therefore$ Average $= \dfrac{B + S}{2} = \dfrac{50 + 20}{2} = 35$ years.

Shortcut:

		B	S	B	S	
Present	→	2.5 : 1 =		5 : 2		3 parts
After 22 years	→			12 : 7		5 parts

Here, the difference between their ages are not equal. So, multiply 1st ratio with 2nd difference and 2nd ratio with 1st difference.

		B	S	
Present	→	(5 : 2) × 5 = 25 : 10		
		↓	↓	11 parts
After 22 years	→	(12 : 7) × 3 = 36 : 21		

11 parts $\longrightarrow$ 22 years $\quad \Rightarrow$ 1 part = 2 years

B = 25 parts = 25 × 2 = 50 years $\quad$ & $\quad$ S = 10 parts = 10 × 2 = 20 years

∴ Average $= \dfrac{B+S}{2} = \dfrac{50+20}{2} = 35$ years.

Q-9 $\quad$ **11 years ago, the ages of A and B was in the ratio of 7 : 4. After 15 years, their ratio will changes to 9 : 7. What are their present ages?**

Sol: $\quad$ Consider, present ages are A and B.

Given that, $\dfrac{A-11}{B-11} = \dfrac{7}{4} \Rightarrow \quad 4A - 44 = 7B - 77 \quad \Rightarrow 7B - 4A = 33$ ...... (1)

Also, $\dfrac{A+15}{B+15} = \dfrac{9}{7} \quad \Rightarrow \quad 7A + 105 = 9B + 135 \Rightarrow 9B - 7A = -30$ ...... (2)

Solve (1) and (2) equations

(7B – 4A = 33) × 7 = 49B – 28A = 231 ...... (3)

(9B – 7A = –30) × 4 = 36B – 28A = –120 ...... (4)

(3) – (4) $\quad\quad\quad\quad \Rightarrow \quad$ 13B = 351 $\quad \Rightarrow \quad$ B = 27 years ..... Sub in (1)

7 × 27 – 4A = 33 $\quad\quad \Rightarrow \quad$ A = 39 years

∴ Present ages of A and B = 39 years and 27 years.

Shortcut:

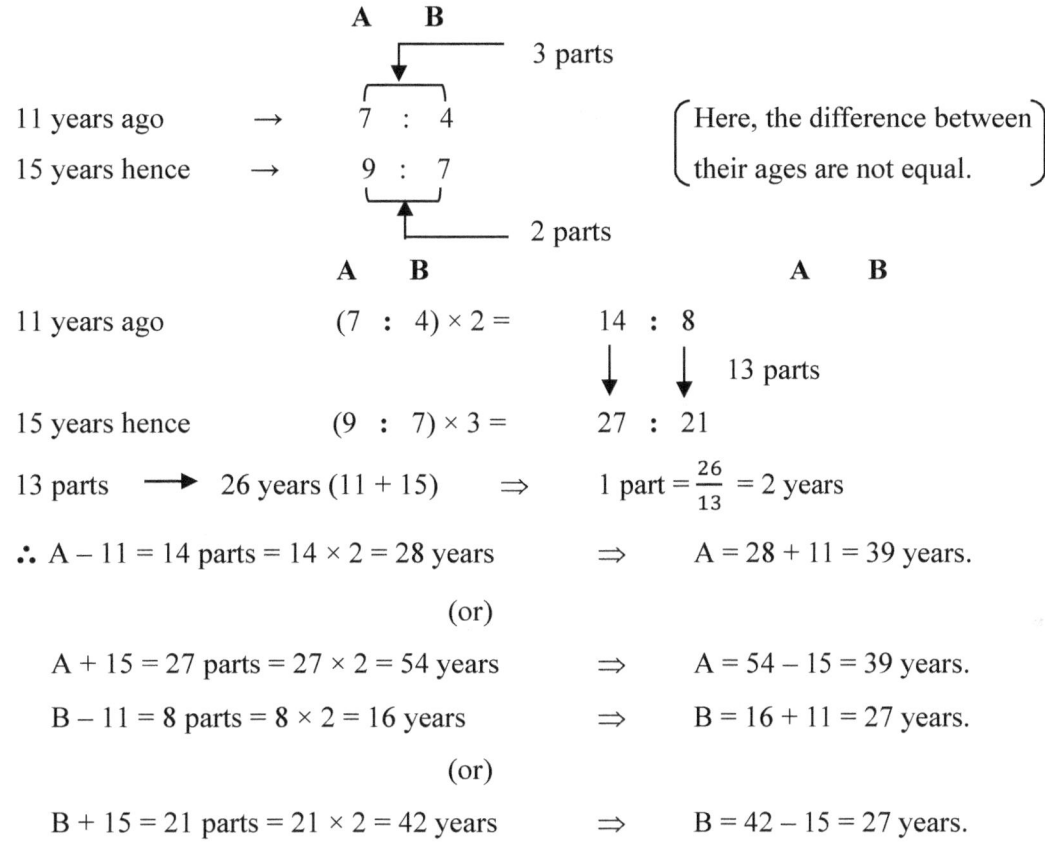

	A	B	
11 years ago	(7 :	4) × 2 =	14 : 8
15 years hence	(9 :	7) × 3 =	27 : 21

13 parts $\longrightarrow$ 26 years (11 + 15) $\Rightarrow$ 1 part $= \dfrac{26}{13} = 2$ years

$\therefore$ A − 11 = 14 parts = 14 × 2 = 28 years $\Rightarrow$ A = 28 + 11 = 39 years.

(or)

A + 15 = 27 parts = 27 × 2 = 54 years $\Rightarrow$ A = 54 − 15 = 39 years.

B − 11 = 8 parts = 8 × 2 = 16 years $\Rightarrow$ B = 16 + 11 = 27 years.

(or)

B + 15 = 21 parts = 21 × 2 = 42 years $\Rightarrow$ B = 42 − 15 = 27 years.

Q - 10 **X's age is thrice the sum of ages of Y and Z. If after 8 years, the age of X will be twice the sum of ages of Y and Z, then find the present age of X.**

Sol: Let, present ages are x, y and z years. According to question,

$x = 3(y + z) = 3y + 3z$ $\longrightarrow$ (1)

$x + 8 = 2[(y + 8) + (z + 8)] = 2(y + z + 16)$

$x = 2y + 2z + 24$ $\longrightarrow$ (2)

$(1) = (2)$ $\Rightarrow$ $3y + 3z = 2y + 2z + 24$ $\Rightarrow$ $y + z = 24$

$\therefore$ Present age of $x = 3(y + z) = 3 \times 24 = 72$ years.

Q - 11 **The ratio of present ages of P and Q is 2 : 3. If the difference between present age of P and age of Q before 6 years is 8, then what is the average age of P and Q?**

Sol: Given that, present P : Q = 2 : 3

Let, P = 2x, Q = 3x (x – common factor)

Also, (Q − 6) − P = 8 $\Rightarrow$ $3x - 6 - 2x = 8$ $\Rightarrow$ $x = 14$

P = 2x = 2 × 14 = 28 years, Q = 3x = 3 × 14 = 42 years

$\therefore$ Average $= \dfrac{P + Q}{2} = \dfrac{28 + 42}{2} = 35$ years.

Q - 12 | The product of the ages of Chandu and Madhu is 320. If twice the age of Madhu is more than Chandu's age by 12 years. What is the age of Chandu?

Sol: Given that, $C \times M = 320$ $\longrightarrow$ (1)

$2M = C + 12 \Rightarrow C = 2M - 12$...... Sub in (1)

Present Ages

$(2M - 12) \times M = 320$

Chandu – C, Madhu – M

$2M^2 - 12M - 320 = 0 \Rightarrow M^2 - 6M - 160 = 0$

$M^2 - 16M + 10M - 160 = 0 \Rightarrow M(M - 16) + 10(M - 16) = 0$

$(M - 16)(M + 10) = 0 \Rightarrow M = 16$ years $(\because M \neq -10)$

$\therefore$ Chandu's age $C = \dfrac{320}{M} = \dfrac{320}{16} = 20$ years.

> **Note:** This problem can be solved through options easily by checking one by one.

Q - 13 | The age of Mohan 15 years ago was $\dfrac{2}{5}^{th}$ of what he is now. What will be the age of Mohan after 8 years?

Sol: Consider, present age of Mohan = 'M' years

Given that, $M - 15 = \dfrac{2}{5}M \Rightarrow M - \dfrac{2}{5}M = 15$

$\dfrac{3M}{5} = 15 \Rightarrow M = 25$ years

$\therefore$ Age of Mohan after 8 years = $M + 8 = 25 + 8 = 33$ years.

Q - 14 | After 11 years, X will be twice as old as Y was 11 years ago. If present age of X is 7 more than Y. What is the present age of X?

Sol: According to question,

$x + 11 = 2(y - 11)$ $\longrightarrow$ (1)

$x = y + 7 \Rightarrow y = x - 7$ ...... Sub in (1)

$x + 11 = 2(x - 7 - 11) \Rightarrow x + 11 = 2(x - 18)$

$x + 11 = 2x - 36 \Rightarrow x = 47$

$\therefore$ Present age of X = 47 years.

Q - 15 | The sum of the ages of A, B and C is 125 years. A is thrice as old as B and C is 10 years younger than B. Find their respective ages.

Sol: Given that, $A + B + C = 125$ years, $A = 3B$ and $C = B - 10$

$3B + B + (B - 10) = 125 \Rightarrow 5B = 135 \Rightarrow B = 27$ years

$A = 3B = 3 \times 27 = 81$ years, $C = B - 10 = 27 - 10 = 17$ years

$\therefore$ A = 81 years, B = 27 years, C = 17 years.

Q - 16 If 9 years are subtracted from present age of Amit and the remainder is divided by 17, then the present age of his grandson Bharath is obtained. If Bharath is 4 years younger to Chandu, whose age is 8 years. What is the age of Amit?

Sol: According to question,

$$\frac{A-9}{17} = B, \; B = C - 4, \; C = 8 \text{ years} \qquad \underline{\textbf{Present Ages}}$$

$$B = 8 - 4 = 4 \text{ years} \qquad\qquad \text{Amit} - A, \text{Bharath} - B, \text{Chandu} - C$$

$$\frac{A-9}{17} = 4 \qquad \Rightarrow \qquad A = 17 \times 4 + 9 = 77 \text{ years}$$

∴ Present age of Amit = 77 years.

Q - 17 If x's age is 1.5 times the average of x, y and z. z's age is one – fourth of the average of x, y, z. If y is 20 years old, then what is the average age of x, y and z?

Sol: Given that, $x = 1.5 \left[\dfrac{x+y+z}{3}\right] = \dfrac{3}{2}\left[\dfrac{x+y+z}{3}\right]$

$$x + y + z = 2x \quad \Rightarrow \quad y + z = x \qquad\longrightarrow \qquad (1)$$

$$z = \frac{1}{4}\left[\frac{x+y+z}{3}\right], y = 20 \text{ years}$$

$$x + y + z = 12z \Rightarrow \quad x + y = 11z \quad \Rightarrow \quad x = 11z - y \quad\longrightarrow\quad (2)$$

$$(1) = (2) \qquad y + z = 11z - y \;\Rightarrow\quad 10z = 2y = 2 \times 20 = 40 \qquad \because y = 20 \text{ years}$$

$$10z = 40 \text{ years} \;\Rightarrow\quad z = 4 \text{ years}$$

From (1), $\qquad x = y + z = 20 + 4 = 24$ years

∴ Average $= \left[\dfrac{x+y+z}{3}\right] = \left[\dfrac{24+20+4}{3}\right] = 16$ years.

Q - 18 Girish is thrice as old as Harish and half as old as Kumar. If the sum of Harish's and Kumar's ages is 84 years, then find the age of Girish.

Sol: Given that, $G = 3H, \; G = \dfrac{K}{2}$ \qquad\qquad\qquad \underline{\textbf{Present Ages}}

$$H + K = 84 \text{ years} \qquad\longrightarrow\qquad (1) \qquad \text{Girish, } - G, \text{Harish} - H, \text{Kumar} - K$$

$$G = 3H = \frac{K}{2} \quad \Rightarrow \qquad K = 6H \; \ldots\ldots \text{ substitute in (1)}$$

$$H + 6H = 84 \quad \Rightarrow \qquad H = 12 \text{ years}$$

∴ Girish's present age $G = 3H = 3 \times 12 = 36$ years.

Q - 19 The sum of the ages of 7 children born at the intervals of 2 years each is 84. What is the age of eldest child?

Sol: Consider, ages of 7 children are

$x - 6, x - 4, x - 2, x, x + 2, x + 4, x + 6$ ∵ Interval gap = 2 years

Sum of their ages = $7x = 84$ ⇒ $x = 12$ years

∴ Age of eldest child = $x + 6 = 12 + 6 = 18$ years.

Shortcut:

Average of children = $\dfrac{Sum\ of\ chilren}{No.of\ chilren} = \dfrac{84}{7} = 12$ years

> **Note:** If no. of observations are odd and the gap between any two consecutive numbers are same, then average is always "middle number."

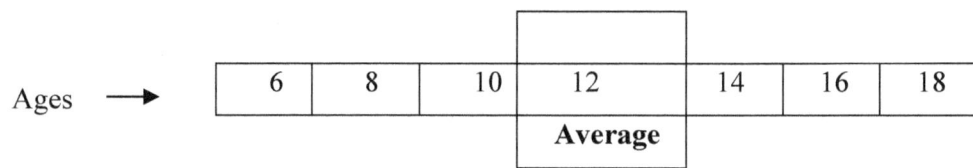

Ages ⟶

6	8	10	12	14	16	18
			Average			

Since the Gap between 2 children = 2 years

∴ Age of eldest child = 18 years.

Q - 20 A man was asked to tell his age in years. He said that, "Take my age 5 years hence, multiply it by 4 and then subtract 4 times my age 2 years ago and you will know how old I am." What was the age of the man?

Sol: Consider, a man's present age = 'M' years

According to question,

$M = (M + 5)\,4 - 4(M - 2)$ ⇒ $M = 4M + 20 - 4M + 8$ ⇒ $M = 28$

∴ Age of the Man = 28 years.

Q - 21 Abhi's present age is twice the age of Gopi three years ago. What is the sum of their present ages, if the ratio of the sum of their present ages to the difference of their present ages is 5 : 1?

Sol: Given that, $A = 2(G - 3)$ ⟶   (1)   **Present Ages**

$(A + G) : (A - G) = 5 : 1$ Abhi – A, Gopi – G

Let, $A + G = 5x$ and $A - G = x$ (x – common factor)

$(A + G) + (A - G) = 5x + x$ ⇒ $2A = 6x$ ⇒ $A = 3x$

$G = 5x - 3x = 2x$ ∵ $A + G = 5x$

Substitute A = 3x, G = 2x in (1)

$3x = 2(2x - 3)$ $\Rightarrow$ $3x = 4x - 6$ $\Rightarrow$ $x = 6$

∴ Sum of their ages A + G = 5x = 5 × 6 = 30 years.

Q - 22 | **Mother's age is equal to the sum of the ages of her son and daughter. The present age of son is 20 years. If 20 years ago, their average age was 25 years, then find the present age of the mother.**

Sol: Given that, M = S + D, S = 20 years   **Present Ages**

Mother – M, Son – S, Daughter – D

20 years ago means at the time of birth of son.

At that time there are only 2 members in the family.

Therefore, Average $= \dfrac{(M - 20) + (D - 20)}{2} = 25$

M + D = 2 × 25 + 40 = 90

M = S + D $\Rightarrow$ M = 20 + (90 – M)   ∵ M + D = 90, S = 20

2M = 110 $\Rightarrow$ M = 55 years

∴ Present age of mother M = 55 years.

Q - 23 | **If the product of ages of father and son is 1792 years and the ratio of their present ages is 7 : 4. What are their present ages?**

Sol: Given that, F : S = 7 : 4   **Present Ages**

Let, F = 7x, S = 4x (x – common factor)   Father – F, Son – S

F × S = 1792 $\Rightarrow$ 7x × 4x = 1792

$x^2 = 64$ $\Rightarrow$ $x = 8$

∴ Present age of Father F = 56 years   &   Present age of Son S = 32 years.

Shortcut:

> **Note:** Product of ages of A and B is 'T' years and the ratio of their present ages is a : b.
>
> Then, Present age of A = a × $\sqrt{\dfrac{T}{ab}}$ years & Present age of B = b × $\sqrt{\dfrac{T}{ab}}$ years

Given that, F : S = 7 : 4, Product = 1792

∴ Present age of Father F = 7 × $\sqrt{\dfrac{1792}{7 \times 4}}$ = 56 years.

Present age of Son S = 4 × $\sqrt{\dfrac{1792}{7 \times 4}}$ = 32 years.

Q - 24 | Deepak's present age is 120% of what he was 15 years ago and 75% of what he will be after 9 years. What is the present age of Deepak?

Sol: Consider, Deepak's present age is 'D' years.

According to question,

$$\frac{120}{100} \times (D - 15) = \frac{75}{100} \times (D + 9)$$

$$\frac{6}{5} \times (D - 15) = \frac{3}{4} \times (D + 9) \qquad \Rightarrow \qquad 24D - 360 = 15D + 135$$

$$9D = 360 + 135 = 495 \qquad \Rightarrow \qquad D = 55 \text{ years}$$

∴ Deepak's present age D = 55 years.

Shortcut:

> **Note:** If a man's age is x% of what he was t_1 years ago and y% of what he will be after t_2 years. Then, the present age of man is $\frac{x t_1 + y t_2}{x - y}$ years $(x > y)$

Here, $x = 120$, $t_1 = 15$, $y = 75$, $t_2 = 9$

∴ Deepak's present age $= \frac{120 \times 15 + 75 \times 9}{120 - 75} = 55$ years.

Q - 25 | The sum of the ages of father and daughter is 51 years. 8 years ago, the product of their ages was 5 times the father's age at that time. What are their present ages?

Sol: Given that, F + D = 51 years   **Present Ages**

(F − 8) (D − 8) = 5(F − 8)   Father – F, Daughter – D

D − 8 = 5   ⇒   D = 13 years

F = 51 − D = 51 − 13 = 38 years

∴ Present age of Father F = 38 years   &   Present age of Daughter D = 13 years.

Q - 26 | Father's present age is 3 times the sum of the ages of his 3 children, but 5 years hence his age will be only twice the sum of their ages. Find the present age of father.

Sol: Consider, Father present age = F and Sum of 3 children = C

Given that, F = 3C and After 5 years F + 5 = 2(C + 3 × 5)

F + 5 = 2(C + 15)   ∵ For every child 5 years increases

3C + 5 = 2C + 30   ⇒   C = 25 years

∴ Present age of father F = 3C = 3 × 25 = 75 years.

Q - 27 A father said to his son, "I was as old as you are at present at the time of your birth". If the father's age is 52 years now, then what will be the age of son after 6 years?

Sol: Consider, Father present age = F and Son present age = S.

According to question,

Father's age was equal to son's present age at the time of birth of son.

∵ Son's birth was taken place 'S' years ago.

'S' years ago, father's age F − S = S $\Rightarrow$ F = 2S

52 = 2S $\Rightarrow$ S = 26 years ∵ F = 52 years

∴ Son's age after 6 years = S + 6 = 26 + 6 = 32 years.

Q - 28 The present age of A is 6 years more than the age of B after 3 years. B's present age is 9 years more than the age of C before 6 years. The present age of C is 23 years. What was the age of A 4 years ago?

Sol: Given that, A = (B + 3) + 6 $\Rightarrow$ A = B + 9

B = (C − 6) + 9 $\Rightarrow$ B = C + 3 ∵ C = 23 years

B = 23 + 3 = 26 years

A = B + 9 = 26 + 9 = 35 years.

∴ Age of A, 4 years ago = A − 4 = 35 − 4 = 31 years.

Q - 29 9 years back, the ratio of the ages of Balu and Sravan was 19 : 12 respectively. The present age of Balu is $4\frac{2}{5}$ times of Ravi's present age. If Ravi's present age is 15 years, then what is Sravan's present age?

Sol: Given that, (B − 9) : (S − 9) = 19 : 12 **Present Ages**

Let, B − 9 = 19x, S − 9 = 12x (x – common factor) Balu – B, Sravan – S, Ravi – R

Also, B = $4\frac{2}{5}$ × R = $\frac{22}{5}$ × 15 = 66 years ∵ R = 15 years

B − 9 = 19x $\Rightarrow$ 19x = 66 − 9 = 57 $\Rightarrow$ x = 3

S − 9 = 12x $\Rightarrow$ S = 12 × 3 + 9 = 45 years

∴ Sravan's present age = 45 years.

Q - 30 The ratio between the present ages of Ajay and Vinay is 7 : 4 respectively. The ratio between Ajay's age 5 years ago and Vinay's age 4 years hence is 3 : 2. What is the ratio between Ajay's age after 3 years and Vinay's age 6 years hence?

Sol: Given that, $\dfrac{A}{V} = \dfrac{7}{4}$ $\quad\Rightarrow\quad$ $A = \dfrac{7}{4}V$ $\qquad\qquad$ **Present Ages**

$\dfrac{A-5}{V+4} = \dfrac{3}{2}$ $\quad\Rightarrow\quad$ $2A - 10 = 3V + 12 \ \ldots\ldots (1)$ $\qquad$ Ajay – A, Vinay – V

Substitute $A = \dfrac{7}{4}V$ in (1)

$2 \times \dfrac{7}{4}V - 10 = 3V + 12$ $\qquad\Rightarrow\qquad$ $\dfrac{7V}{2} - 3V = 22$ $\quad\Rightarrow\quad$ $V = 44$ years

$A = \dfrac{7}{4}V = \dfrac{7}{4} \times 44 = 77$ years

∴ Required ratio = $(A + 3) : (V + 6) = (77 + 3) : (44 + 6) = 8 : 5$.

Q - 31 The average age of a family of 7 members is 31 years. If the present age of younger person is 13 years, then what was the average age of the family at the time of birth of younger person?

Sol: Sum of family at present = $31 \times 7 = 217$ years $\qquad$ ∵ Sum = Avg. × No. of obs.

Sum of family at the time of birth of youngest person (13 years ago) = $217 - 13 \times 7 = 126$

∴ Average at the time of birth of youngest Person = $\dfrac{Sum\ of\ family}{No.of\ persons} = \dfrac{126}{6} = 21$ years.

∵ At the time of birth of youngest person only 6 members were there in the family.

Q - 32 The average age of 24 students and their teacher is 18 years. The average age of first 10 students is 20 years and that of last 14 students is 15 years. Find the age of teacher.

Sol: Sum of 24 students and teacher = Avg. × No. of persons = $18 \times 25 = 450$ years

Sum of 1st 10 students = $20 \times 10 = 200$ years

Sum of last 14 students = $15 \times 14 = 210$ years

Age of teacher = Sum of 24 students and teacher – Sum of 1st 10 – Sum of last 14 students

∴ Age of teacher = $450 - 200 - 210 = 40$ years.

Shortcut:

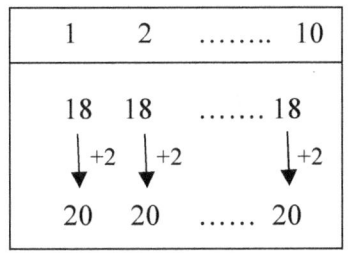

	First 10 students	Last 14 students	Teacher

	1 2 10	11 12 24	
Initial avg.	18 18 18	18 18 18	18
	↓+2 ↓+2 ↓+2	↓−3 ↓−3 ↓−3	
	20 20 20	15 15 15	

Distributed value: $+2 \times 10 = +20$ $-3 \times 14 = -42$

∴ Teacher's age = Initial average + Distributed value = $18 - 20 + 42 = 40$ years.

> **Note:** If teacher's age is not the part of 1st 10 and last 14 values, then
> distributed value sign will be reversed. i.e; '+' becomes '−' & '−' becomes '+'.

Q - 33 Shyam is 4 times his son's age and his wife is $\frac{7}{8}th$ of his present age. If the sum of the ages of all three of them is 119 years, then find the age of Shyam's son.

Sol: Given that, H = 4S   ⇒   S : H = 1 : 4   **Present Ages**

$W = \frac{7}{8} H$ ⇒   H : W = 8 : 7   Husband – H, Wife – W, Son – S

          S   H   W

S : H   =   1 : 4 : $\boxed{4}$

H : W   = $\boxed{8}$: 8 : 7

S : H : W = $1 \times 8 : 4 \times 8 : 4 \times 7 = 2 : 8 : 7$

S + H + W = 17 parts ⟶ 119 years

S = 2 parts ⟶ $? = \frac{119 \times 2}{17} = 14$ years

∴ Shyam's son's age = 14 years.

Q - 34 If Krunal is as much elder than Amit as he is younger to Manish and the sum of the ages of Amit and Manish is 64 years. Find the age of Krunal.

Sol: According to question, A + M = 64 years   Manish (M) → K + x

$(K - x) + (K + x) = 64$ ⇒   K = 32

∴ Krunal's present age K = 32 years.   Krunal   → K

                                                Amit (A)   → K − x

Shortcut:

> **Note:** If A is as much elder than B as he is younger to C and sum of the ages of B and C is 'T' years, then A's present age $= \dfrac{B+C}{2} = \dfrac{T}{2}$

$\therefore$ Krunal's present age $= \dfrac{64}{2} = 32$ years.

Q - 35 At present, Saroj's age is 2 years more than thrice his son's age. After 3 years, his age would be 20 years less than 4 times of his son's age. Find the present age of son.

Sol: Given that, F = 3S + 2   **Present Ages**

F + 3 = 4(S + 3) − 20   Father – F, Son – S

(3S + 2) + 3 = 4S + 12 − 20 $\Rightarrow$ S = 13 years

$\therefore$ Present age of Saroj's son S = 13 years.

Q - 36 Anusha got married 10 years ago. Today her age is $1\frac{1}{3}$ times her age at the time of marriage. Her son's age is $\frac{1}{8}$th of her age. What is the age of the son?

Sol: Given that, A$= 1\frac{1}{3}$ (A − 10)   **Present Ages**

$A = \dfrac{4}{3}$ (A − 10) $\because$ A – got married 10 years ago   Anusha – A, Son – S

$A = \dfrac{4A}{3} - \dfrac{40}{3}$ $\Rightarrow$ A = 40 years

Also, $S = \dfrac{A}{8}$ $\Rightarrow$ $S = \dfrac{40}{8} = 5$ years

$\therefore$ Son's present age S = 5 years.

Q - 37 The ratio of the ages of husband and wife is 9:8. After 6 years, the ratio will be 21:19. If at the time of marriage the ratio was 13:11, then how many years ago were they married?

Sol: Given that, H : W = 9 : 8   **Present Ages**

Let H = 9x, W = 8x   (x – common factor)   Husband – H, Wife – W

After 6 years, $\dfrac{H + 6}{W + 6} = \dfrac{21}{19}$ $\Rightarrow$ $\dfrac{9x + 6}{8x + 6} = \dfrac{21}{19}$

171x + 114 = 168x + 126 $\Rightarrow$ x = 4

H = 9x = 9 × 4 = 36 years, W = 8x = 8 × 4 = 32 years

Consider, 'k' years ago they got married.

$\dfrac{H - K}{W - K} = \dfrac{13}{11}$ $\Rightarrow$ $\dfrac{36 - K}{32 - K} = \dfrac{13}{11}$ $\Rightarrow$ 396 − 11k = 416 − 13k $\Rightarrow$ k = 10 years

$\therefore$ They got married 10 years ago.

Shortcut – 1:

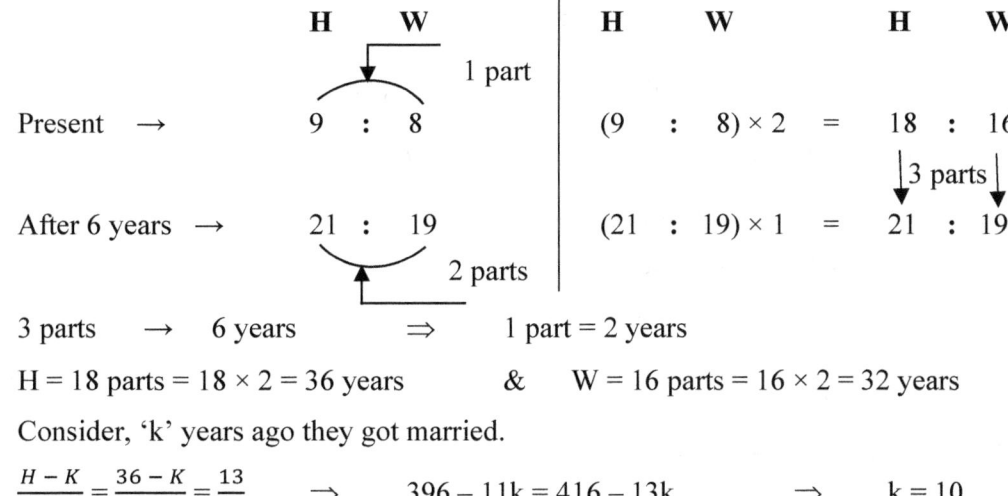

3 parts → 6 years ⇒ 1 part = 2 years

H = 18 parts = 18 × 2 = 36 years   &   W = 16 parts = 16 × 2 = 32 years

Consider, 'k' years ago they got married.

$$\frac{H-K}{W-K} = \frac{36-K}{32-K} = \frac{13}{11} \qquad \Rightarrow \qquad 396 - 11k = 416 - 13k \qquad \Rightarrow \qquad k = 10$$

∴ They got married 10 years ago.

Shortcut – 2:

> **Note:** If the ratio of the present ages of A and B is a : b. After 'T' years, the ratio will be c : d. Then,
>
> Present age of A = $a \times \dfrac{T(c-d)}{ad-bc}$, Present age of B = $b \times \dfrac{T(c-d)}{ad-bc}$

Here a = 9, b = 8, c = 21, d = 19, T = 6 years

Present age of Husband H = $9 \times \dfrac{6(21-19)}{9 \times 19 - 8 \times 21} = 36$ years

Present age of Wife W = $8 \times \dfrac{6(21-19)}{9 \times 19 - 8 \times 21} = 32$ years

Consider, 'k' years ago they got married.

$$\frac{H-K}{W-K} = \frac{36-K}{32-K} = \frac{13}{11} \qquad \Rightarrow \qquad 396 - 11k = 416 - 13k \qquad \Rightarrow \qquad k = 10$$

∴ They got married 10 years ago.

Q - 38   **The present age of Keshav is 4 times the age of Gopi. After 6 years, Keshav will be thrice the age of Gopi, then how many times will Keshav's age be in another 6 years' time with respect to Gopi's age then?**

Sol: Given that, K = 4G   **Present Ages**

After 6 years K + 6 = 3(G + 6)   Keshav – K, Gopi – G

4G + 6 = 3G + 18   ⇒   G = 12 years   ∵ K = 4G

K = 4G = 4 × 12 = 48 years

After 12 years (from 6 years to another 6 years) K + 12 = x(G + 12)

$\Rightarrow$ 48 + 12 = x(12 + 12) $\Rightarrow$ x = 2.5 times

$\therefore$ Keshav's age will be 2.5 times that of Gopi's age after 12 years.

Q - 39   **Akhil got married 5 years back. His present age is $\dfrac{7}{6}$ times his age at the time of his marriage. Akhil's brother was 4 years elder to him at the time of marriage. What will be the age of his brother after 6 years?**

Sol: According to question,   **Present Ages**

$A = \dfrac{7}{6}(A - 5)$ ($\because$ Akhil got married 5 years back)   Akhil – A, Akhil's brother – B

$A = \dfrac{7A}{6} - \dfrac{35}{6} = 35$ years

Also, at the time of marriage B – 5 = (A – 5) + 4

B = 35 – 5 + 4 + 5 = 39 years

$\therefore$ Akhil's brother's age after 6 years = B + 6 = 39 + 6 = 45 years.

Q - 40   **A father said to his son, "I am 5 times as old as you were when I was as old as you are". If the sum of their present ages is 112 years, then find their present ages.**

Sol: Consider, 'k' years ago son's present age and father's past age were equal.

According to question,

S + k = 5(S – k)

S + k = 5S – 5k

4S = 6k $\longrightarrow$ (1)

Sum of present ages = (S + k) + S = 112

k = 112 – 2S   …….. Sub in (1)

4S = 6(112 – 2S) $\Rightarrow$ 4S = 672 – 12S

16S = 672 $\Rightarrow$ S = 42 years

k = 112 – 2S = 112 – 2 × 42 = 28 years

	Father	Son
Present	S + k	S
	$\downarrow$ – k	$\downarrow$ – k
'k' years ago	S	S – k

$\therefore$ Father's present age = S + k = 42 + 28 = 70 years, Son's present age S = 42 years.

Q - 41 | Amar's brother is 4 years elder to him. His father was 34 years of age when his sister was born, while his mother was 31 years of age when he was born. If his sister was 5 years of age when his brother was born. Find the ages of Amar's father and mother respectively when his brother was born.

Sol:　　Amar's father was 34 years when his sister was born.

∴ Amar's father's age, when his brother was born = 34 + 5 = 39 years.

∵ Amar's sister was 5 years elder than his brother

Amar's mother was 31 years when Amar was born.

∴ Amar's mother's age, when his brother was born = 31 – 4 = 27 years.

∵ Amar was 4 years younger than his brother

Q - 42 | The ratio of the present ages of son and his father is 1 : 4 and that of his mother and father is 3 : 4. After 9 years the ratio of the ages of son to that of his mother becomes 4 : 9. What is the present age of father?

Sol:　　Present S : F　=　1 : 4　　　　　　　**Present Ages**

　　　　　　　F : M　=　4 : 3　　　　　　Father – F, Son – S, Mother – M

Present S : F : M = 1 : 4 : 3

Let, S = x, F = $4x$, M = $3x$　　　(x – common factor)

After 9 years　　　　⇒　　$\dfrac{S+9}{M+9} = \dfrac{4}{9}$　　⇒　　$\dfrac{x+9}{3x+9} = \dfrac{4}{9}$

$9x + 81 = 12x + 36$　　⇒　　$x = 15$

∴ Present age of father F = $4x$ = 4 × 15 = 60 years.

Alternative method:

　　　　　　　　　　　　S　　F　　M

Present　→　　　1 : 4

Present　→　　　　　4 : 3

Present ratio, S : F : M = 1 : 4 : 3

		S	M		S	M	S	M
				2 parts				
Present	→	1 :	3		(1 :	3) × 5 = 5 :	15	
							3 parts	
After 9 years	→	4 :	9		(4 :	9) × 2 = 8 :	18	
				5 parts				

3 parts → 9 years ⇒ 1 part = 3 years

∴ Present age of son = 5 parts = 5 × 3 = 15 years.

Present age of son 1 part = 15 years

∴ Father's present age F = 4 parts = 4 × 15 = 60 years.

Q - 43 **Dinesh's father was 45 years of age when he was born, while his mother was 37 years old when his sister 3 years elder to him was born. What is the sum of father and mother at the time of birth of Dinesh?**

Sol: Dinesh's father's age, when Dinesh was born = 45 years

Dinesh's mother's age, when Dinesh's sister was born = 37 years

∴ Dinesh's mother's age, when Dinesh was born = 37 + 3 = 40 years.

∵ Dinesh's sister was 3 years elder than Dinesh

∴ Sum of the ages of father and mother, when Dinesh was born = 45 + 40 = 85 years.

Q - 44 **If the ages of P and R are added to twice the age of Q, the total becomes 59. If the ages of Q and R are added to thrice the age of P, the total becomes 68. If the age of P is added to thrice the age of Q and thrice the age of R, the total becomes 108. What is the age of P?**

Sol: According to question,

$P + R + 2Q = 59$ → (1) $Q + R + 3P = 68$ → (2)

$P + 3Q + 3R = 108$ → (3)

From (2) ⇒ $Q + R = 68 - 3P$, From (3) ⇒ $P + 3(Q + R) = 108$

$P + 3(68 - 3P) = 108$ ⇒ $P + 204 - 9P = 108$ ∵ $Q + R = 68 - 3P$

$8P = 96$ ⇒ $P = 12$

∴ Present age of P = 12 years.

Q - 45 **The average age of Dharma's family consisting of 6 members 5 years ago was 40 years. 3 years ago, a new baby was born in this family. What will be the average age of the family after 4 years?**

Sol: Sum of Dharma's family 5 years ago = 40 × 6 = 240 years ∵ Sum = Avg. × No. of persons

Sum of family at present = 240 + (6 × 5) + 3 = 273 years

∵ 6 members age increased by 5 years and the age of new baby is 3 years

Sum of family after 4 years (7 members) = 273 + (7 × 4) = 301 years

∴ Average after 4 years $= \dfrac{Sum\ after\ 4\ years}{No.of\ persons} = \dfrac{301}{7} = 43$ years.

Q - 46 The average age of 90 employees in an office is 33 years, where $\frac{1}{3}$ employees are males and the ratio of average age of men and women is 5:3. What is the average age of males?

Sol: Sum of all employees = 33 × 90 = 2970 years   ∵ Sum = Avg. × No. of persons

Given that, No. of males = $\frac{1}{3}$ × Total = $\frac{1}{3}$ × 90 = 30 members

No. of females = Total – No. of males = 90 – 30 = 60 members

Average age ratio, Men **:** Women = 5 **:** 3

Let, men average age = 5*x*, women average age = 3*x*

Sum of all employees = Sum of males + Sum of females

2970 = 30 × 5*x* + 60 × 3*x*   ⇒   330*x* = 2970   ⇒   *x* = 9

∴ Average age of males = 5*x* = 5 × 9 = 45 years.

Shortcut:

Sum of all employees = 33 × 90 = 2970 years   ∵ Sum = Avg. × No. of persons

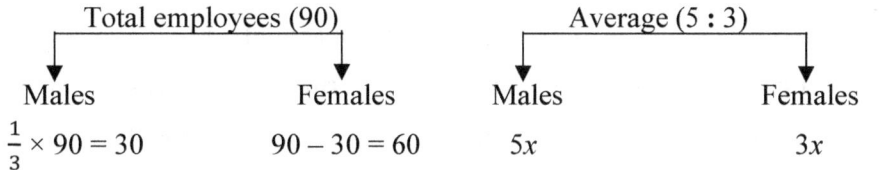

Sum of all employees = Sum of males + sum of females

2970 = 30 × 5*x* + 60 × 3*x*   ⇒   330*x* = 2970   ⇒   *x* = 9

∴ Average age of males = 5*x* = 5 × 9 = 45 years.

Q - 47 The average age of A and B is 34 years. If C replaces A, the average becomes 37 years and if C replaces B, then the average becomes 33 years. If the average age of D, E and F is half of the average age of A, B and C. Find the average age of all 6 members?

Sol: A + B = 34 × 2 = 68 years   →   (1)   ∵ Avg. of A and B is 34

C + B = 37 × 2 = 74 years   →   (2)   ∵ C replaces A

A + C = 33 × 2 = 66 years   →   (3)   ∵ C replaces B

(1) + (2) + (3)   ⇒   2(A + B + C) = 68 + 74 + 66 = 208   ⇒   A + B + C = 104

Also, given that

Average of D, E and F = $\frac{1}{2}$ × [Average of A, B and C]

$\frac{D + E + F}{3} = \frac{1}{2} \times \frac{A + B + C}{3}$ ⇒   D + E + F = $\frac{1}{2}$ × 104

∴ Average of all 6 members = $\frac{(A + B + C) + (D + E + F)}{6} = \frac{104 + 52}{6} = 26$ years.

Q - 48 In a joint family the average age of 5 males and 8 females is 38 years, 25 years respectively. If two persons whose average age is 16 years have left the family and three more members joined the family whose ages are 36, 22 and 32 years, then how much is the average age of the new family is increased?

Sol: Sum of family of 13 members (5 males and 8 females) = 5 × 38 + 8 × 25 = 390 years

∵ Sum = Avg. × No. of persons

Average of 13 members $= \dfrac{390}{13} = 30$ years

Also, avg. of males = 38, average of females = 25

New sum of family 14 members = 390 – (2 × 16) + (36 + 22 + 32) = 448

New avg. of family $= \dfrac{448}{14} = 32$ years

∵ 2 members left with an average of 16 and 3 members join the family

∴ New average is increased by 2 years.

Q - 49 The average age of all 30 teachers of a school 8 years ago was 42 years. 5 years ago, the principal has retired from his post at the age of 64 years and after 3 years a new principal whose age was 52 years recruited. What is the average age of all the teachers, if principal is also considered as a teacher?

Sol: Sum of all 30 teachers 8 years ago = 42 × 30 = 1260 years

∵ Sum = Avg. × No. of persons

Without any replacement, present sum = 1260 + (30 × 8) = 1500 years

But, the principal was retired 5 years ago at the age of 64 years and after 3 years a new principal recruited at the age of 52 years

New present sum = 1500 – 5 – 64 + 2 + 52 = 1485 years

∴ New average age $= \dfrac{New\ sum}{No.of\ persons} = \dfrac{1485}{30} = 49\dfrac{1}{2}$ years.

Q - 50 The average age of officers of a public sector bank, having 15 officers was 38 years. 2 officers whose ages were 56 years and 55 years transferred to other branches and on the same day one officer resigned from his duties. So 2 new officers aged 32 years and 30 years joined in bank. After 2 years on the same date the average age of all 14 officers was found to be 35 years. What was the age of the officer who resigned from his duties?

Sol: Sum of 15 officers = 38 × 15 = 570 years   ∵ Sum = Avg. × No. of persons

Here, 2 officers transferred, 1 officer resigned and 2 officers joined.

Present sum of 14 officers = 570 – 56 – 55 – x + 32 + 30   →   (1)

where 'x' is age of resigned officer

Sum of 14 officers after 2 years = 35 × 14 = 490 years

Sum of 14 officers at present = 490 – (2 × 14) = 462 years   →   (2)

(1) = (2)   ⇒   570 – 56 – 55 – x + 32 + 30 = 462   ⇒ x = 59 years

∴ Age of officer who resigned x = 59 years.

ASSESSMENT TEST

1. The ratio of the present ages of A and B is 11 : 9. If the difference between A and B is 8 years, then what is the age of A?

2. Srikanth was 4 times as old as Vinay, 8 years ago. What is the age of Vinay, if Srikanth will be 50 years old 14 years hence?

3. The sum of the ages of 4 children born at the intervals of 3 years each is 46. What is the age of the youngest child?

4. If the product of ages of A and B is 884 years and the ratio of their present ages is 17 : 13. What are their present ages?

5. The ratio of present ages of A and B is 6 : 5. If the difference between present age of A and age of B after 4 years is 5, then what is the sum of their present ages?

6. The product of the ages of Karan and Anand is 308. If twice the age of Anand is 6 years more than Karan. What is the age of Anand?

7. 24 years ago, Sunitha was one – fourth of what she is now. What was the age of Sunitha before 5 years?

8. After 16 years, A will be thrice as old as B was 7 years ago. If the present age of A is 9 years more than B. Find the age of B.

9. The sum of the ages of x, y and z is 143 years. x is 15 years older than z, y is twice as old as z. Find their respective ages.

10. A's age is twice the sum of ages of B and C. If after 15 years the age of A is equal to sum of ages of B and C. What is the present age of A?

11. Sandeep's age is 140% of what he was 9 years ago and 80% of what he will be after 12 years. What is his present age?

12. Present age of father is equal to the sum of the ages of his daughter and son. The age of daughter is 24 years. If 24 years ago, their average age was 16 years, then find the present age of son.

13. If B's age is twice the average age of A, B and C. C's age is one – third of the average of A, B and C. If A is 10 years old, then find the average age of all 3 members A, B and C.

14. At present ratio of the ages of mother and daughter is 2 : 1 and the product of their ages is 800. What will be the ratio of their ages after 5 years?

15. If 12 years are subtracted from the present age of grandfather and the remainder is divided by 7, then the present age of his granddaughter is obtained. If granddaughter is 2 years more than his grandson's age who is 6 years old. What is the age of grandfather?

16. The sum of the ages of mother and son is 46 years. 6 years, ago the product of their ages was 3 times the mother's age at that time. What are their present ages?

17. A father said to his son, "I was as old as you are at present at the time of your birth". If the father's age is 48 years, then what will be the age of son before 7 years?

18. Sudhir is 2.5 times that of Dayakar's age and 1.5 times that of Ramu's age. If the sum of the ages of Dayakar and Ramu is 64 years, then what is the age of Sudhir?

19. A father's age is 4 times the sum of the ages of his two children, after 24 years his age will be equal to the sum of their ages. What is the age of the father?

20. 17 years ago, the ages of A and B was in the ratio of 5 : 2. 16 years hence, their ratio will changes to 7 : 5. What is the present age of B?

21. The average age of a family of 6 members is 28 years. If the present age of youngest member is 8 years, then what was the average age of the family at the time of birth of youngest member?

22. The present age of A is 7 years more than the age of B before 5 years. B's present age is 3 years more than the age of C after 2 years. The present age of C is 34 years. What will be the age of A after 8 years?

23. 15 years ago, the ratio of the ages of Prabhu and Vineet was 5 : 7 respectively. The present age of Prabhu is $3\frac{1}{8}$ times of Abbas present age. If Abbas present age is 16 years, then what is Vineet's present age?

24. At present, the ratio of the ages of Karthik and Rishab is 8 : 17. Rishab is 12 years younger than Prasad. Prasad's age before 6 years was 74 years. What is the present age of Karthik's father, who is 27 years elder than Karthik?

25. Venky's grandmother was 6 times older to him 12 years ago. She would be two times of his age 6 years hence. What was the ratio of the ages of Venky and his grandmother 6 years ago?

26. The ratio of the ages of husband and wife is 6 : 5. 8 years hence, the ratio will be 7 : 6. If at the time of marriage the ratio was 7 : 5, then how many years ago were they married?

27. A woman was asked to tell her age in years. She said that "Take my age 6 years hence, multiply it by 5 and then subtract 5 times my age 3 years back and you will know how old I am". What was the age of the woman?

28. The average age of 35 students and their teacher is 15 years. The average age of first 17 students is 18 years and that of last 18 students is 11 years. Find the age of teacher.

29. Venkat present age is 2.5 times his son's age and $\frac{4}{7}$th of his mother's present age. The sum of the present ages of all three of them is 126 years. What is the difference between the present age of Venkat's mother and Venkat's son?

30. Ramu's present age is thrice the age of Srinu 5 years ago. What is the age of Ramu, if the ratio of the sum of their present ages to the difference of their present ages is 11 : 3?

31. If Bhanu is as much elder than Swathi as she is younger than Prasanna and the sum of the ages of Swathi and Prasanna is 76 years. What is the age of Bhanu?

32. A father said to his daughter, "I am 3 times as old as you were when I was as old as you are". If the sum of their present ages is 150 years, then find their present ages.

33. Presently, a man's age is 5 years more than twice his daughter's age. After 4 years, his age will be 18 years less than thrice the daughter's age. Find the father's present age.

34. Swetha's father was 32 years of age when she was born while her mother was 29 years old when her brother 5 years younger to her was born. What is the difference between the ages of her parents?

35. The ratio of the present ages of son and his father is 2 : 5 and that of his father and mother is 5 : 4. After 12 years the ratio of the age of son to that of his mother becomes 5 : 8. What is the present age of father?

36. The average age of the family of 4 members, 7 years ago was 36 years. 3 years ago a new baby was born in this family. What will be the average age of the family after 6 years?

37. The average age of a family of 6 members, 5 years ago was 32 years. During this period a child was born in this family and still the average age of the family is same today. What is the present age of the child?

38. The average age of 70 employees in an office is 41 years, where $\frac{4}{7}$th employees are females and the ratio of average age of men and women is 7 : 5. What is the average age of females?

39. The average age of 28 teachers of a school 11 years ago was 34 years. 3 years ago, the principal has retired from his post at the age of 55 years and after one year a new principal whose age was 49 years recruited. What is the average age of all the teachers, if principal is also considered as a teacher?

40. The average age of officers of an organization, having 12 officers was 45 years. One officer whose age was 70 years transferred and on the same day one officer resigned from his duties. So one new officer aged 36 years joined in organization. After 3 years on the same date the average age of all 11 officers was found to be 43 years. What was the age of the officer who resigned from his duties?

KEY

1. 44 years

2. 15 years

3. 7 years

4. 34 years, 26 years

5. 99 years

6. 14 years

7. 27 years

8. 23 years

9. 47 years, 64 years and 32 years

10. 30 years

11. 37 years

12. 28 years

13. 15 years

14. 9 : 5

15. 68 years

16. 37 years, 9 years

17. 17 years

18. 60 years

19. 32 years

20. 29 years

21. 24 years

22. 49 years

23. 64 years

24. 59 years

25. 5 : 16

26. 10 years ago

27. 45 years

28. 36 years

29. 54 years

30. 21 years

31. 38 years

32. 90 years, 60 years

33. 35 years

34. 8 years

35. 45 years

36. 41 years

37. 2 years

38. 35 years

39. 44.75 years

40. 66 years

PROFIT AND LOSS

- ✓ Basically, the terms profit and loss come in the field of business.
- ✓ The terms profit and loss depend on two factors, one is cost price (CP) and the other is selling price (SP).
- ✓ Cost price (CP) means the price at which an article is purchased.
- ✓ Selling price (SP) means the price at which an article is sold.
- ✓ If SP more than CP, then we are getting profit. If SP is less than CP, then we are getting loss.
- ✓ Profit percentage and loss percentage both are calculated over cost price (CP).

Profit (SP > CP)	Loss (SP < CP)
Profit = SP − CP	Loss = CP − SP
$\text{Profit\%} = \dfrac{Profit}{Cost\ price} \times 100\%$	$\text{Loss\%} = \dfrac{Loss}{Cost\ price} \times 100\%$
$\text{Profit\%} = \dfrac{SP - CP}{CP} \times 100\%$	$\text{Loss\%} = \dfrac{CP - SP}{CP} \times 100\%$
$SP = CP \times \dfrac{100 + P\%}{100}$	$SP = CP \times \dfrac{100 - L\%}{100}$
$CP = SP \times \dfrac{100}{100 + P\%}$	$CP = SP \times \dfrac{100}{100 - L\%}$

Some important formulae to be remember:

- ✓ When a person sells two similar articles, one at a gain of $x\%$ and another at a loss of $x\%$, then the seller will always incurs a loss. Then,

$$\text{Loss percentage} = \frac{x^2}{100}\%$$

- ✓ If a merchant buys x articles for Rs. y and sold z articles for Rs. w, then

$$\text{Profit (or) Loss percentage} = \left(\frac{xw}{yz} - 1\right) \times 100\%$$

+ve answer indicates profit & −ve answer indicates loss

- ✓ If a dishonest trader professes to sell his goods at Cost Price (CP) but uses false weights, then

$$\text{Gain percent} = \frac{Error}{True\ value - Error} \times 100\%$$

$$\text{Gain percent} = \frac{True\ weight - False\ weight}{False\ weight} \times 100\%$$

✓ If a dishonest trader professes to sell his goods at x% profit (or) loss on cost price and uses y% less weight, then

$$\text{Profit (or) Loss percentage} = \frac{y \pm x}{100 - y} \times 100\%$$

Use, +ve sign for profit & –ve sign for loss

✓ If a person purchased two articles for Rs. A. He sold first article at a profit (or) loss of x% and the second article at a profit (or) loss of y%. If selling price (SP) of two articles are same, then

$$\text{CP of first article} = \frac{A(100 \pm y)}{(100 \pm x) + (100 \pm y)}$$

$$\text{CP of second article} = \frac{A(100 \pm x)}{(100 \pm x) + (100 \pm y)}$$

Use, +ve sign for profit & –ve sign for loss

✓ 1. If a shopkeeper sells his goods at a gain of x% on selling price (SP), then

$$\text{Actual gain percent on CP} = \frac{x}{100 - x} \times 100\%$$

2. If a shopkeeper sells his goods at a loss of x% on selling price (SP), then

$$\text{Actual loss percent on CP} = \frac{x}{100 + x} \times 100\%$$

✓ If a person sells x articles gains (or) loses the cost price (CP) of y articles, then

$$\text{Gain (or) loss percentage} = \frac{y}{x} \times 100\%$$

✓ 1. If a person sells x articles gains the selling price (SP) of y articles, then

$$\text{Gain percentage} = \frac{y}{x - y} \times 100\%$$

2. If a person sells x articles loses the selling price (SP) of y articles, then

$$\text{Loss percentage} = \frac{y}{x + y} \times 100\%$$

SOLVED EXAMPLES

Q-1 **The cost price of an article is Rs. 800 and selling price is Rs. 960, then find the profit percentage.**

Sol: $\text{Profit\%} = \dfrac{SP - CP}{CP} \times 100\% = \dfrac{960 - 800}{800} \times 100\% = 20\%$.

Shortcut:

Consider, cost price (CP) is 100%

CP = Rs. 800 $\longrightarrow$ 100%

SP = Rs. 960 $\longrightarrow$ $? = \dfrac{960 \times 100}{800} = 120\%$

∴ Profit percentage = SP – CP = 120% – 100% = 20%.

Q-2 **The cost price of an article is Rs. 1200 and selling price is Rs. 900, then find the loss percentage.**

Sol: $\text{Loss\%} = \dfrac{CP - SP}{CP} \times 100\% = \dfrac{1200 - 900}{1200} \times 100\% = 25\%$.

Shortcut:

Consider, cost price (CP) is 100%

CP = Rs. 1200 $\longrightarrow$ 100%

SP = Rs. 900 $\longrightarrow$ $? = \dfrac{900 \times 100}{1200} = 75\%$

∴ Loss percentage = CP – SP = 100% – 75% = 25%.

Q-3 **By selling a TV set for Rs. 17000, Varun loses 15%. Find the cost price of the TV set.**

Sol: Given that, SP of a TV set = Rs. 17000, Loss = 15%

Cost price CP = $SP \times \dfrac{100}{100 - L\%} = 17000 \times \dfrac{100}{100 - 15} = \text{Rs. } 20000$

∴ Cost price of a TV set is Rs. 20000.

Shortcut:

Consider, cost price (CP) of a TV set is 100%   ∵ Loss = 15%,

SP = 85% $\longrightarrow$ Rs. 17000   SP = CP – L = 100 – 15 = 85%

CP = 100% $\longrightarrow$ $? = \dfrac{17000 \times 100}{85} = \text{Rs. } 20000$.

Q - 4 | **Shiva buys a cooler for Rs.7800. How much should he sell in order to gain 13%?**

Sol: Given that, CP of a cooler = Rs. 7800, Gain = 13%

$$SP = CP \times \frac{100 + G\%}{100} = 7800 \times \frac{100 + 13}{100} = Rs.\ 8814$$

∴ SP of a cooler is Rs. 8814.

Shortcut:

Consider, cost price (CP) of a cooler is 100%   ∵ Gain = 13%

CP = 100%   ⟶   Rs. 7800   SP = CP + G = 100 + 13 = 113%

SP = 113%   ⟶ $? = \frac{7800 \times 113}{100} = Rs.\ 8814.$

Q - 5 | **If a person sold an article at Rs. 2500, there is 15% loss. Find the profit (or) loss percentage when he sells at Rs. 4000.**

Sol: Given that, initial SP = Rs. 2500, Loss = 15%

$$CP = SP \times \frac{100}{100 - L\%} = 2500 \times \frac{100}{100 - 15} = Rs.\ \frac{50000}{17}$$

New SP = Rs. 4000, which is more than CP. Therefore we will get profit.

$$\text{Profit percentage} = \frac{SP - CP}{CP} \times 100\% = \frac{4000 - \frac{50000}{17}}{\frac{50000}{17}} \times 100\% = 36\%.$$

Shortcut:

Consider, cost price (CP) is 100%.

SP_1 = Rs. 2500   ⟶   85%   ∵ Loss = 15%

SP_2 = Rs. 4000   ⟶ $? = \frac{85 \times 4000}{2500} = 136\%$

Since $SP_2 > CP$. Therefore, Profit percentage = 136% − 100% = 36%.

Q - 6 | **If loss is $\frac{1}{4}$ of SP, then find the loss percentage.**

Sol: Loss = CP − SP   ⟹ $\frac{SP}{4} = CP - SP$ ∵ Loss $= \frac{Sp}{4}$

$\frac{SP}{4} + SP = CP$ ⟹ $\frac{5SP}{4} = CP$

⟹ $\frac{SP}{CP} = \frac{4}{5}$ ⟹   SP = 4, CP = 5

∴ Loss percentage $= \frac{CP - SP}{CP} \times 100\% = \frac{5 - 4}{4} \times 100\% = 20\%.$

Shortcut:

$$\text{If loss is } \frac{x}{y} \text{ of SP, then Loss percentage} = \frac{x}{x+y} \times 100\%$$

Here, $x = 1$ and $y = 4$

$$\therefore \text{Loss percentage} = \frac{1}{1+4} \times 100\% = 20\%.$$

Q-7 The cost price of 15 articles is same as selling price of 9 articles, then find the profit (or) loss percentage.

Sol: Given that, $15CP = 9SP$ $\Rightarrow$ $5CP = 3SP$

$\frac{SP}{CP} = \frac{5}{3}$ $\Rightarrow$ $SP = 5, CP = 3$ $\qquad$ (SP > CP = Profit)

$$\therefore \text{Profit percentage} = \frac{SP - CP}{CP} \times 100\% = \frac{5-3}{3} \times 100\% = 66\frac{2}{3}\%.$$

Shortcut:

From the given statement, it is clear that there is a profit of 6 articles by selling 9 articles

$$\therefore \text{Profit percentage} = \frac{6}{9} \times 100\% = 66\frac{2}{3}\%.$$

Q-8 The cost price of 20 articles is same as selling price of 25 articles, then find the profit (or) loss percentage.

Sol: Given that, $20CP = 25SP$ $\Rightarrow$ $4CP = 5SP$

$\frac{SP}{CP} = \frac{4}{5}$ $\Rightarrow$ $SP = 4, CP = 5$ $\qquad$ (SP < CP = Loss)

$$\therefore \text{Loss percentage} = \frac{CP - SP}{CP} \times 100\% = \frac{5-4}{5} \times 100\% = 20\%.$$

Shortcut:

From the given statement, it is clear that there is a loss of 5 articles by selling 25 articles

$$\therefore \text{Loss percentage} = \frac{5}{25} \times 100\% = 20\%.$$

Q-9 The loss made on selling 8 meters of cloth is equal to the cost price of 3 meters of that cloth. Find the loss percentage.

Sol: Total loss = Total CP – Total SP

$3CP = 8CP - 8SP$ $\Rightarrow$ $8SP = 5CP$ $\qquad\qquad$ $\because$ Loss = 3CP

$\frac{SP}{CP} = \frac{5}{8}$ $\Rightarrow$ $SP = 5, CP = 8$ $\qquad$ (SP < CP = Loss)

$$\therefore \text{Loss percentage} = \frac{CP - SP}{CP} \times 100\% = \frac{8-5}{8} \times 100\% = 37.5\%.$$

Shortcut:

If a person sells x articles gains (or) loses the cost price (CP) of y articles. Then,

$$\text{Gain (or) loss percentage} = \frac{y}{x} \times 100\%$$

Here, $x = 8$ and $y = 3$

$\therefore$ Loss percentage $= \frac{3}{8} \times 100\% = 37.5\%$.

Q - 10 **Ramesh Sold a DVD player to Vijay at a gain of 20% and Vijay sold it to Devan at a gain of 10%. If Devan pays Rs.8580 to Vijay, then what is the cost price of DVD player to Ramesh?**

Sol: Consider, Ramesh cost price CP is 100%.

	CP		**SP**	
Ramesh	100%	20% gain	120%	$\because$ Ramesh SP = Vijay CP
Vijay	120%	10% gain	132%	$\because$ Vijay SP = Devan CP
Devan	132%			

Devan CP = 132% $\longrightarrow$ Rs. 8580

Ramesh CP = 100% $\longrightarrow$? $= \dfrac{8580 \times 100}{132} = $ Rs. 6500

$\therefore$ Ramesh's cost price of the DVD player is Rs. 6500.

Shortcut:

Consider, Ramesh cost price CP = Rs. x

	Ramesh		**Vijay**		**Devan**
Devan payment =	x	$\times$	$\dfrac{100 + 20}{100}$	$\times$	$\dfrac{100 + 10}{100}$

$8580 = x \times \dfrac{120}{100} \times \dfrac{110}{100}$ $\qquad \Rightarrow \qquad x = $ Rs. 6500

$\therefore$ Cost price of the DVD player to Ramesh is Rs. 6500.

Q - 11 | P sold an article to Q at 40% profit. Q sold it to R at 25% profit. If P's profit is Rs. 560 more than Q's profit, then find the cost price of P.

Sol: Consider, cost price of P is 100%

	CP		**SP**	**Profit = SP – CP**
P	100%	40% gain →	140%	140% – 100% = 40%
Q	140%	25% gain →	175%	175% – 140% = 35%
R	175%			

Given that, P's profit = Q's profit + Rs. 560

$40\% = 35\% + 560 \quad \Rightarrow \quad 5\% = 560 \quad \Rightarrow \quad 1\% = 112$

∴ Cost price of P = 100% = 100 × 112 = Rs. 11200.

Q - 12 | If 12 chocolates are bought for Rs. 16 and sold at 16 for Rs. 12. Find the profit (or) loss percentage.

Sol: CP of each chocolate = Rs. $\frac{16}{12}$ = Rs. $\frac{4}{3}$

SP of each chocolate = Rs. $\frac{12}{16}$ = Rs. $\frac{3}{4}$ ∵ SP < CP = Loss

∴ Loss percentage = $\frac{CP - SP}{CP} \times 100\% = \frac{\frac{4}{3} - \frac{3}{4}}{\frac{4}{3}} \times 100\% = 43\frac{3}{4}\%.$

Shortcut:

If a shopkeeper buys x articles for Rs. y and sold z articles for Rs. w, then

$$\text{Profit (or) Loss percentage} = \left(\frac{xw}{yz} - 1\right) \times 100\%$$

Here $x = 12$, $y = 16$, $z = 16$ and $w = 12$

Profit (or) Loss percentage = $\left(\frac{12 \times 12}{16 \times 16} - 1\right) \times 100\% = \left(\frac{9}{16} - 1\right) \times 100\% = -43\frac{3}{4}\%$

Here, –ve sign indicates that there is a loss of $43\frac{3}{4}\%.$

Q - 13 | When the selling price of an article is increased by Rs. 168, then the loss of 12% is converted into a profit of 9%. Find the cost price of the article.

Sol: Consider, Cost price (CP) of an article is Rs. x

$SP_1 = 88\%$ of $x = \frac{88}{100}x$ ∵ Loss = 12%

$SP_2 = SP_1 + 168 \quad \Rightarrow \quad 109\%$ of $x = 88\%$ of $x + 168$ ∵ Profit = 9%

$\frac{109}{100}x - \frac{88}{100}x = 168 \quad \Rightarrow \quad 21x = 16800 \quad \Rightarrow \quad x = Rs. 800$

∴ Cost price of an article is Rs. 800.

Shortcut:

Consider, CP = 100%

$SP_2 = SP_1 + 168$

109% − 88% = 168

$SP_2 − SP_1 = 21\% \longrightarrow$ Rs. 168

CP = 100% $\longrightarrow$? = $\dfrac{168 \times 100}{21}$ = Rs. 800

∴ Cost price of an article is Rs. 800.

109% — SP$_2$	
	P = 9%
21%	100% — CP
	L = 12%
88% — SP$_1$	

Q - 14 | On selling 60 articles, a shopkeeper got a profit equal to the selling price of 15 articles. **What is the profit percentage?**

Sol: We know that, profit = SP − CP

15SP = 60SP − 60CP   ⇒   60CP = 45SP

$\dfrac{SP}{CP} = \dfrac{60}{45}$ ⇒   SP = 60, CP = 45   (SP > CP = Profit)

∴ Profit percentage = $\dfrac{SP - CP}{SP} \times 100\% = \dfrac{60 - 45}{45} \times 100\% = 33\dfrac{1}{3}\%$.

Shortcut:

If a person sells *x* articles, gains the selling price (SP) of *y* articles. Then

$$\text{Gain percentage} = \dfrac{y}{x - y} \times 100\%$$

Here, *x* = 60 and *y* = 15

∴ Profit percentage = $\dfrac{15}{60 - 15} \times 100\% = 33\dfrac{1}{3}\%$.

Q - 15 | **If selling price is tripled, the profit is four times. Find the profit percentage.**

Sol: We know that, profit P = SP − CP ……. (1)

According to question, 4P = 3SP − CP …….. (2)

Substitute (1) in (2)

4(SP − CP) = 3SP − CP   ⇒   4SP − 4CP = 3SP − CP

⇒   SP = 3CP   ⇒ $\dfrac{SP}{CP} = \dfrac{3}{1}$ ⇒   SP = 3, CP = 1

∴ Profit percentage = $\dfrac{3 - 1}{1} \times 100\% = 200\%$.

Q - 16   **Satya purchased 150 chairs at a price of Rs. 120 per chair. He sold 40 chairs at a profit of Rs. 15 per chair and 85 chairs at a profit of Rs. 12 per chair and the rest at a loss of Rs. 9 per chair. What is the average profit per chair?**

Sol: Total CP of all chairs = 150 × 120 = Rs. 18000

SP of 40 chairs = 40 × (120 + 15) = Rs. 5400

SP of 85 chairs = 85 × (120 + 12) = Rs. 11220

SP of remaining 25 chairs = 25 × (120 – 9) = Rs. 2775

Total SP of all chairs = 5400 + 11220 + 2775 = Rs. 19395

Total Profit = Total SP – Total CP = 19395 – 18000 = Rs. 1395

∴ Average profit per chair = $\frac{Total\ profit}{No.of\ Chairs} = \frac{1395}{150}$ = Rs. 9.30.

Q - 17   **If Naveen sold an article at five-sixth of its actual selling price, he would have incurred a loss of 30%. Find his actual profit (or) loss percent.**

Sol: Consider, CP of an article = 100% and SP is Rs. x

Given that, SP = $\frac{5}{6}x$ ⟶   70%   ∵ Loss = 30%

Actual SP = x ⟶   ? = $\frac{70x \times 6}{5x}$ = 84%

Since, actual SP < CP. Therefore Loss = CP – SP = 100% – 84% = 16%.

Q - 18   **If a shopkeeper sells an article at a gain of 30%. If he had bought it at 25% less and sold it for Rs. 240 less, he would have earned 20% Profit. Find the cost price of the article.**

Sol: Consider, Initial CP of an article = 100%

Initial SP = 130%   ∵ Gain = 30%

New CP = 100% – 25% = 75%, New SP = $\frac{120}{100}$ × 75% = 90%   ∵ CP – 25% less

Initial SP – New SP = Rs. 240

130% – 90% =   40%   ⟶   Rs. 240   ∵ SP – Rs. 240 less

CP of an article = 100%   ⟶   ? = $\frac{240 \times 100}{40}$ = Rs. 600.

Q - 19 **A person bought 10 kgs of sugar for certain price. He sold 5 kgs at a profit of 20% and 2 kgs with neither profit nor loss and 3 kgs at 10% loss. What is the total profit percentage?**

Sol:

$$\boxed{\text{Total cost} = \text{Quantity} \times \text{Price per kg}}$$

Consider, CP of sugar per kg = Rs. 100

Total CP of sugar = 10 × 100 = Rs. 1000

SP of 5 kgs = 5 × 120 = Rs. 600   ∵ Gain = 20%

SP of 2 kgs = 2 × 100 = Rs. 200   ∵ No profit, No Loss

SP of 3 kgs = 3 × 90 = Rs. 270   ∵ Loss = 10%

Total SP of sugar = 600 + 200 + 270 = Rs. 1070

∴ Profit percentage $= \dfrac{SP - CP}{CP} \times 100\% = \dfrac{1070 - 1000}{1000} \times 100\% = 7\%$.

Q - 20 **If the selling price of an article is increased by 20%, then the profit is 56%. Find the profit percentage, if the selling price is decreased by 15%.**

Sol: Consider, CP of an article = 100% and SP = Rs. x

$\text{SP} = \dfrac{120}{100} \times x = \dfrac{6x}{5} \quad \longrightarrow \quad 156\%$ ∵ Profit = 56% & SP increased by 20%

$\text{SP} = \dfrac{85}{100} \times x = \dfrac{17x}{5} \quad \longrightarrow \quad ? = 156 \times \dfrac{17x}{5} \times \dfrac{5}{6x} = 110.5\%$ ∵ SP decreased by 15%

∴ If SP is decreased by 15%, then profit is 10.5%.

Q - 21 **A calculator X which when sold for Rs. 480 gives a profit of 20%. Another calculator Y of the same cost price was sold at less cost due to customer bargain. By what percent was the selling price of Y less than that of X, if there was neither profit nor loss for the seller?**

Sol: Consider, CP of each calculator = 100%

For calculator X, SP = 120% $\longrightarrow$ Rs. 480   ∵ Profit = 20%

For calculator X, CP = 100% $\longrightarrow$ $? = \dfrac{480 \times 100}{120} = \text{Rs. } 400$

Total CP of both calculators = 400 + 400 = Rs. 800

Given that there was neither profit nor loss for seller means, Total CP = Total SP

800 = 480 + SP of Y $\Rightarrow$ SP of Y = Rs. 320

Selling price of calculator Y = Rs. 320

	X	Y
SP	480	320

$$\boxed{\text{Percentage less than} = \dfrac{Difference}{High\ value} \times 100\%}$$

∴ Percentage less than $= \dfrac{480 - 320}{480} \times 100\% = 33\dfrac{1}{3}\%$.

Q - 22 A mobile phone worth Rs. 12000 is sold by X to Y at 20% loss and Y sells the mobile phone back to X at a gain of 20%. Who gains and who loses? Find also values.

Sol: SP of X = CP of Y = 80% of 12000 = $\frac{80}{100} \times 12000$ = Rs. 9600 $\because$ Loss = 20%

SP of Y = CP of X = 120% of 9600 = $\frac{120}{100} \times 9600$ = Rs. 11520 $\because$ Gain = 20%

In this transaction Y's CP = Rs. 9600, Y's SP = Rs. 11520

$\therefore$ Y gains and X loses = 11520 – 9600 = Rs. 1920

X loss percentage = $\frac{Loss}{CP} \times 100\%$ = $\frac{1920}{12000} \times 100\%$ = 16%.

Q - 23 A milkman buys some milk. If he sells it at Rs. 4 a liter he loses Rs. 300, but when he sells it at Rs. 6 at a liter, he gains Rs. 250. How much milk did he purchase?

Sol: Consider, quantity of milk purchased = x liters

SP of milk, when sold at Rs. 4 per liter = $4x$

SP of milk, when sold at Rs.6 per liter = $6x$

Difference in selling prices = 250 – (–300) = Rs. 550

$6x - 4x = 550 \quad \Rightarrow \quad x = 275$ liters.

Shortcut:

$$\text{Quantity purchased} = \frac{Difference\ in\ selling\ prices}{Difference\ in\ SP\ per\ liter}$$

Use, +ve sign for profit & –ve sign for loss

$\therefore$ Quantity of milk purchased = $\frac{250 - (-300)}{6 - 4}$ = 275 liters.

Q - 24 A fruit vendor makes a profit of 20% by selling watermelon at a certain price. If he charges Rs. 6 more on watermelon, he would gain 40%. Find the cost price and selling price of watermelon.

Sol: Consider, CP of watermelon = Rs. x

SP_1 = 120% of $x = \frac{120}{100}x = \frac{6x}{5}$ $\because$ Profit = 20%

$SP_2 = SP_1 + Rs. 6 \quad \Rightarrow \quad 140\%$ of $x = \frac{6x}{5} + 6$ $\because$ Profit = 40%

$\frac{140x}{100} = \frac{6x}{5} + 6 \quad \Rightarrow \quad \frac{7x}{5} - \frac{6x}{5} = 6 \quad \Rightarrow \quad x = Rs. 30$

$\therefore$ Cost price of watermelon is Rs. 30 & First selling price = $\frac{6x}{5} = \frac{6}{5} \times 30$ = Rs. 36.

Shortcut:

Consider, CP of watermelon = 100%

According to question, SP_1 = 120% and SP_2 = 140%

Difference in SP's = Rs. 6 $\Rightarrow$ 140% − 120% = Rs. 6

      Difference = 20% $\longrightarrow$ Rs. 6

CP of watermelon = 100% $\longrightarrow$ $? = \frac{6 \times 100}{20}$ = Rs. 30.

First selling price = 120% $\longrightarrow$ $? = \frac{6 \times 120}{20}$ = Rs. 36.

Q - 25 **4 chairs cost as much as 7 stools, 5 stools as much as 2 tables, 6 tables as much as 4 fans, if the cost of 7 fans is Rs. 9450, then find the price of 6 chairs.**

Sol: Given that, 4 Chairs = 7 Stools, 5 Stools = 2 Tables, 6 Tables = 4 Fans,

7 Fans cost = Rs. 9450 $\Rightarrow$ Each fan cost = $\frac{9450}{7}$ = Rs. 1350

6 Tables cost = 4 × 1350 $\Rightarrow$ 1 Table cost = Rs. 900

5 Stools cost = 2 × 900 $\Rightarrow$ 1 Stool cost = Rs. 360

4 Chairs cost = 7 × 360 $\Rightarrow$ 1 Chair cost = Rs. 630

$\therefore$ Cost of 6 Chairs = 6 × 630 = Rs. 3780.

Shortcut:

Consider cost of 6 Chairs = Rs. x

    Given that, Rs. x = 6 Chairs

    4 Chairs = 7 Stools

    5 Stools = 2 Tables

    6 Tables = 4 Fans

    7 Fans = Rs. 9450

$\therefore$ 6 Chairs cost $x = \frac{6 \times 7 \times 2 \times 4 \times 9450}{4 \times 5 \times 6 \times 7}$ = Rs. 3780.

Q - 26 **What will be the profit percentage after selling an article at a certain price, if there is a loss of $16\frac{2}{3}\%$ when the same article is sold at half of the previous selling price?**

Sol: Consider, CP of an article is 100% and SP is Rs. x.

If SP is half, then loss = $16\frac{2}{3}\%$ $\Rightarrow$ SP = CP − L = 100% − $16\frac{2}{3}\%$ = $83\frac{1}{3}\%$

Half of the SP = $\frac{x}{2}$ $\longrightarrow$ $83\frac{1}{3}\%$

Actual SP = x $\longrightarrow$ $? = \frac{83\frac{1}{3} \times x}{\frac{x}{2}}$ = $166\frac{2}{3}\%$

$\therefore$ Profit percentage = SP − CP = $166\frac{2}{3}\%$ − 100% = $66\frac{2}{3}\%$.

Q - 27 A shopkeeper sells eggs at 25 for a rupee gaining 36%. How many eggs did he buy for a rupee?

Sol: Let, CP of each egg = 100%

SP of each egg = Rs. $\frac{1}{25}$ and SP of each egg = 136%   ∵ Gain = 36%

SP of each egg = 136% ⟶ Rs. $\frac{1}{25}$

CP of each egg = 100% ⟶ $? = \frac{1}{25} \times \frac{100}{136} = $ Rs. $\frac{1}{34}$

For Rs. $\frac{1}{34}$ ⟶ 1 egg

For 1 rupee ⟶ ? = 34 eggs

∴ No. of eggs bought for a rupee = 34.

Shortcut:

Consider, cost price of each egg = 100

SP of each egg = 136   ∵ Gain = 36%

According to question,

SP of 25 eggs = CP of x eggs

$25 \times 136 = x \times 100$ ⇒ $x = 34$ eggs

∴ No. of eggs bought for a rupee = 34.

Q - 28 The price of a land passing through three hands, rises on the whole by 61%. If the first and second sellers earned a profit of 15% and 20% respectively, then find the profit earned by the third seller.

Sol: Consider, CP of the 1st seller = 100 and profit of 3rd seller = x%

	CP	SP	
1st seller	100	$100 \times \left(\frac{100 + 15}{100}\right) = 115$	∵ Profit = 15%
2nd seller	115	$115 \times \left(\frac{100 + 20}{100}\right) = 138$	∵ Profit = 20%
3rd seller	138	$138 \times \left(\frac{100 + x}{100}\right) = 100 + 61$	∵ Overall profit = 61%

$138 \times \left(\frac{100 + x}{100}\right) = 161$ ⇒ $x = 16\frac{2}{3}$%

∴ Profit of the 3rd seller is $16\frac{2}{3}$%.

Shortcut:

Consider, initial value = 100 and third seller profit = x%

New value		Initial value	1st seller		2nd seller		3rd seller
100 + 61	=	100	$\times \dfrac{100 + 15}{100}$	$\times$	$\dfrac{100 + 20}{100}$	$\times$	$\dfrac{100 + x}{100}$

$$161 = 100 \times \frac{115}{100} \times \frac{120}{100} \times \frac{100+x}{100} \quad \Rightarrow \quad x = 16\frac{2}{3}\%$$

$\therefore$ Profit of the 3rd seller is $16\frac{2}{3}$%.

Q - 29   A shopkeeper sells a drafter at the rate of Rs. 300 each. So that he will get a profit of 20%. What amount of profit will he earn in 23 days, if he sells 9 drafters per day?

Sol: Consider, CP of each drafter = 100%

Given that, SP of each drafter = 120% $\longrightarrow$ Rs. 300 $\because$ Profit = 20%

CP of each drafter = 100% $\longrightarrow$? $= \dfrac{300 \times 100}{120} = $ Rs. 250

Profit on each drafter = SP – CP = 300 – 250 = Rs. 50

No. of drafters sold in 23 days = 23 × 9 = 207

$\therefore$ Total profit for 207 drafters = 207 × 50 = Rs. 10350.

Q - 30   An article is sold at 25% profit. If its cost price and selling price are increased by Rs. 50 and Rs. 30 respectively, then the percentage of profit decreases by 5%. Find the cost price of an article.

Sol: Consider, CP of an article = Rs. x

Initial SP = 125% of $x = \dfrac{125}{100} \times x = \dfrac{5}{4}x$ $\because$ Profit = 25%

New CP = x + 50, New SP $= \dfrac{5}{4}x + 30$

New SP = 120% of New CP $\because$ Profit decreases by 5% (25 – 5 = 20%)

$$\frac{5}{4}x + 30 = \frac{120}{100} \times (x + 50)$$

$$\frac{5x + 120}{4} = \frac{6}{5}(x + 50) \quad \Rightarrow \quad 25x + 600 = 24x + 1200 \quad \Rightarrow \quad x = 600$$

$\therefore$ Cost price of an article is Rs. 600.

Q - 31   Rakesh bought 30 chairs at Rs. 150 per chair. He sold 14 chairs at 20% profit and the remaining at 25% profit. What is his profit percentage in this transaction?

Sol: Total CP of 30 Chairs = 30 × 150 = Rs. 4500

SP of first 14 chairs = 14 × 120% of 150 $\because$ Profit = 20%

$$= 14 \times \frac{120}{100} \times 150 = \text{Rs. } 2520$$

SP of remaining chairs = 16 × 125% of 150 ∵ Profit = 25%

$$= 16 \times \frac{125}{100} \times 150 = \text{Rs. } 3000$$

Total SP = 2520 + 3000 = Rs. 5520

∴ Profit percentage $= \frac{SP - CP}{CP} \times 100\% = \frac{5520 - 4500}{4500} \times 100\% = 22\frac{2}{3}\%.$

Shortcut:

∴ Overall profit percentage $= \frac{14 \times 20 + 16 \times 25}{14 + 16} = 22\frac{2}{3}\%.$

Q - 32 **A man purchases a certain number of toffees at 30 a rupee and the same number of toffees at 40 a rupee. He mixes them and sells at 70 for 3 rupees. What is his gain (or) loss percent in the whole transaction?**

Sol: Consider, initial number of toffees = x

CP of initial x toffees $= \frac{x}{30}$ ∵ CP of 1 toffee $= \frac{1}{30}$

New no. of toffees = x

CP of new x toffees $= \frac{x}{40}$ ∵ CP of 1 toffee $= \frac{1}{40}$

Total CP of $2x$ toffees $= \frac{x}{30} + \frac{x}{40} = \frac{7x}{120}$

SP of $2x$ toffees $= 2x \times \left(\frac{3}{70}\right) = \frac{6x}{70}$ ∵ SP of 1 toffee $= \frac{3}{70}$

∴ Profit% $= \frac{SP - CP}{CP} \times 100\% = \frac{\frac{6x}{70} - \frac{7x}{120}}{\frac{7x}{120}} \times 100\% = 46\frac{46}{49}\%.$ (SP > CP = Profit)

Shortcut:

Consider, total CP of all toffees = 100%

Overall percentage $= \frac{2 \times \frac{3}{70}}{1 \times \frac{1}{30} + 1 \times \frac{1}{40}} \times 100\% = 146\frac{46}{49}\%$

Since overall percentage is more than 100% by $46\frac{46}{49}\%$

∴ Profit percentage is $46\frac{46}{49}\%.$

> **Note:** If overall percentage is more than 100 we will get profit, less than 100 we will get loss.

Q - 33 | If chocolates are bought at 30 for a rupee, how many chocolates must be sold for a rupee so as to gain 20%?

Sol: Consider, CP of each chocolate = 100%

SP of each chocolate = 120%   ∵ Gain = 20%

CP of each chocolate = 100%   ⟶ $\frac{1}{30}$ rupees

SP of each chocolate = 120%   ⟶ $? = \frac{1}{30} \times \frac{120}{100} = \frac{1}{25}$ rupees

∴ No. of chocolates sold for a rupee = 25.

Shortcut:

Let CP of each chocolate = 100

SP of each chocolate = 120   ∵ Gain = 20%

According to question,

CP of 30 chocolates = SP of x chocolates

$30 \times 100 = x \times 120$ ⇒ $x = 25$

∴ No. of chocolates sold for a rupee = 25.

Q - 34 | A man buys 4 bags and 5 toys for Rs. 6600. He sells the bags at a loss of 14% and toys at a loss of 12% and his whole loss is Rs. 860. Find the price of each bag and each toy.

Sol: Consider, CP of 4 bags = Rs. x and CP of 5 toys = $(6600 - x)$.

Given that, Loss on bags + Loss on toys = Rs. 860

$\frac{x \times 14}{100} + \frac{(6600 - x) \times 12}{100} = 860$ ⇒ $14x + 79200 - 12x = 86000$ ⇒ $x =$ Rs. 3400

∴ CP of 4 bags $x =$ Rs. 3400   ⇒   CP of each bag $= \frac{3400}{4} =$ Rs. 850.

CP of 5 toys $= 6600 - x = 6600 - 3400 =$ Rs. 3200

CP of each toy $= \frac{3200}{5} =$ Rs. 640.

Q - 35 | A man buys two articles for Rs. 2610. He sells one at a loss of 7% and the other at a gain of 8% on the whole he neither gains nor loses. What is the cost of each article?

Sol: Consider, CP of article sold at loss = Rs. x & CP of article sold at gain = Rs. $(2610 - x)$

According to question, on the whole neither gains nor loses means

Total SP = Total CP

$x \times \left(\frac{100 - 7}{100}\right) + (2610 - x) \times \left(\frac{100 + 8}{100}\right) = 2610$

$93x + 281880 - 108x = 261000$ ⇒ $x =$ Rs. 1392

∴ CP of article sold at loss $x =$ Rs. 1392.

CP of article sold at gain $= 2610 - x = 2610 - 1392 =$ Rs. 1218

Shortcut:

Consider, CP of article sold at loss = Rs. x & CP of article sold at gain = Rs. $(2610 - x)$

According to question, on the whole neither gains nor loses means

Loss on 1st article = Gain on 2nd article $\Rightarrow$ 7% of x = 8% of $(2610 - x)$

$\frac{7}{100} \times x = \frac{8}{100} \times (2610 - x)$ $\Rightarrow$ x = Rs. 1392

$\therefore$ CP of article sold at loss x = Rs. 1392.

CP of article sold at gain = $2610 - x = 2610 - 1392$ = Rs. 1218.

Q - 36

Sagar purchased 220 reams of paper at Rs. 70 per ream. He spent Rs. 390 on transportation, paid octroi at the rate of 50 paisa per ream and paid Rs. 200 to the coolie. If he wants to have a gain of 10%, then find the selling price of paper per ream.

Sol: CP of 220 reams of paper for sagar = 220×70 = Rs. 15400

Given that, Transport charges = Rs. 390, coolie = Rs. 200

Octroi charges for 220 reams = 220×0.5 = Rs. 110

Total cost spent by Sagar = $15400 + 390 + 200 + 110$ = Rs. 16100

Total SP of 220 reams of paper = $110\% \times 16100$ = Rs. 17710 $\because$ Gain = 10%

$\therefore$ SP of paper per ream = $\frac{17710}{220}$ = Rs. 80.5.

Q - 37

The profit earned by selling an article for Rs. 560 is equal to loss incurred when the same article is sold for Rs. 280. What is the cost price of an article?

Sol: According to question,

Profit when sold at Rs. 560 = Loss when sold at Rs. 280

$560 - CP = CP - 280$ $\because$ Profit = SP – CP

$2CP = 840$ $\Rightarrow$ CP = Rs. 420 $\because$ Loss = CP – SP

$\therefore$ Cost price of an article is Rs. 420.

Shortcut:

> **Note:** If profit when sold at Rs. x is equal to loss when sold at Rs. y.
>
> Then,   Cost price $CP = \frac{x + y}{2}$

$\therefore$ Cost price of an article = $\frac{560 + 280}{2}$ = Rs. 420.

Q - 38 **A dishonest fruit vendor professes to sell his goods at cost price but he uses a weight of 850 grams for every kg. What is his gain percent?**

Sol: Actual weight = 1kg = 1000 grams

Let, CP of each gram = 1 rupee $\Rightarrow$ 1000 grams CP = Rs. 1000

Fruit vendor announces to sell his goods at CP

$\Rightarrow$ Selling price for fruit vendor = Rs. 1000

But fruit vendor uses false weights, giving only 850 grams

$\Rightarrow$ CP of 850 grams for vendor = Rs. 850

$\therefore$ Gain percentage $= \dfrac{SP - CP}{CP} \times 100\% = \dfrac{1000 - 850}{850} \times 100\% = 17\dfrac{11}{17}\%$.

Shortcut:

$$\text{Gain}\% = \dfrac{True\ weight - False\ weight}{False\ weight} \times 100\%$$

$\therefore$ Gain percentage $= \dfrac{1000 - 850}{850} \times 100\% = 17\dfrac{11}{17}\%$.

Q - 39 **A shopkeeper sells sugar at a profit of 20% and uses a weight which is 30% less. Find his total percentage gain.**

Sol: Consider, actual weight of sugar = 1000 grams and also CP of 1gm = 1rupee

CP of 1000 grams = Rs. 1000

SP of 1000 gm of sugar = 120% of 1000 = Rs. 1200 $\because$ Profit = 20%

But, shopkeeper uses 30% less weight means he is giving only 700 gm for every 1000 gm

CP of 700 gm of sugar for shopkeeper = Rs. 700

$\therefore$ Profit percentage $= \dfrac{SP - CP}{CP} \times 100\% = \dfrac{1200 - 700}{700} \times 100\% = 71\dfrac{3}{7}\%$.

Shortcut:

Note: If a shopkeeper sells his goods at $x\%$ profit (or) loss on CP and uses $y\%$ less weight.

Then, Profit (or) Loss % $= \dfrac{y \pm x}{100 - y} \times 100\%$

Use, +ve sign for profit & –ve sign for loss

Here, $x = 20\%$ and $y = 30\%$

Profit (or) Loss % $= \dfrac{30 + 20}{100 - 30} \times 100\% = 71\dfrac{3}{7}\%$

Here, +ve answer indicates there is a profit of $71\dfrac{3}{7}\%$.

Q - 40   A dishonest dealer sells goods at 12% loss on cost price, but uses 25 gm instead of 30 gm. What is his profit (or) loss percentage?

Sol: Actual weight = 30 gm, Consider CP of 1 gm = 1 rupee

⇒   CP of 30 gm = Rs. 30

But dealer announces to sell his goods at 12%loss

⇒   SP for the dealer = 88% of 30 = Rs. 26.4

But, the dealer uses false weights giving 25 gm instead of 30 gm

CP of 25gm for the dealer = Rs. 25   (SP > CP = profit)

∴ Profit percentage $= \frac{SP - CP}{CP} \times 100\% = \frac{26.4 - 25}{25} \times 100\% = 5.6\%.$

Shortcut:

Percent less than of weight $= \frac{Difference\ of\ weights}{Actual\ weight} \times 100\% = \frac{30 - 25}{30} \times 100\% = \frac{50}{3}\%$

> **Note:** If a dealer sells his goods at x% profit (or) loss on CP and uses y% less weight.
>
> Then, Profit (or) Loss % $= \frac{y \pm x}{100 - y} \times 100\%$

Use, +ve sign for profit & –ve sign for loss

Here, $x = 12\%$ and $y = \frac{50}{3}\%$

∴ Profit (or) Loss % $= \frac{\frac{50}{3} - 12}{100 - \frac{50}{3}} \times 100\% = \frac{14}{250} \times 100\% = 5.6\%.$

Here, +ve answer indicates there is a profit of 5.6%.

Q - 41   A dishonest dealer sells goods at 30% loss on cost price but uses 20% less weight. What is his profit (or) loss percentage?

Sol: Consider, Actual weight = 1000 gm, CP of 1 gm = 1 rupee

CP of 1000 gm = Rs. 1000

SP of goods = 70% of 1000 = Rs. 700   ∵ Loss = 30%

But, dealer uses 20% less weight which means giving only 800 gm

Therefore, CP of 800 gm for the dealer = Rs. 800 (SP < CP = Loss)

∴ Loss percentage $= \frac{CP - SP}{CP} \times 100\% = \frac{800 - 700}{800} \times 100\% = 12.5\%.$

Shortcut:

> **Note:** If a dealer sells his goods at x% profit (or) loss on CP and uses y% less weight.
>
> Then, Profit or Loss % $= \dfrac{y \pm x}{100 - y} \times 100\%$

Use, +ve sign for profit & –ve sign for loss

Here, $x = 30\%$ and $y = 20\%$

∴ Profit (or) Loss % $= \dfrac{20 - 30}{100 - 20} \times 100\% = -12.5\%.$

Here, –ve answer indicates there is a loss of 12.5%.

Q - 42 A shopkeeper uses 1200 grams for every 1 kg to sell his goods. Find his actual profit (or) loss percentage, when he sells his goods at 20% profit.

Sol: Actual weight = 1 kg = 1000 grams

Consider, CP of 1 gm = 1 rupee   ⇒   CP of 1000 grams = Rs. 1000

SP of goods = 120% of 1000 = Rs. 1200   ∵ Profit = 20%

But, shopkeeper giving 1200 grams for every 1 kg

CP of 1200 grams for shopkeeper = Rs. 1200

It is clear that, CP of goods and SP of goods are equal.

∴ There is neither profit nor loss in the whole transaction.

Q - 43 A person had a fan to sell. I offered him a sum of money for the fan which he refused as being 8% below the value of the fan. I then offered Rs. 720 more, the second offer was 10% more than the estimated value. Find the value of the fan.

Sol: Consider, a value of fan = Rs. x

According to question,

1st offer $= (100 - 8)\%$ of $x = \dfrac{92}{100}x$ ∵ 8% less

2nd offer $= 1^{st}$ offer $+ 720$

$\dfrac{(100 + 10)}{100}x = \dfrac{92}{100}x + 720$ ⇒ $\dfrac{110}{100}x - \dfrac{92}{100}x = 720$ ⇒ $x = $ Rs. 4000

∴ Value of fan $x = $ Rs. 4000.

Shortcut:

Consider, value of fan = 100%

From the diagram, it is clear that

Difference between 1st and 2nd offer = 18% ⟶ Rs. 720

∴ Value of fan = 100% ⟶ ? $= \dfrac{720 \times 100}{18} = 4000.$

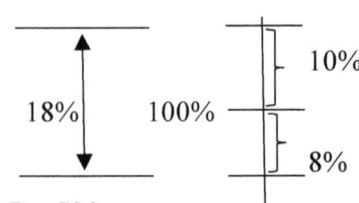

Q - 44 | **A man bought an article and sold it at a gain of 12%. If he had bought it at 10% less and sold it for Rs. 280 more, then he would have gained 30%. Find the cost price of an article.**

Sol: Consider, initial CP = 100%

Initial SP = 112% ∵ Gain = 12%

New cost price = 100% – 10% = 90% and New SP = Initial SP + Rs. 280

$\frac{130}{100} \times 90\% = 112\% + 280$ ⇒ 117% – 112% = 280 ∵ Gain = 30%

Difference of SP's = 5% ⟶ Rs. 280

∴ CP of an article = 100% ⟶ $? = \frac{280 \times 100}{5} = $ Rs. 5600.

Shortcut:

	First	**Second**
CP	10% less 100% ⟶	90%
	12% Profit	30% Profit
SP	112%	$90 \times \frac{130}{100} = 117\%$

According to question,

Difference between two SP's = Rs. 280

Difference of SP's = 5% ⟶ Rs. 280

∴ CP of an article = 100% ⟶ $? = \frac{280 \times 100}{5} = $ Rs. 5600.

Q - 45 | **An article is sold at a profit of 25%. If both the cost price and selling prices are Rs. 125 less, then the profit would be 5% more. Find the cost price.**

Sol: Consider, CP of an article = 100%

SP of an article = 125% ∵ Profit = 25%

Given that, if CP and SP are Rs. 125 less, then the profit is 5% more (25% + 5% = 30%)

New CP = 100% – 125, New SP = 125% – 125, Profit = 30%

New SP = 130% of New CP ⇒ $125\% - 125 = \frac{130}{100}(100\% - 125)$

1250% – 1250 = 1300% – 1625 ⇒ 50% = Rs. 375

∴ CP of an article = 100% = 50% × 2 = 375 × 2 = Rs. 750.

Q - 46 The sale price of an article including the sales tax is Rs. 11760. The rate of sales tax is 12%. If the shopkeeper has made a profit of 40%, then find the cost price of the article.

Sol: Consider, CP of an article = Rs. x

SP of an article = 140% of $x = \dfrac{140}{100} x$ $\because$ Profit = 40%

Sales tax = 12% of $\dfrac{140}{100}x = \dfrac{12}{100} \times \dfrac{140}{100} \times x = \dfrac{1680}{10000}x$

Total SP including tax $= \dfrac{140}{100}x + \dfrac{1680}{10000}x = \dfrac{15680}{10000}x$

$11760 = \dfrac{15680}{10000}x \qquad \Rightarrow \qquad x = $ Rs. 7500

$\therefore$ Cost price of an article is Rs. 7500.

Q - 47 The cost price of two articles are same. On one article there is 35% profit, on the other article there is 15% loss. Find the overall profit (or) loss percentage.

Sol: Consider, CP of first article = CP of second article = Rs. 100

SP of 1st article = Rs. 135 $\because$ Profit = 35%

SP of 2nd article = Rs. 85 $\because$ Loss = 15%

Total CP of 2 articles = 100 + 100 = Rs. 200

Total SP of 2 articles = 135 + 85 = Rs. 220   (SP > CP = Profit)

$\therefore$ Profit percentage $= \dfrac{Sp - Cp}{CP} \times 100\% = \dfrac{220 - 200}{200} \times 100\% = 10\%$.

Alternate Method:

	First		Second	
CP	100	+	100	= 200
	$\downarrow$ 35% Profit		$\downarrow$ 15% Loss	
SP	135	+	85	= 220

$\therefore$ Profit percentage $= \dfrac{Sp - Cp}{CP} \times 100\% = \dfrac{220 - 200}{200} \times 100\% = 10\%$.

Shortcut:

> **Note:** If CP of 2 (or) more than 2 articles are same. On first article $x_1\%$ profit (or) loss, on second article $x_2\%$ profit (or) loss and so on.
>
> Then overall Profit (or) Loss $\% = \dfrac{\pm x_1\% \pm x_2\% \pm .... \pm x_n\%}{n}$
>
> Where 'n' is number of articles

Use +ve sign for profit & –ve sign for loss

Profit (or) Loss $\% = \dfrac{+35 - 15}{2} = +10\%$

Here, +ve sign indicates there is a profit of 10%.

Q - 48 **The cost price of two articles are same. On first article there is 12% profit, on the second article there is 25% loss. Find the overall profit (or) loss percent.**

Sol: Consider, CP of 1st article = CP of 2nd article = Rs. 100

SP of 1st article = Rs. 112 $\quad\quad\quad\quad$ $\because$ Profit = 12%

SP of 2nd article = Rs. 75 $\quad\quad\quad\quad$ $\because$ Loss = 25%

Total CP of two articles = 100 + 100 = Rs. 200

Total SP of two articles = 112 + 75 = Rs. 187 $\quad\quad\quad$ (SP < CP = Loss)

$\therefore$ Loss percentage $= \dfrac{CP - SP}{CP} \times 100\% = \dfrac{200 - 187}{200} \times 100\% = 6.5\%$.

Alternate Method:

	First		**Second**	
CP	100	+	100	= 200
	$\downarrow$ 12% Profit		$\downarrow$ 25% Loss	
SP	112	+	75	= 187

$\therefore$ Loss percentage $= \dfrac{CP - SP}{CP} \times 100\% = \dfrac{200 - 187}{200} \times 100\% = 6.5\%$.

Shortcut:

> **Note:** If CP of 2 (or) more than 2 articles are same. On first article $x_1\%$ profit (or) loss, on second article $x_2\%$ profit (or) loss and so on.
>
> Then overall Profit (or) Loss $\% = \dfrac{\pm x_1\% \pm x_2\% \pm \cdots \pm x_n\%}{n}$
>
> Where 'n' is number of articles

Profit (or) Loss $\% = \dfrac{+12 - 25}{2} = -6.5\%$

Here, –ve sign indicates there is a loss of 6.5%.

Q - 49 | **The selling price of two articles are same. On one article there is 70% profit, on the other there is 30% profit. Find the overall profit percent.**

Sol: Let, CP of two articles are $100x$ and $100y$ respectively.

SP of 1st article $= 170x$ & SP of 2nd article $= 130y$ $\because$ Profit on $1^{st} = 70\%$

Given that, SP of 2 articles are same $\because$ Profit on $2^{nd} = 30\%$

$170x = 130y \implies \dfrac{x}{y} = \dfrac{13}{17} \implies x = 13, y = 17$

CP of 1st article $= 100x = 100 \times 13 = 1300$

CP of 2nd article $= 100y = 100 \times 17 = 1700$

Total CP of 2 articles $= 1300 + 1700 =$ Rs. 3000

SP of 1st article $= 170 \times 13 = 2210$ & SP of 2nd article $= 130 \times 17 = 2210$

Total SP of 2 articles $= 2210 + 2210 =$ Rs. 4420

$\therefore$ Profit percentage $= \dfrac{SP - CP}{CP} \times 100\% = \dfrac{4420 - 3000}{3000} \times 100\% = 47\dfrac{1}{3}\%.$

Shortcut:

If SP of two articles are same. On one article there is $x\%$ profit (or) loss, on the other article there is $y\%$ profit (or) loss, then

> Required percentage $= \dfrac{2(100 \pm x)(100 \pm y)}{(100 \pm x) + (100 \pm y)}$

Use +ve sign for profit & –ve sign for loss

> **Note:** If the value is more than 100, then we will get profit.
> If the value is less than 100, then we will get loss.

Required percentage $= \dfrac{2(100+70)(100+30)}{(100+70)+(100+30)} = \dfrac{2 \times 170 \times 130}{300} = 147\dfrac{1}{3}\%.$

Here, the answer is more than 100. Therefore, profit percentage $= 47\dfrac{1}{3}\%.$

Q - 50   **Laxman sold two TV sets, each for Rs. 25000. If he makes 40% profit on the first and 30% loss on the second, then find his gain (or) loss percentage in the whole transaction.**

Sol: Consider, CP of 2 TV sets are $100x$ and $100y$ respectively.

	First	**Second**
CP	$100x$	$100y$
	$\downarrow$ 40% Profit	$\downarrow$ 30% Loss
SP	$140x$	$70y$

Given that, SP of 2 TV sets are same $140x = 70y$

$\Rightarrow$ $x : y = 1 : 2$ $\Rightarrow$ $x = 1, y = 2$

$CP_1 = 100x = 100$, $CP_2 = 100y = 100 \times 2 = 200$

$SP_1 = 140x = 140$, $SP_2 = 70y = 70 \times 2 = 140$

Total CP = $100 + 200 = 300$, Total SP = $140 + 140 = 280$ (SP < CP = Loss)

$\therefore$ Loss percentage = $\dfrac{Cp - Sp}{CP} \times 100\% = \dfrac{300 - 280}{300} \times 100\% = 6\frac{2}{3}\%$.

> **Note:** While we are calculating profit (or) loss percentages for these kinds of problems, there is no requirement of given selling price values.

Shortcut:

If SP of 2 articles are same. On one article there is $x\%$ profit (or) loss, on the other article there is $y\%$ profit (or) loss, then

$$\text{Required percentage} = \dfrac{2(100 \pm x)(100 \pm y)}{(100 \pm x) + (100 \pm y)}$$

Use +ve sign for profit & –ve sign for loss

$$\text{Required percentage} = \dfrac{2(100 + 40)(100 - 30)}{(100 + 40) + (100 - 30)} = \dfrac{2 \times 140 \times 70}{210} = 93\frac{1}{3}\%.$$

Here, the answer is less than 100. Therefore, loss percentage = $100\% - 93\frac{1}{3}\% = 6\frac{2}{3}\%$.

Q - 51 | **Vinesh sold two articles for Rs. 2500 each. If he makes 40% profit on one and 40% loss on the other, then find the overall profit (or) loss percentage.**

Sol: Consider, CP of 2 articles are $100x$ and $100y$ respectively.

	First	**Second**
CP	$100x$	$100y$
	$\downarrow$ 40% Profit	$\downarrow$ 40% Loss
SP	$140x$	$60y$

Given that, SP of 2 articles are same

$140x = 60y \quad \Rightarrow \quad x : y = 3 : 7 \quad \Rightarrow \quad x = 3, y = 7$

$CP_1 = 100x = 100 \times 3 = 300, CP_2 = 100y = 100 \times 7 = 700$

$SP_1 = SP_2 = 140x \text{ (or) } 60y = 420$

Total CP = 300 + 700 = 1000, Total SP = 420 + 420 = 840   (SP < CP = Loss)

$\therefore$ Loss percentage $= \dfrac{CP - SP}{CP} \times 100\% = \dfrac{1000 - 840}{1000} \times 100\% = 16\%$.

Shortcut-1:

    ✓   If SP of 2 articles are same. On one article there is $x\%$ profit (or) loss, on the other article there is $y\%$ profit (or) loss, then

$$\text{Required percentage} = \frac{2(100 \pm x)(100 \pm y)}{(100 \pm x) + (100 \pm y)}$$

Required percentage $= \dfrac{2(100 + 40)(100 - 40)}{(100 + 40) + (100 - 40)} = 84\%$.

Here, the answer is less than 100. Therefore, loss percentage = 100% – 84% = 16%.

Shortcut-2:

    ✓   When a person sells two similar articles, one at a gain of $x\%$ and another at a loss of $x\%$, then the seller will always incurs a loss. Then,

$$\text{Loss percentage} = \frac{x^2}{100}\%$$ Here, $x = 40\%$

$\therefore$ Loss percentage $= \dfrac{40^2}{100}\% = 16\%$.

Q - 52 | **By mistake a person calculated the profit percentage over selling price and it is found as 20%, then find the actual profit percentage.**

Sol: Given that, a person mistakenly calculated profit percentage over SP and it is 20%.

So, consider SP = 100 $\Rightarrow$ CP = 80 $\because$ Profit = 20%

$\therefore$ Actual profit % = $\frac{SP - CP}{CP} \times 100\% = \frac{100 - 80}{80} \times 100\% = 25\%$.

> **Note:** Actual profit (or) loss percentage both are must calculate over CP only.

Q - 53 | **By mistake a person calculated the loss percentage over selling price and it is found as 40%, then find the actual loss percentage.**

Sol: Given that, a person mistakenly calculated loss percentage over SP and it is 40%.

So, consider SP = 100 $\Rightarrow$ CP = 140 $\because$ Loss = 40%

$\therefore$ Actual loss % = $\frac{CP - SP}{CP} \times 100\% = \frac{140 - 100}{140} \times 100\% = 28\frac{4}{7}\%$.

> **Note:** Actual profit (or) loss percentage both are must calculate over CP only.

Q - 54 | **Rohan purchased an article for Rs. 10000 and sold it at a gain of 20%. From that amount he purchased another article and sold it at a loss of 10%. What is her overall gain (or) loss percentage?**

Sol: Consider, CP of first article = 100%

	CP	SP
1st article	100% $\longrightarrow$ 20% Gain	120%
2nd article	120% $\longrightarrow$ 10% Loss	120% × $\frac{90}{100}$ = 108%

From the above table, it is clear that SP of 1st article is same as CP of 2nd article.

So, compare CP of 1st article and SP of 2nd article.

CP of 1st article = 100%, SP of 2nd article = 108%

$\therefore$ Profit = SP – CP = 108% – 100% = 8%.

Q - 55 | **Venky purchased an item for Rs. 8500 and sold it at a gain 12% with that amount he purchased another item and sold it at a loss of 5%. What was his overall gain (or) loss?**

Sol:

	CP	12% Gain	**SP**
First item	8500	$\longrightarrow$	$\frac{112}{100} \times 8500 = 9520$
Second item	9520	$\longrightarrow$ 5% Loss	$\frac{95}{100} \times 9520 = 9044$

Total CP = 8500 + 9520 = Rs. 18020

Total SP = 9520 + 9044 = Rs. 18564   (SP > CP = Profit)

∴ Profit = SP – CP = 18564 – 18020 = Rs. 544.

Q - 56 | **A shopkeeper has 560 kgs of rice, part of which he sells at 4% profit and the rest at 12% profit. He gains 7% on the whole. Find the quantity sold at 4% profit and 12% profit respectively.**

Sol:

Given that, total quantity of rice = 560 kgs, total profit = 7%

Consider, quantity sold at 4% profit = 'x' kgs

Quantity sold at 12% profit = $(560 – x)$ kgs

Total SP of rice = Rice sold at 4% profit + Rice sold at 12% profit

$$560 \times \left(\frac{100 + 7}{100}\right) = x \left(\frac{100 + 4}{100}\right) + (560 - x) \left(\frac{100 + 12}{100}\right)$$

$59920 = 104x + 62720 - 112x \quad \Rightarrow \quad 8x = 2800 \quad \Rightarrow \quad x = 350$ kgs

∴ Quantity sold at 4% profit 'x' = 350 kgs.

Quantity sold at 12% profit = $(560 - x) = 560 - 350 = 210$ kgs.

Shortcut:

This problem can be solved easily by using alligation principle.

According to alligation principle,

Use, +ve sign for Profit

–ve sign for Loss

$12 - 7 = 5 : 7 - 4 = 3$

Quantity ratio = 5 : 3

Total quantity = 8 parts = 560 kgs $\Rightarrow$ 1 part = 70 kgs

∴ Quantity sold at 4% profit = 5 parts = 5 × 70 = 350 kgs

Quantity sold at 12% profit = 3 parts = 3 × 70 = 210 kgs.

Q-57 A shopkeeper has 750 kgs of wheat, part of which he sells at 5% loss and the rest at 10% profit. He gains 4% on the whole. Find the quantity sold at 5% loss and 10% profit.

Sol: Given that, total quantity = 750 kgs, total profit = 4%

Consider, quantity sold at 5% loss = 'x' kgs

Quantity sold at 10% profit = $(750 - x)$ kgs

Total SP of wheat = Wheat sold at 5% of loss + Wheat sold at 10% profit

$$750 \times \left(\frac{100 + 4}{100}\right) = x\left(\frac{100 - 5}{100}\right) + (750 - x)\left(\frac{100 + 10}{100}\right)$$

$$78000 = 95x + 82500 - 110x \quad \Rightarrow \quad x = 300 \text{ kgs}$$

∴ Quantity sold at 5% loss 'x' = 300 kgs.

Quantity sold at 10% profit = $(750 - x) = 750 - 300 = 450$ kgs.

Shortcut:

This problem can be solved easily by using alligation principle.

According to alligation principle,

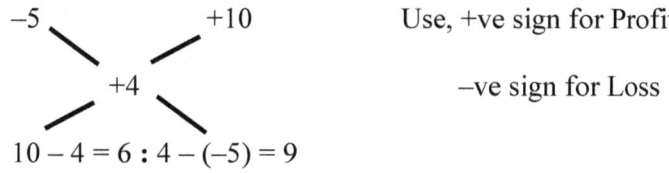

Use, +ve sign for Profit

−ve sign for Loss

$$10 - 4 = 6 : 4 - (-5) = 9$$

Quantity ratio = 6 : 9 = 2 : 3

Total quantity = 5 parts = 750 kgs $\quad \Rightarrow \quad$ 1 part = 150 kgs

∴ Quantity sold at 5% loss = 2 parts = 2 × 150 = 300 kgs.

Quantity sold at 10% profit = 3 parts = 3 × 150 = 450 kgs.

Q-58 The cost of a coffee powder packet and a liter of milk are Rs. 30 and Rs. 50 respectively. 20 cups of coffee is made with one packet coffee powder and for each cup 150 ml of milk is used. Find the selling price of each cup of coffee, if coffee is sold at $16\frac{2}{3}$% profit.

Sol: Given that, CP of coffee powder packet = Rs. 30 and milk per liter = Rs. 50

Milk required for 20 cups of coffee = 20 × 150 = 3000 ml = 3 liters

Cost of 3 liters of milk = 3 × 50 = Rs. 150

Cost of 1 packet coffee powder = Rs. 30

Total CP of 20 cups of coffee = 150 + 30 = Rs. 180

CP of coffee per cup $= \frac{180}{20}$ = Rs. 9

SP of coffee per cup $= 116\frac{2}{3}$% of CP $= \frac{350}{3} \times \frac{1}{100} \times 9$ = Rs. 10.5. ∵ Profit $= 16\frac{2}{3}$%

Q - 59   **A fruit vendor earns a profit of 40% by selling 50 mangoes for Rs. 350, but he gives one – fifth of it to his friend at cost price and sells the remaining mangoes. In order to earn the same profit at what price must he sell each mango?**

Sol: Given that, SP of 50 mangoes = Rs. 350 and Profit = 40%

SP of each mango $= \dfrac{350}{50} =$ Rs. 7

CP of each mango $=$ SP $\times \dfrac{100}{100 + P\%} = 7 \times \dfrac{100}{100 + 40} =$ Rs. 5

Total SP of mangoes = SP of $\dfrac{1}{5}$ of mangoes + SP of remaining $\dfrac{4}{5}$ of mangoes

$350 = 10 \times 5 + 40 \times x \quad \Rightarrow \quad 40x = 300 \quad \Rightarrow \quad x =$ Rs. 7.5

∴ In order to gain the same 40%, he must sell each mango for Rs. 7.5.

Q - 60   **A fruit seller brought some apples. He sold 30% of them at a gain of 40% and 30% of them at a loss of 40%. The rest being rotten were not sold. What is the loss percentage on the whole?**

Sol: Consider, no. of apples brought = 100 and cost per apple = 1 rupee

Total CP of 100 apples = 100 × 1 = Rs. 100

No. of apples sold at gain = No. of apples sold at loss = 30% of 100 = 30

SP of apples (30) sold at gain $= 30 \times \dfrac{(100 + 40)}{100} =$ Rs. 42

SP of apples (30) sold at loss $= 30 \times \dfrac{(100 - 40)}{100} =$ Rs. 18

Remaining 40 apples being rotten were not sold

Total SP of 100 apples = 42 + 18 = Rs. 60.

∴ Loss percentage $= \dfrac{CP - SP}{CP} \times 100\% = \dfrac{100 - 60}{100} \times 100\% = 40\%.$

Q - 61   **A manufacturer buys a second-hand machine for Rs. 60000 and incurs Rs. 15000 on installation and repairs. After 1 year he sells the machine for Rs. 70000. What is his profit (or) loss percentage, if 20% is to be deducted on account of depreciation?**

Sol: Given that, CP of machine = Rs. 60000 and installation and repairs = Rs. 15000

Total CP of manufacturer = Rs. 60000 + Rs. 15000 = Rs. 75000

SP of machine after one year = Rs. 70000

Reduction of cost of machine after one year = 20% of 75000 = Rs. 15000

Total CP of machine after one year = 75000 – 15000 = Rs. 60000

∴ Profit percentage $= \dfrac{SP - CP}{CP} \times 100\% = \dfrac{70000 - 60000}{60000} \times 100\% = 16\dfrac{2}{3}\%.$

Q - 62 The cost of production of a car which is sold at 25% profit went up by 30%. What should be the percent increase in the selling price to maintain the profit percentage the same, even at the new cost of production?

Sol: Consider, cost of production of a car = Rs. 100

SP of a car = Rs. 125 ∵ Profit = 25%

New cost of production of a car = Rs. 130 ∵ Cost increased by 30%

New SP of a car = 125% of new cost = $\frac{125}{100} \times 130$ = Rs. 162.5 ∵ Profit = 25%

$$\text{Percentage increase} = \frac{Difference}{Less\ value} \times 100\%$$

Percentage increase in SP = $\frac{162.5 - 125}{125} \times 100\% = 30\%$

∴ To maintain the same profit (25%), SP is increased by 30%.

Alternate method:

	CP		SP
		Profit 25%	
Initial	100	⟶	125
	↓ 30% increase		
New	130	⟶	162.5
		Profit 25%	

∴ Percentage increase in SP = $\frac{162.5 - 125}{125} \times 100\% = 30\%$.

Q - 63 A shopkeeper bought 60 TV sets at Rs. 15000 each. For every set purchased from him, he gave one set free. The loss made by him is equal to selling price of 20 sets. What is the selling price of each set, that is bought?

Sol: CP of each TV set = Rs. 15000

Shopkeeper sold only 30 TV sets out of 60 TV sets because for every TV set he sold, he gave one set free.

Total CP = Total SP + Loss

60 × 15000 = 30SP + 20SP = 50SP ∵ Loss = 20SP

∴ SP of each TV set = Rs. 18000.

Q - 64 | **If an article A is sold for Rs. 5000, there is a loss of 20%. An article B should be sold in such a way that there is a profit of 20% on both A and B articles. If the cost price of article B is 30% more than that of A, then what is selling price of B?**

Sol: Consider, CP of article A = 100%

SP of A = 80% $\longrightarrow$ Rs. 5000 $\qquad$ ∵ Loss = 20%

CP of A = 100% $\longrightarrow$ $? = \dfrac{5000 \times 100}{80} = $ Rs. 6250

CP of article B = 130% of CP of A = $\dfrac{130}{100} \times 6250 = $ Rs. 8125 $\qquad$ ∵ B – 30% more than A

Total CP of both A and B = 6250 + 8125 = Rs. 14375

Total SP of both A and B = 120% of 14375 = $\dfrac{120}{100} \times 14375 = $ Rs. 17250 $\quad$ ∵ Profit = 20%

SP of A + SP of B = Rs. 17250

∴ SP of article B = 17250 – 5000 = Rs. 12250.

Shortcut:

	SP	CP
Article A	5000	Loss 20% $\longrightarrow$ $5000 \times \dfrac{100}{80} = 6250$
		30% more ↓
Article B	x	$6250 \times \dfrac{130}{100} = 8125$

Total CP = 6250 + 8125 = 14375

Total SP = $14375 \times \dfrac{120}{100} = 17250$ $\qquad$ ∵ Profit = 20%

$5000 + x = 17250$ $\qquad \Rightarrow \qquad x = $ Rs. 12250

∴ SP of article B = Rs. 12250.

Q - 65 | **A person sells his table at a profit of 25% and the chair at a loss of 15% but on the whole he gains Rs. 72. On the other hand if he sells the table at a loss of 15% and chair at a profit of 25%, then he neither gains nor loses. Find the cost price of the table and the chair.**

Sol: Consider, CP of table = T and CP of chair = C

According to question,

$$\frac{25}{100}T - \frac{15}{100}C = 72$$

$$5(5T - 3C) = 7200 \qquad \Rightarrow \qquad 5T - 3C = 1440 \quad \ldots\ldots\ldots (1)$$

$$\frac{-15}{100}T + \frac{25}{100}C = 0 \qquad \Rightarrow \qquad \frac{25}{100}C = \frac{15}{100}T \qquad \Rightarrow \qquad C = \frac{3}{5}T \ldots\ldots\ldots (2)$$

Substitute (2) in (1)

$$5T - 3 \times \frac{3}{5}T = 1440 \qquad \Rightarrow \qquad T = \text{Rs. } 450$$

$$\therefore \text{ CP of chair } C = \frac{3}{5}T = \frac{3}{5} \times 450 = \text{Rs. } 270.$$

Q - 66 $\frac{1}{5}$ of goods are sold at 25% profit, $\frac{1}{2}$ is sold at 12% profit and the rest at 15% profit. If the total profit earned is Rs. 124, then find the value of goods.

Sol: Consider, value of goods = Rs. x

According to question,

$$\frac{1}{5} \times \frac{25}{100} \times x + \frac{1}{2} \times \frac{12}{100} \times x + \left(1 - \frac{1}{5} - \frac{1}{2}\right) \times \frac{15}{100} \times x = 124$$

$$\frac{25x}{100} + \frac{12x}{100} + \frac{3x}{10} \times \frac{15}{100} = 124 \qquad \Rightarrow \qquad x = \text{Rs. } 800$$

$$\therefore \text{ Value of goods } x = \text{Rs. } 800.$$

Q - 67   Three-fourth of a commodity is sold at 8% profit and the rest at a loss of 7%. If there was an overall profit of Rs. 1105. Find the value of commodity.

Sol: Consider, value of commodity = Rs. x

According to question,

$$\frac{3}{4} \times \frac{8}{100} \times x - \left(1 - \frac{3}{4}\right) \times \frac{7}{100} \times x = 1105$$

$$24x - 7x = 1105 \times 400 \qquad \Rightarrow \qquad x = \text{Rs. } 26000$$

$$\therefore \text{ Value of commodity } x = \text{Rs. } 26000.$$

Q - 68   A person bought two books A and B for Rs. 515. He sold book A at a loss of 18% and book B at a profit of 24% and he found that each book was sold at the same price. Find the cost price of each book.

Sol: Consider, CP of book A = Rs. x and CP of book B = $(515 - x)$.

According to question, SP of two books are same.

$$\left(\frac{100 - 18}{100}\right) \times x = \left(\frac{100 + 24}{100}\right) \times (515 - x)$$

$$82x = 63860 - 124x \qquad \Rightarrow \qquad 206x = 63860 \qquad \Rightarrow \qquad x = \text{Rs. } 310$$

$$\therefore \text{ CP of book A, } x = \text{Rs. } 310 \text{ \& CP of book B} = 515 - x = 515 - 310 = \text{Rs. } 205.$$

Alternate method:

Given that, SP of two books are same.

Total CP of two books A + B = Rs. 515

$$\frac{100-18}{100} \text{ of A} = \frac{100+24}{100} \text{ of B} \qquad \Rightarrow \qquad 82A = 124B \qquad \Rightarrow \qquad A : B = 62 : 41$$

Total CP of 2 books = 103 parts ⟶ Rs. 515 ⟹ 1 part = Rs. 5

∴ CP of book A = 62 parts = 62 × 5 = Rs. 310.

CP of book B = 41 parts = 41 × 5 = Rs. 205.

Shortcut:

✓ If a person purchased two articles for Rs. A. He sold first article at a profit (or) loss of x% and the second article at a profit (or) loss of y%. If selling price (SP) of two articles are same, then

$$CP \text{ of first article} = \frac{A(100 \pm y)}{(100 \pm x) + (100 \pm y)}$$

$$CP \text{ of second article} = \frac{A(100 \pm x)}{(100 \pm x) + (100 \pm y)}$$

Use, +ve sign for Profit & –ve sign for Loss

Here, A = Rs. 515, x = 18% Loss, y = 24% Profit

$$\therefore CP \text{ of book A} = \frac{515(100+24)}{(100-18) + (100+24)} = Rs. 310.$$

$$CP \text{ of book B} = \frac{515(100-18)}{(100-18) + (100+24)} = Rs. 205.$$

Q - 69 A man bought two articles x and y for Rs. 705. He sold article x at a gain of 20% and article y at a gain of 15% and he found that each article was sold at the same price. Find the cost price of each article.

Sol: Given that SP of two articles are same.

Total CP of two articles x + y = Rs. 705

$$SP_1 = SP_2 \qquad \Rightarrow \qquad \left(\frac{100+20}{100}\right) \text{ of } x = \left(\frac{100+15}{100}\right) \text{ of } y \qquad \Rightarrow \qquad x : y = 23 : 24$$

Total CP of two articles = 47 parts ⟶ Rs. 705 ⟹ 1 part = Rs. 15

∴ CP of article x = 23 parts = 23 × 15 = Rs. 345.

CP of article y = 24 parts = 24 × 15 = Rs. 360.

Shortcut:

✓ If a person purchased two articles for Rs. A. He sold first article at a profit (or) loss of $x\%$ and the second article at a profit (or) loss of $y\%$. If selling price (SP) of two articles are same, then

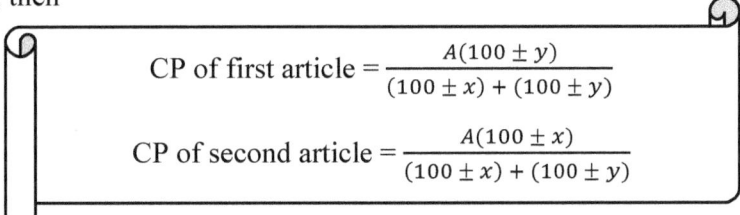

$$\text{CP of first article} = \frac{A(100 \pm y)}{(100 \pm x) + (100 \pm y)}$$

$$\text{CP of second article} = \frac{A(100 \pm x)}{(100 \pm x) + (100 \pm y)}$$

Use, +ve sign for Profit & –ve sign for Loss

Here, A = Rs. 705, $x = 20\%$ Profit and $y = 15\%$ Profit

$$\therefore \text{CP of article } x = \frac{705(100 + 15)}{(100 + 20) + (100 + 15)} = \text{Rs. } 345.$$

$$\text{CP of article } y = \frac{705(100 + 20)}{(100 + 20) + (100 + 15)} = \text{Rs. } 360.$$

Q - 70 **Vinay calculates his profit percentage on selling price whereas Ajay calculates his profit percentage on cost price. They find that the difference of their profits is Rs.150. If the selling price of both of them are same and both will get 20% profit, then find their selling price.**

Sol: Consider, SP of both the persons = Rs. x

$$\text{CP of Vinay} = x\left(\frac{100 - 20}{100}\right) = \frac{4}{5}x \text{ and CP of Ajay} = x\left(\frac{100}{100 + 20}\right) = \frac{5}{6}x$$

Vinay's profit = SP – CP = $x - \frac{4}{5}x = \frac{x}{5}$

Ajay's profit = SP – CP = $x - \frac{5}{6}x = \frac{x}{6}$

Now, difference of their profits $= \frac{x}{5} - \frac{x}{6} = 150$

$\frac{x}{30} = 150 \qquad \Rightarrow \qquad x = \text{Rs. } 4500$

∴ SP of both the persons is Rs. 4500.

Shortcut:

Consider, CP is 100%

According to question,

$$\frac{20}{100} \times \text{SP} - \frac{20}{100} \times \text{CP} = 150 \qquad \Rightarrow \qquad \text{SP} - \text{CP} = 750$$

SP – CP = profit = 20% ⟶ Rs. 750

∴ Selling price = 120% ⟶ $? = \frac{750 \times 100}{20} = \text{Rs. } 4500.$

Q - 71 A man bought two watches for Rs. 2500 and sold the first at 25% profit. If he sold the first at 25% profit and the second at 15% profit, he would get Rs. 10 less. Find the difference in cost prices of the two watches.

Sol: Consider, CP of 1st watch = Rs. x and CP of 2nd watch = 2500 – x

According to question,

Actual profit = 15% of x + 25% of (2500 – x)

Assumption profit = 25% of x + 15% of (2500 – x)

Given that, Actual profit – Assumption profit = Rs. 10

15% of x + 25% of (2500 – x) – 25% of x – 15% of (2500 – x) = 10

$\frac{10}{100}(2500 - x) - \frac{10}{100}x = 10 \implies 25000 - 10x - 10x = 1000 \implies x = 1200$

∴ CP of two watches are x = Rs. 1200 and 2500 – x = Rs. 1300.

Q - 72 A manufacturer undertakes to supply 6000 pieces of a particular component at a rate of Rs. 50 per piece. According to his estimation, even if 12% fail to pass the quality tests, then he will make a profit of 10%. But it is turned out, 40% of the components were rejected. What is the loss to the manufacturer?

Sol: According to manufacturer estimation,

No. of pieces = 88% of 6000 = $\frac{88}{100}$ × 6000 = 5280   ∵ 12% – Fail

Total SP of 5280 pieces = 5280 × 50 = Rs. 264000

Total SP = 110% ⟶ Rs. 264000

Total CP = 100% ⟶ $? = \frac{264000 \times 100}{110}$ = Rs. 240000

But, actually 40% components were rejected.

SP of 60% of 6000 pieces = $50 \times \frac{60}{100} \times 6000$ = Rs. 180000

∴ Loss to the manufacturer = CP – SP = 240000 – 180000 = Rs. 60000.

Q - 73 A business man buys 18 coolers for a total of Rs. 60000. 8 of them are first grade quality and the rest of them are second grade quality. At what price should he sell the first quality coolers, so that if he sells them at 2.5 times the price of second quality, he would make a profit of 40%?

Sol: Given that, total CP of 18 coolers = Rs. 60000

Total SP of 18 coolers = 140% of 60000 = Rs. 84000   ∵ Profit = 40%

Consider, SP of 2nd grade quality = x & SP of 1st grade quality = 2.5x

Total SP = SP of 1st quality + SP of 2nd quality

84000 = 8 × 2.5x + 10x = 30x ⟹ x = Rs. 2800

∴ SP of first grade quality = 2.5x = 2.5 × 2800 = Rs. 7000.

| Q - 74 | A fruit merchant purchased 200 kgs of apples at Rs. 25 per kg, 30% of them are spoilt in transportation and he sold them at Rs. 16 per kg. What should be the selling price of the remaining apples per kg to get a total profit of 20% in the transaction? |

Sol:

> Total cost = Quantity × Price per kg

Total CP of apples = 200 × 25 = Rs. 5000

Total SP of apples = 120% of 5000 = Rs. 6000   ∵ Profit = 20%

Consider, SP of remaining apples per kg = Rs. x

Total SP of apples = SP of 30% of apples + SP of remaining 70% of apples

$6000 = (\frac{30}{100} \times 200) \times 16 + (\frac{70}{100} \times 200) x$ ⇒ x = Rs. 36

∴ SP of remaining apples per kg is Rs. 36.

| Q - 75 | A sold an article to B at a profit of 30% and B sold it to C at a profit of 10%. D sold the similar article to E at a loss of 10% and E sold it to F at a loss of 20%. The sum of the prices that C and F paid for their respective articles is Rs. 34500 more than that what A paid. If A and D bought the article at same price, then find the sum of the prices paid by A and D for their respective articles. |

Sol: Given that, CP of A and D are same. So, consider CP of article for A and D = 100%.

	CP	SP		CP	SP
A	100%	30% Profit ⟶ 130%	D	100%	10% Loss ⟶ 90%
B	130%	10% Profit ⟶ $130 \times \frac{110}{100} = 143\%$	E	90%	20% Loss ⟶ $90 \times \frac{80}{100} = 72\%$
C	143%		F	72%	

Given that, C + F = A + 34500

143% + 72% = 100% + 34500   ⇒   115% = 34500   ⇒   1% = Rs. 300

∴ Sum of the prices paid by A & D = 100% + 100% = 200% = 200 × 300 = Rs. 60000.

ASSESSMENT TEST

1. If a person sold an article at Rs. 680, there is 20% loss. Find the profit (or) loss percentage, when he sells at Rs. 1105.

2. If Venu sold an article at seven-eighth of its actual selling price, he would have incurred a loss of 16%. Find his actual profit (or) loss percentage.

3. If the manufacturer gains 10%, the wholesale dealer gains 15% and the retailer gains 20%. If the retail price is Rs. 18216, then find the cost of production.

4. Joseph bought a machine for Rs. 60000 and spent Rs. 8000 on repairs and Rs. 2000 on transport and sold it with 30% profit. What is the selling price of machine?

5. The profit made on selling 14 meters of cloth is equal to the cost price of four meters of that cloth. Find the profit percentage.

6. Anu sold an article to Bhanu at 30% profit. Bhanu sold it to Jhanu at 15% profit. If Anu's profit is Rs. 1260 more than Bhanu's profit, then what is the cost price of Anu?

7. A sold a tape recorder to B at a gain of 10% and B sold it to C at a gain of 15% and C sold it to D at a loss of 10%. If D pays Rs. 9108, then what is the cost price of A?

8. If 20 chocolates are bought for Rs. 15 and at 15 for Rs. 20. Find the profit (or) loss percentage.

9. Suma purchased a second-hand cycle for Rs. 1340 and spent Rs. 360 on its repairs. She sold it for Rs. 2210. Find her profit (or) loss percentage.

10. The profit earned by selling an article for Rs. 750 is equal to the loss incurred when the same article is sold for Rs. 460. What should be the selling price of the article, if the profit earned is 40%?

11. If goods are purchased for Rs. 560 and one-third be sold at a loss of 10% at what gain percent should the remainder be sold, so as to gain 10%, on the whole transaction?

12. Pravalika purchased 25 tables at Rs. 250 per table. She sold 12 tables at 15% profit and the remaining at 10% profit. What is her profit percentage in this transaction?

13. A person buys a certain number of oranges at 15 a rupee and the same number of oranges at 20 a rupee. He mixes them and sells at 35 for 2 rupees. What is his gain (or) loss percentage in the whole transaction?

14. A man buys 4 chairs and 6 tables for Rs. 5440. He sells the chairs at a loss of 15% and tables at a loss of 10% and his whole loss is Rs. 672. Find the price of each chair and each table.

15. A cooler worth Rs. 8500 is sold by A to B at 15% loss and B sells the cooler back to A at a loss of 8%. Who gains and who loses? Find also values.

16. A milkman buys some milk. If he sells it at Rs. 3 a liter, he loses Rs. 560, but when he sells it at Rs. 7 a liter, he gains Rs. 420. How much milk did he purchase?

17. A fruit merchant makes a profit of 14% by selling watermelon at a certain price. If he charges Rs. 2.7 more on watermelon, he would gain 32%. Find the cost price of watermelon and first selling price.

18. 5 cups cost as much as 3 plates, 10 plates as much as 4 kettles, 3 kettles as much as 7 dishes, if the price of 4 dishes is Rs. 450, then find the price of 3 cups.

19. What will be the profit percentage after selling an article at a certain price, if there is a loss of 35% when the same article is sold at one-third of the previous selling price?

20. A shopkeeper purchases some goods for Rs. 2600. If the overhead expenses are 13% of the cost price. What is the selling price of goods to gain 20%?

21. The price of a land passing through three hands, rises on the whole by 50.15%. If the first and second sellers earned a profit of 10% and 26% respectively, then find the profit earned by the third seller.

22. A shopkeeper sells a calculator at the rate of Rs. 483 each. So that he will get a profit of 40%. What amount of profit will he earn in 12 days, if he sells 5 calculators per day?

23. By selling 24 chocolates for Rs. 18 a man loses 13%. How many chocolates should be sold for Rs. 36, so as to gain 16% in the transaction?

24. A dealer fixed the selling price of his articles at Rs. 924 after adding 32% profit to the cost price. As the sale was low at this price level, then he decided to fix the selling price at 15% profit. Find the new selling price.

25. If the selling price of an article is increased by 12%, then the profit is 68%. Find the profit percent, if the selling price is decreased by 30%.

26. An article A which when sold for Rs. 1540 gives a profit of 10%. Another article B of the same cost price was sold at less cost due to customer bargain. By what percent was the selling price of B less than that of A, if there was neither profit nor loss for the seller?

27. The cost of production of a bike which is sold at 40% profit went up by 50%. What should be the percent increase in the selling price to maintain the profit percentage the same even at the new cost of production?

28. A shopkeeper bought 70 mobile phones at Rs. 20000 each. For every mobile phone purchased from him, he gave one mobile phone free. The loss made by him is equal to selling price of 15 mobile phones. What is the selling price of each mobile phone that is bought?

29. The cost price of two articles are same. On one article there is 40% profit, on the other article there is 25% loss. Find the overall profit (or) loss percentage.

30. A milk dealer announces to sell the milk at 24% profit but gives 800 ml for every liter. Find the actual profit percentage.

31. Ramana sold two articles for Rs. 30000 each. If he makes 30% profit on one and 50% profit on the other, then find the overall profit percentage.

32. By mistake a person calculated the profit percentage over selling price and it is found as 40%, then find the actual profit percentage.

33. By mistake a person calculated the loss percentage over selling price and it is found as 20%, then find the actual loss percentage.

34. Nani purchased an article for Rs. 15000 and sold it at a gain of 30%. From that amount he purchased another article and sold it at a loss of 25%. What is his overall gain (or) loss percentage?

35. Revanth purchased an item for Rs. 14800 and sold it at a gain of 26%. With that amount he purchased another item and sold it at a loss of 30%. What was his overall gain (or) loss?

36. Manoj purchased 150 reams of paper at Rs. 90 per ream. He spent Rs. 500 on transportation, paid octroi at the rate of 60 paise per ream and paid Rs. 110 to the coolie. If he wants to have a gain of 20%, then what is the selling price of paper per ream?

37. A dishonest fruit vendor professes to sell his fruits at cost price but he uses a weight of   840 grams for every kg. What is his gain percent?

38. A shopkeeper sells wheat at a profit of 15% and uses a weight which is 25% less. Find his total percentage gain.

39. A merchant professes to lose $8\frac{1}{3}$% on a certain rice, but uses a weight equal to 750 grams instead of 1 kg. Find his real loss (or) profit percentage.

40. A dishonest dealer sells his goods at 25% loss on cost price, but uses 40% less weight. What is his percentage profit (or) loss?

41. A seller uses 900 grams for every one kg to sell his goods. Find his actual profit (or) loss percent, when he sells his article at 40% profit.

42. A man bought an article and sold it at a gain of 28%. If he had bought it at 16% less and sold it for Rs. 140 less, then he would have gained 50%. What is the cost price of an article?

43. An article is sold at a profit of 30%. If both the cost price and selling prices are Rs. 480 less, then the profit would be 8% more. Find the cost price.

44. An article is sold at 30% profit. If its cost price and selling prices are increased by Rs. 75 and Rs. 60 respectively, then the percentage of profit decreases by 15%. Find the cost price of an article.

45. A shopkeeper has 920 kg of sugar, part of which he sells at 7% loss and the rest at 16% profit. He gains 5% on the whole. Find the quantity sold at 7% loss.

46. The cost of coffee powder packet and a liter of milk are Rs. 25 and Rs. 40 respectively. 22 cups of coffee is made with two packets of coffee powder and for each cup 250 ml of milk is used. Find the selling price of each cup of coffee, if coffee is sold at 32% profit.

47. A fruit vendor earns a profit of 30% by selling 60 oranges for Rs. 468, but he gives one-third of it to his friend at cost price and sells the remaining oranges. In order to earn the same profit, at what price must he sell each orange?

48. A fruit seller bought some apples. He sold 40% of them at a gain of 25% and 25% of them at a loss of 20%. The rest being rotten were not sold. What is the loss percentage on the whole?

49. If an article A is sold for Rs. 4200, there is a gain of 20%. An article B should be sold in such a way that there is a profit of 30% on both A and B articles. If the cost price of B is 10% more than that of A, then what is the selling price of B?

50. A man sells a fan at a profit of 36% and the light at a loss of 16% but on the whole he gains Rs. 91. On the other hand if he sells the fan at a loss of 16% and light at a profit of 36%, then he neither gains nor loses. Find the cost price of fan and light.

51. One-fourth of goods are sold at 20% profit, one-sixth of goods are sold at 30% profit and the rest at 10% profit. If the total profit earned is Rs. 171, then find the value of goods.

52. A man bought two books A and B for Rs. 1218. He sold book A at a loss of 14% and book B at a profit of 17% and he found that each book was sold at the same price. Find the cost prices of each book.

53. Nandu calculates his profit percentage on selling price. Rahul calculates his profit percentage on cost price. They find that the difference of their profits is Rs. 180. If the selling price of both of them are same and both will get 15% profit, then find their selling price.

54. A man bought 2 bags for Rs. 3200 and sold the first at 12% profit and the second at 20% profit. If he sold the first at 20% profit and the second at 12% profit, he would get Rs. 16 more. Find the difference in cost prices of the 2 bags.

55. A business man purchases 14 washing machines for a total of Rs. 59520. 6 of them are first grade quality and the rest of them are second grade quality. At what price should he sell the first quality washing machines so that if he sells them at 2.8 times the price of second quality, he would make a profit of 25%?

56. A fruit vendor purchased 150 kgs of oranges at Rs. 30 per kg. 20% of them are spoilt in transportation and he sold them at Rs. 15 per kg. What should be the selling price of the remaining oranges per kg to get a total profit of 40% in the transaction?

57. A manufacturer buys a second-hand machine for Rs. 80000 and incurs Rs. 20000 on installation and repairs. After one year he sells the machine for Rs. 120000. What is his profit (or) loss percentage, if 15% is to be deducted on account of depreciation?

58. The sale price of an article including the sales tax is Rs. 6900. The rate of sales tax is 15%. If the shopkeeper has made a profit of 25%, then find the cost price of an article.

59. A manufacturer undertakes to supply 4000 pieces of a particular component at a rate of Rs. 40 per piece. According to his estimation, even if 10% fail to pass the quality tests, then he will make a profit of 20%. But it is turned out, 35% of the components were rejected. What is the loss to the manufacturer?

60. A sold an article to B at a profit of 15% and B sold it to C at a profit of 20%. D sold the similar article to E at a loss of 20% and E sold it to F at a loss of 15%. The sum of the prices that C and F paid for their respective articles is Rs. 37100 more than that of what A paid. If A and D bought the article at same price, then find the sum of the prices paid by A and D for their respective articles.

KEY

1. 30% profit
2. 4% loss
3. Rs. 12000
4. Rs. 91000
5. $28\frac{4}{7}\%$
6. Rs. 12000
7. Rs. 8000
8. $77\frac{7}{9}\%$ profit
9. 30% profit
10. Rs. 847
11. 20%
12. $22\frac{2}{3}\%$
13. $2\frac{2}{49}\%$ loss
14. Rs. 640, Rs. 480
15. A gains, B loses, A gain% = 6.8%
16. 245 liters
17. CP = Rs. 15, SP_1 = Rs. 17.10
18. Rs. 189
19. 95%
20. Rs. 3525.60
21. $16\frac{2}{3}\%$
22. Rs. 8280
23. 36
24. Rs. 805
25. 5%
26. $18\frac{2}{11}\%$
27. 50%
28. Rs. 28000
29. 7.5% profit
30. 55%

31. $39\frac{2}{7}\%$

32. $66\frac{2}{3}\%$

33. $16\frac{2}{3}\%$

34. 2.5% loss

35. Loss = Rs. 1764.40

36. Rs. 113.60

37. $19\frac{1}{21}\%$

38. $53\frac{1}{3}\%$

39. $22\frac{2}{9}\%$ profit

40. 25% profit

41. $55\frac{5}{9}\%$ profit

42. Rs. 7000

43. Rs. 2280

44. Rs. 175

45. 440 kg

46. Rs. 16.20

47. Rs. 8.70

48. 30%

49. Rs. 5355

50. Fan = Rs. 315, Light = Rs. 140

51. Rs. 1080

52. Book A = Rs. 702, Book B = Rs. 516

53. Rs. 9200

54. Rs. 200

55. Rs. 8400

56. Rs. 48.75

57. $41\frac{3}{17}\%$ profit

58. Rs. 4800

59. Rs. 16000

60. Rs. 70000

DISCOUNTS

✓ **DISCOUNT:** Discount is defined as the reduction of amount on a fixed price (Marked Price) of an article.

✓ **Marked Price:** Marked Price means the price on the label of an article.

✓ Discount percentage will always calculate over marked price of an article.

Discount D = Marked Price (MP) – Selling Price (SP)

Discount percentage = $\dfrac{Discount}{Marked\ Price} \times 100\%$

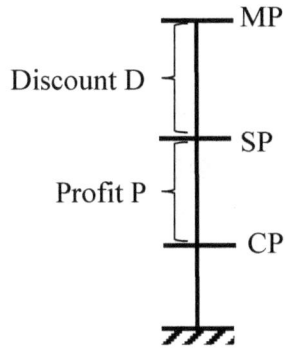

$$\text{Discount \%} = \frac{MP - SP}{MP} \times 100\%$$

$$SP = MP \times \frac{100 - D\%}{100}$$

$$MP = SP \times \frac{100}{100 - D\%}$$

Successive discounts:

✓ If a series of discounts (one after the other) are given on Marked Price (MP), then those discounts are called as successive discounts.

✓ Let $D_1\%$, $D_2\%$, $D_3\% \ldots$ be the series of discounts on marked price, then

$$SP = MP \times \frac{100 - D_1\%}{100} \times \frac{100 - D_2\%}{100} \times \frac{100 - D_3\%}{100} \times \ldots$$

✓ If two successive discounts $D_1\%$, $D_2\%$ are equivalent to single discount of

$D_1 + D_2 - \dfrac{D_1 \times D_2}{100}\%$

SOLVED EXAMPLES

Q - 1   **Find a single discount equivalent to two successive discounts of 15% and 25%.**

Sol: Consider, Marked price = 100%

Given that, $D_1 = 15\%$, $D_2 = 25\%$

We know that, $SP = MP \times \dfrac{100 - D_1\%,}{100} \times \dfrac{100 - D_2\%,}{100}$

$SP = 100 \times \dfrac{100 - 15}{100} \times \dfrac{100 - 25}{100} = 100 \times \dfrac{85}{100} \times \dfrac{75}{100} = 63.75\%$

∴ Single discount = MP – SP = 100% – 63.75% = 36.25%.

Shortcut:

✓   If two successive discounts $D_1\%$, $D_2\%$ are equivalent to single discount of

$$D_1 + D_2 - \frac{D_1 \times D_2}{100} \%$$

Here, $D_1 = 15\%$ and $D_2 = 25\%$

∴ Single discount = $15 + 25 - \dfrac{15 \times 25}{100}\% = 36.25\%$.

Q - 2   **Find a single discount which is equivalent to three successive discounts of 10%, 20% and 25%.**

Sol: Consider, Marked price = 100%

Given that, $D_1 = 10\%$, $D_2 = 20\%$ and $D_3 = 25\%$

$SP = MP \times \dfrac{100 - D_1\%,}{100} \times \dfrac{100 - D_2\%,}{100} \times \dfrac{100 - D_3\%}{100}$

$SP = 100 \times \dfrac{100 - 10}{100} \times \dfrac{100 - 20}{100} \times \dfrac{100 - 25}{100} = 100 \times \dfrac{90}{100} \times \dfrac{80}{100} \times \dfrac{75}{100} = 54\%$

∴ Single discount = MP – SP = 100% – 54% = 46%.

Q - 3   **A tradesman marks his goods at 30% above his cost price and allows customers a discount of 15% for cash payment. Find his profit percentage.**

Sol: Consider, CP of goods = 100%

Marked Price MP = 100% + 30% = 130%   ∵ MP = 30% above CP

Discount D = 15%

$SP = MP \times \dfrac{100 - D\%}{100} = 130 \times \dfrac{100 - 15}{100} = 130 \times \dfrac{85}{100} = 110.5\%$

∴ Profit percentage = SP – CP = 110.5% – 100% = 10.5%.

Q-4 If a discount of 10% is given on marked price of an article, the shopkeeper gets a profit of 25%. Find his profit percent, if he offers a discount of 19% on the same article.

Sol: Given that, Discount = 10% and Profit = 25%

Consider, CP of an article = 100% $\Rightarrow$ SP of an article = 125%

$SP = MP \times \dfrac{100 - D\%}{100}$ $\Rightarrow$ $125 = MP \times \dfrac{100 - 10}{100}$ $\Rightarrow$ $MP = 125 \times \dfrac{100}{90} = \dfrac{1250}{9}$

If shopkeeper offers a discount of 19%, then

$SP = \dfrac{1250}{9} \times \dfrac{100 - 19}{100} = 112.5\%$

∴ Profit percentage = SP – CP = 112.5% – 100% = 12.5%.

Q-5 The marked price of DVD player is Rs. 7500. A shopkeeper allows a discount of 28% and still gets a profit of 20%. Find the cost price of DVD player.

Sol: Consider, CP of DVD player = 100%

$SP = MP \times \dfrac{100 - D\%}{100} = 7500 \times \dfrac{100 - 28}{100} = Rs.\ 5400$

SP of DVD player = 120% $\longrightarrow$ Rs. 5400 ∵ SP = CP + P, Profit P = 20%

∴ CP of DVD player = 100% $\longrightarrow$ $? = \dfrac{5400 \times 100}{120} = Rs.\ 4500.$

Q-6 While selling an article, a merchant allows 30% discount on marked price and there is a loss of 16%. What is the profit percent, if it is sold at the marked price?

Sol: Consider, CP of an article = 100%

SP of an article = 100% – 16% = 84% ∵ Loss = 16%

$SP = MP \times \dfrac{100 - D\%}{100}$ $\Rightarrow$ $MP = SP \times \dfrac{100}{100 - D\%}$

$MP = 84 \times \dfrac{100}{100 - 30} = 120\%$ ∵ Discount D = 30%

If the article is sold at marked price = 120%, then

∴ Profit % = SP – CP = 120% – 100% = 20%.

Q-7 The cost price of an article is Rs. 9000. After allowing a discount of 13%, a gain of 16% was made. Find the marked price of an article.

Sol: Consider, CP of an article = 100%

Given that, CP of an article = 100% $\longrightarrow$ Rs. 9000

SP of an article = 116% $\longrightarrow$ $? = \dfrac{9000 \times 116}{100} = Rs.\ 10440$ ∵ Gain = 16%

$MP = SP \times \dfrac{100}{100 - D\%} = 10440 \times \dfrac{100}{100 - 13} = Rs.\ 12000$ ∵ D = 13%

∴ Marked Price of an article is Rs. 12000.

Q - 8 A man purchased an article with a discount of 20% on its marked price. He sold it at a price 35% more than the price at which he bought it. How much percent is the new selling price to its marked price?

Sol: Consider, MP of an article = Rs. 100

CP of an article for a man = $100 \times \dfrac{100 - 20}{100}$ = Rs. 80   ∵ D = 20%

SP of an article for a man = 135% of CP = $\dfrac{135}{100} \times 80$ = Rs. 108

∴ Required percentage = $\dfrac{108 - 100}{100} \times 100\% = 8\%$.

Q - 9 Three successive discounts of x% each is equal to a single discount of Rs. 13664. If the selling price after all the discounts was Rs. 14336, then find the value of x.

Sol: MP = SP + D = 14336 + 13664 = Rs. 28000

$SP = MP \times \dfrac{100 - D_1\%}{100} \times \dfrac{100 - D_2\%}{100} \times \dfrac{100 - D_3\%}{100}$

$14336 = 28000 \times \dfrac{100 - x}{100} \times \dfrac{100 - x}{100} \times \dfrac{100 - x}{100}$

$(100 - x)^3 = \dfrac{14336 \times 100 \times 100 \times 100}{28000}$ = Rs. 512000

$100 - x = \sqrt[3]{512000} = 80$ ⇒ $x = 100 - 80$ ⇒ $x = 20\%$

Therefore, each discount is 20%.

Q - 10 An article is marked 40% above its cost price and sold at a profit of 26% after a certain discount. Find the new profit percent, if the discount is doubled.

Sol: Consider, CP of an article = 100%

MP of an article = 100% + 40% = 140%   ∵ MP – 40% above CP

SP of an article = 100% + 26% = 126%   ∵ Profit = 26%

$SP = MP \times \dfrac{100 - D\%}{100}$ ⇒ $126 = 140 \times \dfrac{100 - D\%}{100}$ ⇒   D = 10%

If discount is double, then new discount = 20%

$SP = 140 \times \dfrac{100 - 20}{100} = 112\%$

∴ Profit percentage = 112% – 100% = 12%.

Q - 11 At first a mobile phone was being offered at a discount of 30%. The dealer reduced this discounted price further by 10% because the customer bargained. If the selling price of mobile phone is Rs. 12600, then what is its marked price?

Sol: Here, the dealer giving two discounts because of customer bargain.

$SP = MP \times \dfrac{100 - D_1\%}{100} \times \dfrac{100 - D_2\%}{100}$ ⇒ $12600 = MP \times \dfrac{100 - 30}{100} \times \dfrac{100 - 10}{100}$

∴ MP of mobile phone = $\dfrac{12600 \times 100 \times 100}{70 \times 90}$ = Rs. 20000.

Q - 12   After allowing a discount of 20% an article was sold for Rs. 7200. If no discount is offered on the article, there would be a profit of 50%, then find the cost price of the article.

Sol: $SP = MP \times \dfrac{100 - D\%}{100}$ $\Rightarrow$ $7200 = MP \times \dfrac{100 - 20}{100}$ $\Rightarrow$ $MP = Rs.\ 9000$

If there is no discount offered, then MP = SP = Rs. 9000

Consider, CP of an article = 100%

SP of an article = 150% $\longrightarrow$ Rs. 9000 $\because$ Profit = 50%

$\therefore$ CP of an article = 100% $\longrightarrow$ $? = \dfrac{9000 \times 100}{150} = Rs.\ 6000.$

Q - 13   If a company sells a bike with a marked price of Rs. 156000 and gives a discount of 8% on Rs. 90000 and 5% on the remaining amount, then find the amount charged by the company for the bike.

Sol: Given that, MP of a bike = Rs. 156000

Company is giving 8% discount on Rs. 90000 and 5% discount on remaining amount (156000 – 90000) Rs. 66000

Total discount offered $= \dfrac{8}{100} \times 90000 + \dfrac{5}{100} \times 66000 = Rs.\ 10500$

$\therefore$ Amount charged by company for the bike = 156000 – 10500 = Rs. 145500.

Q - 14   What is the maximum discount percent that a merchant can offer on his marked price, so that he ends up selling at no profit (or) no loss, if he initially marked his goods up at 50%?

Sol: Consider, CP of goods = 100%

MP of goods = 150% $\because$ MP – 50% more than CP

Merchant ends up "neither profit nor loss" which means,

CP of goods = SP of goods = 100%

$SP = MP \times \dfrac{100 - D\%}{100}$ $\Rightarrow$ $100 = 150 \times \dfrac{100 - D\%}{100}$

$100 - D = 66\dfrac{2}{3}$ $\Rightarrow$ $D = 33\dfrac{1}{3}\%$

$\therefore$ Maximum discount offered by merchant is $33\dfrac{1}{3}\%$.

Q - 15 A shopkeeper marked the price of an item $\frac{4}{5}$th of its cost price. If he sold it at a price 30% more than marked price, then what will be the percent of profit (or) loss?

Sol: Consider, CP of an item = 100%

MP of an item $= \frac{4}{5} \times CP = \frac{4}{5} \times 100\% = 80\%$

SP of an item $= 130\%$ of MP $= \frac{130}{100} \times 80\% = 104\%$ ∵ SP – 30% more than MP

∴ Profit % = SP – CP = 104% – 100% = 4%.

Q - 16 Two shopkeepers sell machines at the same list price. The first one allows two successive discounts of 32% and 15% and the second one allows 22% and 25%. Which discount series is more advantage to the customers?

Sol: Consider, list price of two shopkeepers (MP) = 100

First Shopkeeper	Second Shopkeeper
$SP = MP \times \frac{100 - D_1\%}{100} \times \frac{100 - D_2\%}{100}\%$	$SP = MP \times \frac{100 - D_1\%}{100} \times \frac{100 - D_2\%}{100}\%$
$D_1 = 32\%, \ D_2 = 15\%$	$D_1 = 22\%, \ D_2 = 25\%$
$SP = 100 \times \frac{100 - 32}{100} \times \frac{100 - 15}{100}$	$SP = 100 \times \frac{100 - 22}{100} \times \frac{100 - 25}{100}$
$SP = 100 \times \frac{68}{100} \times \frac{85}{100} = 57.8\%$	$SP = 100 \times \frac{78}{100} \times \frac{75}{100} = 58.5\%$
Single discount = MP – SP	Single discount = MP – SP
$\Rightarrow 100\% - 57.8\% = 42.2\%$	$\Rightarrow 100\% - 58.5\% = 41.5\%$

∴ By comparing two discount series,

first discount series is more advantage to the customers.

> **Note:**
> ✓ For **customer advantage** DISCOUNT should be more (or) SP should be less.
> ✓ For **shopkeeper advantage** DISCOUNT should be less (or) SP should be more.

Shortcut:

First Shopkeeper:

Single Discount $= D_1 + D_2 - \frac{D_1 \times D_2}{100}\% = 32 + 15 - \frac{32 \times 15}{100}\% = 42.2\%$.

Second Shopkeeper:

Single Discount $= D_1 + D_2 - \frac{D_1 \times D_2}{100}\% = 22 + 25 - \frac{22 \times 25}{100}\% = 41.5\%$.

∴ Among two discount series, first discount series is more advantage to the customers.

Q - 17 The cost of mobile phone is marked at Rs. 24000. If successive discounts of 10%, 15% and 20% be allowed, then at what price does a customer buy it?

Sol: $SP = MP \times \dfrac{100 - D_1\%}{100} \times \dfrac{100 - D_2\%}{100} \times \dfrac{100 - D_3\%}{100}$

Given that, MP = Rs. 24000, D_1 = 10%, D_2 = 15%, D_3 = 20%

$SP = 24000 \times \dfrac{100 - 10}{100} \times \dfrac{100 - 15}{100} \times \dfrac{100 - 20}{100} = $ Rs. 14688

∴ Customer bought the mobile phone for Rs. 14688.

Q - 18 Krishna marked an article at 50% above its cost price and then sold it after a discount of 12%. He made a profit of Rs. 192, then find his cost price.

Sol: Consider, CP of an article = 100%

MP of an article = 150%   ∵ MP – 50% above CP

$SP = MP \times \dfrac{100 - D\%}{100} = 150 \times \dfrac{100 - 12}{100} = 132\%$ ∵ D = 12%

Profit = SP – CP = 132% – 100% = 32%

Given that,   Profit   = 32%   ⟶   Rs. 192

∴ CP of an article = 100%   ⟶ $? = \dfrac{192 \times 100}{32} = $ Rs. 600.

Q - 19 A shopkeeper sells a mobile phone, with a marked price of Rs. 7000 at a discount of 15% and gives a headphone costing Rs. 400 free with each mobile phone. Even then, he makes a profit of 20%. Find the cost price of each mobile phone.

Sol: $SP = MP \times \dfrac{100 - D\%}{100} = 7000 \times \dfrac{100 - 15}{100} = $ Rs. 5950

But, he is giving Rs. 400 headphones free for every mobile phone.

SP of mobile phone = 5950 – 400 = Rs. 5550

SP of mobile phone = 120%   ⟶   Rs. 5550   ∵ Profit = 20%

∴ CP of mobile phone = 100%   ⟶ $? = \dfrac{5550 \times 100}{120} = $ Rs. 4625.

Q - 20 An uneducated shopkeeper marks his goods at 30% above the cost price and thinking that he will get a profit of 15%, after offering a discount of 20% on the marked price. What is his actual profit percentage?

Sol: Consider, CP of goods = 100%

MP of goods = 130%   ∵ MP – 30% above CP

$SP = MP \times \dfrac{100 - D\%}{100} = 130 \times \dfrac{100 - 20}{100} = 104\%$ ∵ D = 20%

∴ Actual profit percentage = SP – CP = 104% – 100% = 4%.

Q-21 | If a shopkeeper wants to earn 32% profit on an item, after giving a discount of 25% to the customer. By what percent should he increase his marked price to arrive at the label price?

Sol: Consider, CP of an item = 100%

SP of an item = 132%   ∵ Profit = 32%

$$SP = MP \times \frac{100 - D\%}{100} \quad \Rightarrow MP = SP \times \frac{100}{100 - D\%} = 132 \times \frac{100}{100 - 25} = 176 \quad \because D = 25\%$$

MP of an item = 176%

∴ MP is more than the CP by 76% (176% − 100%).

Q-22 | Vikas bought an article with 10% discount on marked price. He sold it with 30% profit on marked price. What was his profit percent on the price he bought?

Sol: Consider, MP of an article = Rs. 100

Vikas bought an article with 10% discount

CP of Vikas = 100 − 10 = Rs. 90

Vikas SP = Rs. 130   ∵ Profit = 30% on MP

∴ Profit percentage $= \frac{SP - CP}{CP} \times 100\% = \frac{130 - 90}{90} \times 100\% = 44\frac{4}{9}\%.$

Q-23 | By selling an article for Rs. 45, a shopkeeper gains 25%. During a clearance sale, the shopkeeper allows a discount of $13\frac{1}{3}\%$ on previous selling price. What is his gain percentage during the sale season?

Sol: Let CP of an article = 100%

SP of an article = 125% ⟶ Rs. 45   ∵ Profit = 25%

CP of an article = 100% ⟶ $? = \frac{45 \times 100}{125} = $ Rs. 36

During a clearance sale, discount is $13\frac{1}{3}\%$ on SP

Discount $= 13\frac{1}{3}\%$ of $45 = \frac{40}{3} \times \frac{1}{100} \times 45 = $ Rs. 6

New SP of an article = 45 − 6 = Rs. 39

∴ Profit percentage $= \frac{SP - CP}{CP} \times 100\% = \frac{39 - 36}{36} \times 100\% = 8\frac{1}{3}\%.$

Q - 24 A person bought an article marked at Rs. 3200 with a discount of 15% offered on the marked price. What additional discount must be offered to that person to bring the net price to Rs. 2448?

Sol: Given that, MP = Rs. 3200, D_1 = 15%, SP = Rs. 2448, D_2 = ?

$SP = MP \times \frac{100 - D_1\%}{100} \times \frac{100 - D_2\%}{100}$ $\Rightarrow$ $2448 = 3200 \times \frac{100 - 15}{100} \times \frac{100 - D_2\%}{100}$

$100 - D_2 = \frac{2448 \times 100 \times 100}{3200 \times 85} = 90$ $\Rightarrow$ $D_2 = 10\%$

$\therefore$ Additional discount offered is 10%.

Q - 25 The marked price of a TV set is Rs. 24000. It is to be sold at Rs. 18768 after giving two successive discounts. If the first discount is 15%, then find the second discount.

Sol: Given that, MP of TV set = Rs. 24000

SP of TV set = Rs. 18768 and D_1 = 15%

$SP = MP \times \frac{100 - D_1\%}{100} \times \frac{100 - D_2\%}{100}$ $\Rightarrow$ $18768 = 24000 \times \frac{100 - 15}{100} \times \frac{100 - D_2\%}{100}$

$100 - D_2 = \frac{18768 \times 100 \times 100}{24000 \times 85} = 92$ $\Rightarrow$ $D_2 = 8\%$

$\therefore$ Second discount is 8%.

Q - 26 While selling a dining table, a shopkeeper gives a discount of 5%. If he gives a discount of 8%, he earns Rs. 210 less as profit. Find the marked price of dining table.

Sol: Consider, MP of dining table = 100%

If discount is 5%, then $SP_1 = MP - D_1 = 100\% - 5\% = 95\%$

If discount is 8%, then $SP_2 = MP - D_2 = 100\% - 8\% = 92\%$

Given that, Difference = 3% $\longrightarrow$ Rs. 210

$\therefore$ Marked price MP = 100% $\longrightarrow$? $= \frac{210 \times 100}{3} = $ Rs. 7000.

Q - 27 Karan purchased an article after two successive discounts of 20% and 25%. Charan purchased the same article after three successive discounts of 20%, 15% and 10% respectively. If the marked price is Rs. 5000 for each of them, then find the difference between the price, at which Karan and Charan purchased the articles.

Sol: Given that, Karan purchased an article after two successive discounts of 20% and 25%

Also, MP = Rs. 5000

Cost of Karan $= MP \times \frac{100 - D_1\%}{100} \times \frac{100 - D_2\%}{100} = 5000 \times \frac{100 - 20}{100} \times \frac{100 - 25}{100} = $ Rs. 3000

Charan purchased the same article after 3 successive discounts of 20%, 15%, 10%

Cost of Charan $= MP \times \frac{100 - D_1\%}{100} \times \frac{100 - D_2\%}{100} \times \frac{100 - D_3\%}{100}$

Cost of Charan $= 5000 \times \frac{100 - 20}{100} \times \frac{100 - 15}{100} \times \frac{100 - 10}{100} = $ Rs. 3060

$\therefore$ Difference of prices of Karan and Charan = 3060 - 3000 = Rs. 60.

Q - 28 A dealer bought 120 drafters at Rs. 250 each. He spent Rs. 2000 on packing and transportation. He marked each drafter at Rs. 350 and offered a discount of 20%. What is his profit percent?

Sol: CP of drafters for dealer = 120×250 = Rs. 30000

Packing and transportation = Rs. 2000

Total CP of dealer = $30000 + 2000$ = Rs. 32000

SP of each drafter = $MP \times \dfrac{100 - D\%}{100} = 350 \times \dfrac{100 - 20}{100}$ = Rs. 280 $\because$ MP = Rs. 350

Total SP of drafters for dealer = 120×280 = Rs. 33600

$\therefore$ Profit percentage = $\dfrac{SP - CP}{CP} \times 100\% = \dfrac{33600 - 32000}{32000} \times 100\% = 5\%$.

Q - 29 A shopkeeper earns a profit of 25% on selling a toy at 15% discount on the marked price. Find the ratio of the cost price and marked price of the toy.

Sol: Consider, CP of a toy = Rs. 100

SP of a toy = Rs. 125 $\because$ Profit = 25%

$SP = MP \times \dfrac{100 - D\%}{100} \Rightarrow MP = SP \times \dfrac{100}{100 - D\%} = \dfrac{125 \times 100}{100 - 15} = $ Rs. $\dfrac{12500}{85}$ $\because$ D = 15%

$\therefore$ Required ratio CP : MP = $100 : \dfrac{12500}{85} = 85 : 125 = 17 : 25$.

Q - 30 A shopkeeper allows a discount of 15% on the marked price of an article but charges a sales tax of 12% on the discounted price. If the customer pays Rs. 1142.40 as the price including the sales tax, then what is the marked price of an article?

Sol: Consider, MP of an article = Rs. x

Given that, discount = 15% and sales tax = 12%

Customer payment including tax = $(100 + 12)\%$ of $(100 - 15)\%$ of x

$1142.40 = \dfrac{112}{100} \times \dfrac{85}{100} \times x \qquad \Rightarrow \qquad x =$ Rs. 1200

$\therefore$ Marked price of an article is Rs. 1200.

Q - 31 A shopkeeper buys two articles x and y for Rs. 500 each. He marks them at same list price, he sells x after two successive discounts of 15% and 10% and still gains Rs. 418. If he sells y at 20% discount, then what is his profit on article y?

Sol: CP of x and y = Rs. 500 each

Profit on x = Rs. 418 $\Rightarrow$ SP of x = CP + P = 500 + 418 = Rs. 918

On article x, two successive discounts are 15% and 10%

$$SP = MP \times \frac{100 - D_1\%}{100} \times \frac{100 - D_2\%}{100} \qquad \Rightarrow \qquad 918 = MP \times \frac{100 - 15}{100} \times \frac{100 - 10}{100}$$

$\therefore$ MP of an article $= \dfrac{918 \times 100 \times 100}{(100 - 15)\,(100 - 10)} = $ Rs. 1200.

On article y, discount = 20%

$$SP \text{ of } y = MP \times \frac{100 - D\%}{100} = 1200 \times \frac{100 - 20}{100} = \text{Rs. } 960$$

$\therefore$ Profit on article y = SP − CP = 960 − 500 = Rs. 460.

Q - 32   **The amount of loss incurred on selling an article at Rs. 480 is equal to the amount of profit made on selling the same article at 20% profit. What is the profit percentage, if $6\frac{2}{3}\%$ discount is allowed on the marked price of Rs. 900?**

Sol: Consider, CP of an article = Rs. x

According to question,   Loss amount = profit amount

$$CP - SP_1 = SP_2 - CP \quad \Rightarrow \quad x - 480 = \frac{120}{100}x - x$$

$$x - 480 = \frac{20}{100}x \qquad \Rightarrow \qquad \frac{80x}{100} = 480 \qquad \Rightarrow \qquad x = \text{Rs. } 600$$

$\therefore$ Cost price of an article is Rs. 600

Given that, MP = Rs. 900 and discount $= 6\frac{2}{3}\%$

$$SP = MP \times \frac{100 - D\%}{100} = 900 \times \frac{100 - 6\frac{2}{3}}{100} = 9\left(100 - 6\frac{2}{3}\right) = \text{Rs. } 840$$

$\therefore$ Profit percentage $= \dfrac{SP - CP}{CP} \times 100\% = \dfrac{840 - 600}{600} \times 100\% = 40\%$.

Q - 33   **A merchant purchases an article marked at Rs. 32000 with 25% and 10% off. He spends Rs. 2400 on repairs and sells it for Rs. 32000. What is his gain (or) loss percentage?**

Sol: Merchant purchases an article after two successive discounts 25% and 10%

Cost price of an article $= 32000 \times \dfrac{100 - 25}{100} \times \dfrac{100 - 10}{100} = $ Rs. 21600

Expenses on repairs = Rs. 2400

Total CP of an article = 21600 + 2400 = Rs. 24000 and SP = Rs. 32000

$\therefore$ Profit percentage $= \dfrac{SP - CP}{CP} \times 100\% = \dfrac{32000 - 24000}{24000} \times 100\% = 33\frac{1}{3}\%$.

Q - 34 **Anitha marks up an article by 40% and sells it at a discount of 10% to Sunitha. Sunitha marks up the price of the article to a certain amount which happens to be 50% more than Anitha's cost price. What is the maximum discount Sunitha can offer without going into loss?**

Sol: Consider, Anitha CP of an article = 100

MP of an article = 140 $\because$ MP – 40% above CP

Anitha SP = Sunitha CP = $140 \times \dfrac{100 - 10}{100} = 126$ $\because$ Discount = 10%

Sunitha MP = 150% of Anitha CP = 150

CP of Sunitha = SP of Sunitha = 126 $\because$ No profit, No loss

$SP = MP \times \dfrac{100 - D\%}{100}$ $\Rightarrow$ $126 = 150 \times \dfrac{100 - D\%}{100}$ $\Rightarrow$ D = 16%

$\therefore$ Maximum discount Sunitha can offer is 16%.

Q - 35 **Karuna bought an article at Rs. 300 and marked it at Rs. 500. She offered a discount and then sold it, her profit (or) loss percentage and discount percentage are in the ratio of 5 : 3. Find her profit (or) loss percentage.**

Sol: According to question, Profit (or) Loss % : Discount % = 5 : 3

Case I	Case II
$\dfrac{Profit\ \%}{Discount\ \%} = \dfrac{5}{3}$	$\dfrac{Loss\ \%}{Discount\ \%} = \dfrac{5}{3}$
$\dfrac{\frac{SP - CP}{CP} \times 100\%}{\frac{MP - SP}{MP} \times 100\%} = \dfrac{5}{3}$	$\dfrac{\frac{CP - SP}{CP} \times 100\%}{\frac{MP - SP}{MP} \times 100\%} = \dfrac{5}{3}$
$\dfrac{(SP - 300)500}{300(500 - SP)} = \dfrac{5}{3}$	$300 - SP = 500 - SP$
	Solution is not possible if it is loss%
$SP - 300 = 500 - SP \Rightarrow SP = 400$	

$\therefore$ Profit percentage $= \dfrac{400 - 300}{300} \times 100\% = 33\frac{1}{3}\%.$

ASSESSMENT TEST

1. If an article is sold for Rs. 615 after giving a discount of 18%, then find the marked price of the article.

2. The cost price of an article is Rs. 7250. After allowing a discount of 26%, a gain of 11% was made. Find the marked price of an article.

3. While selling an article, a shopkeeper allows 35% discount on marked price and there is a loss of $15\frac{1}{2}\%$. What is the profit percent, if it is sold at marked price?

4. If a dealer wants to earn 25% profit on an item after giving a discount of 15% to the customer, by what percent should he increase his marked price to arrive at the label price?

5. If a discount of 16% is given on marked price of an article, the shopkeeper gets a profit of 40%. Find his profit percent, if he offers a discount of 28% on the same article.

6. The marked price of a watch is Rs. 4550. A shopkeeper allows a discount of 20% and still gets a profit of 30%. Find the cost price of watch.

7. A boy purchased an article with a discount of 30% on its marked price. He sold it at a price 25% more than the price at which he bought it. How much percent is the new selling price to its marked price?

8. Three successive discounts of $x\%$ each is equal to a single discount of Rs. 7569. If the selling price after all the discounts was Rs. 28431, then find the value of 'x'.

9. An article is marked 56% above its cost price and sold at a profit of 32.6% after a certain discount. Find the new profit percent, if the discount percentage is doubled.

10. While selling a TV set, a shopkeeper gives a discount of 11%. If he gives a discount of 15%, he earns Rs. 960 less as profit. Find the marked price of TV set.

11. After allowing a discount of 18%, an article was sold for Rs. 7380. If no discount is offered on the article, there would be a profit of 60%, then find the cost price of the article.

12. At first a laptop was being offered at a discount of 30%. The dealer reduced this discounted price further by 16% because the customer bargained. If the selling price of a laptop is Rs. 26460, then what is its marked price?

13. If a company sells a car with a marked price of Rs. 348000 and gives a discount of 6% on Rs. 232000 and 3.5% on the remaining amount, then find the amount charged by the company for the car.

14. Ramu bought a cooler with 15% discount on marked price. Had he bought it with 22% discount, he would have saved Rs. 910. At what price did he buy the cooler?

15. Agarwal marked an article at 60% above its cost price and then sold it after a discount of 24%. He made a profit of Rs. 108, then find his cost price.

16. A shopkeeper sells a cricket bat, with a marked price of Rs. 500, at a discount of 10% and gives a ball costing Rs. 30 free with each cricket bat. Even then he makes a profit of 25%. Find the cost price of each cricket bat.

17. Two shopkeepers sell machines at the same list price. The first one allows two successive discounts of 24% and 16% and the second one allows 17% and 23%. Which discount series is more advantage to the shopkeeper?

18. A shopkeeper marked the price of an article $\frac{5}{8}$th of its cost price. If he sold it at a price 40% more than marked price, then what will be the percentage of profit (or) loss?

19. What is the maximum discount percentage that a shopkeeper can offer on his marked price, so that he ends up selling at no profit and no loss, if he initially marked his goods up at 40%?

20. If on a marked price, the difference of selling prices with a discount of 40% and two successive discounts of 24% and 16% is Rs. 192. What is the marked price?

21. An illiterate shopkeeper marks his goods at 60% above the cost price and thinking that he will get a profit of 25%, after offering a discount of 30% on the marked price. What is his actual profit percentage?

22. Sanjay bought an article with 20% discount on marked price. He sold it with 10% profit on marked price. What was his profit percent on the price he bought?

23. By selling an article for Rs. 52, a shopkeeper gains 30%. During a clearance sale, the shopkeeper allows a discount of $15\frac{5}{13}$%, on previous selling price. What is his gain percent during sale season?

24. A man bought an article marked at Rs. 5000 with a discount of 18% offered on the marked price. What additional discount must be offered to that person to bring the net price to Rs. 3608?

25. The marked price of a mobile phone is Rs. 15000. It is to be sold at Rs. 10560 after giving two successive discounts. If the first discount is 20%, then find the second discount.

26. A shopkeeper earns a profit of 8% on selling an article at 24% discount on the marked price. Find the ratio of the cost price and marked price of the article.

27. Venkat purchased an article after two successive discounts of 12% and 15%. Uday purchased the same article after three successive discounts of 15%, 10% and 5% respectively. If the marked price is Rs. 8000 for each of them, then find the difference between the prices, at which Venkat and Uday purchased the articles.

28. An article was sold after allowing three successive discounts of 5%, 20% and 25% respectively. If a single discount of 48% was allowed, the seller would have got Rs. 150 less. Find the marked price of the article.

29. A shopkeeper bought 80 calculators at Rs. 130 each. He spent Rs. 2600 on packing and transportation. He marked each calculator at Rs. 200 and offered a discount of 15%. What is his profit percent?

30. A merchant buys two articles A and B for Rs. 1000 each. He marks them at same list price. He sells A after two successive discounts of 15% and 30% and still gains Rs. 666. If he sells B at 25% discount, then what is his profit on article B?

31. The amount of loss incurred on selling an article at Rs. 600 is equal to the amount of profit made on selling the same article at 25% profit. What is the profit percentage, if 15% discount is allowed on the marked price of Rs. 1200?

32. A shopkeeper allows a discount of 20% on the marked price of an article but charges a sales tax of 15% on the discounted price. If the customer pays Rs. 2392 as the price including the sales tax, then what is the marked price of an article?

33. A shopkeeper purchases an article marked at Rs. 40000 with 15% and 8% off. He spends Rs. 3720 on repairs and sells it for Rs. 40000. What is his gain (or) loss percent?

34. Seema marks up an article by 50% and sells it at a discount of 20% to Hema. Hema marks up the price of the article to a certain amount which happens to be 60% more than Seema's cost price. What is the maximum discount Hema can offer without going into loss?

35. Durga bought an article at Rs. 400 and marked it at Rs. 700. She offered a discount and then sold it, her profit (or) loss percentage and discount percentage are in the ratio of 7 : 4. Find her profit (or) loss percentage.

KEY

1. Rs. 750
2. Rs. 10875
3. 30%
4. $47\frac{1}{17}\%$
5. 20%
6. Rs. 2800
7. 12.5%
8. 10%
9. 9.2%
10. Rs. 24000
11. Rs. 5625
12. Rs. 42000
13. Rs. 330020
14. Rs. 11050
15. Rs. 500
16. Rs. 336
17. Second discount series
18. 12.5% Loss
19. $28\frac{4}{7}\%$
20. Rs. 5000
21. 12%
22. 37.5%
23. 10%
24. 12%
25. 12%
26. 19 : 27
27. Rs. 170
28. Rs. 3000
29. $4\frac{8}{13}\%$
30. Rs. 1100
31. 27.5%

32. Rs. 2600

33. $14\frac{2}{7}$% Gain

34. 25%

35. 37.5% Profit

PARTNERSHIPS

✓ In general partnerships are in the field of business. The persons who invests money in the business are called as "**Partners**".

❖ Basically there are two types of partners.

  1. Active partner   2. Sleeping partner

<u>Active partner:</u>

The person who invests the money in the business and also works in that business (or) manages the business are called as "Active partner".

<u>Sleeping partner:</u>

The person who invests the money in the business but doesn't works in that business is called as "Sleeping partner".

✓ In the business, profits are divided among the partners based on investment (capital) and time period.

> Profit (P) = Investment (I) × Time Period (T)

✓ As active partner is works in the business, so that he will get some extra amount (called as salary) which is from the profit and the remaining profit is divided between active partner and sleeping partner based on their profit ratio.

Profit ratio = $P_1 : P_2 : P_3 = I_1T_1 : I_2T_2 : I_3T_3$

Where, P_1, P_2, P_3 are Profits

I_1, I_2, I_3 are Investments

T_1, T_2, T_3 are Time periods

Investment ratio = $I_1 : I_2 : I_3 = \dfrac{P_1}{T_1} : \dfrac{P_2}{T_2} : \dfrac{P_3}{T_3}$

Time period ratio = $T_1 : T_2 : T_3 = \dfrac{P_1}{I_1} : \dfrac{P_2}{I_2} : \dfrac{P_3}{I_3}$

> **Note:** If time periods of all the partners are equal, then
>
> profit ratio is equal to investment ratio.

SOLVED EXAMPLES

Q-1 **A, B and C started a business with Rs. 10000, Rs. 15000 and Rs. 2000 respectively for a period of 2 years. Then find the ratio of their profits at the end of 2 years.**

Sol: Profit ratio $P_1 : P_2 : P_3 = I_1T_1 : I_2T_2 : I_3T_3$

Profit ratio of A, B and C = $10000 \times 2 : 15000 \times 2 : 20000 \times 2 = 2 : 3 : 4$.

Q-2 **There are three partners x, y and z in a business x invests Rs. 12000 for 8 months, y invests Rs. 18000 for 6 months and z invests Rs. 15000 for 10 months. Find the ratio of their profits.**

Sol: Profit ratio $P_1 : P_2 : P_3 = I_1T_1 : I_2T_2 : I_3T_3$

Profit ratio of x, y and z = $12000 \times 8 : 18000 \times 6 : 15000 \times 10 = 16 : 18 : 15$.

Q-3 **A, B and C entered in to a business. A invested Rs. 24000 for 12 months, B invested Rs. 30000 for 9 months and C invested Rs. 36000 for 6 months. If at the end of the year there was a profit of Rs. 34400, then find the shares of A, B and C.**

Sol: Profit ratio $P_1 : P_2 : P_3 = I_1T_1 : I_2T_2 : I_3T_3$

Profit ratio of A, B and C = $24000 \times 12 : 30000 \times 9 : 36000 \times 6 = 16 : 15 : 12$

Total profit = 43 parts $\longrightarrow$ Rs. 34400 $\Rightarrow$ 1 part = Rs. 800

A's share = 16 parts = 16×800 = Rs. 12800.

B's share = 15 parts = 15×800 = Rs. 12000.

C's share = 12 parts = 12×800 = Rs. 9600.

Q-4 **Bannu starts a business with Rs. 63000 and after 3 months Prasad joins with Bannu as his partner. After a year the profits are divided in the ratio of 4 : 5. How much did Prasad contribute?**

Sol: Consider, Prasad's investment = Rs. x

Profit ratio $P_1 : P_2 = I_1T_1 : I_2T_2$

$4 : 5 = 63000 \times 12 : x \times 9$ ∵ Prasad joined after 3 months, his time period is 9 months

∴ Prasad's investment x = Rs. 10500.

Q-5 **P, Q and R invested capitals in the ratio of 2 : 4 : 5, the timing of their investment being in the ratio of 6 : 5 : 3. In what ratio would be their profits be distributed?**

Sol: Given that, Investment ratio $I_1 : I_2 : I_3 = 2 : 4 : 5$

Time period ratio $T_1 : T_2 : T_3 = 6 : 5 : 3$

Profit ratio $P_1 : P_2 : P_3 = I_1T_1 : I_2T_2 : I_3T_3$

∴ $P_1 : P_2 : P_3 = 2 \times 6 : 4 \times 5 : 5 \times 3 = 12 : 20 : 15$.

Q-6   Nikhil, Akhil and Shakil invested capitals in the ratio of 6 : 8 : 9. At the end of the business they receive the profits in the ratio 3 : 4 : 6. Find the ratio of time for which they invested their capitals.

Sol: Time period ratio = $T_1 : T_2 : T_3 = \frac{P_1}{I_1} : \frac{P_2}{I_2} : \frac{P_3}{I_3}$

Given that, Investments ratio = $I_1 : I_2 : I_3 = \frac{P_1}{T_1} : \frac{P_2}{T_2} : \frac{P_3}{T_3} = 6 : 8 : 9$

Profit ratio $P_1 : P_2 : P_3 = 3 : 4 : 6$

$\therefore T_1 : T_2 : T_3 = \frac{3}{6} : \frac{4}{8} : \frac{6}{9} = \frac{1}{2} : \frac{1}{2} : \frac{2}{3} = \frac{1}{2} \times 6 : \frac{1}{2} \times 6 : \frac{2}{3} \times 6 = 3 : 3 : 4.$

> **Note:** To convert any fraction ratio into normal ratio multiply every fraction with LCM.

Q-7   Ankith and Riya enter into a business with capitals in the ratio 8 : 9. At the end of 6 months, Ankith withdraws his capital. If they receive the profits in the ratio of 4 : 9. How long Riya's capital was used?

Sol: $\frac{Ankith's\ profit}{Riya's\ profit} = \frac{I_1 T_1}{I_2 T_2}$ $\Rightarrow$ $\frac{4}{9} = \frac{8 \times 6}{9 \times T_2}$ $\Rightarrow$ $T_2 = 12$ months

$\therefore$ Riya's capital was used for 12 months.

Q-8   A began a business with Rs. 42000 and B joined afterwards with Rs. 56000. When did B joined, if the profits at the end of the year were divided equally?

Sol: Given that, A's profit = B's profit

$I_1 T_1 = I_2 T_2$ $\Rightarrow$ $42000 \times 12 = 56000 \times T_2$ $\Rightarrow$ $T_2 = 9$ months

As, B will be in the business for 9 months. Therefore, B joined after 3 months.

Q-9   P started a business with an investment of Rs. 40000. After 6 months Q joined the business with an investment of Rs. 70000. After how many years, from the beginning will the profit shares of both the persons are equal?

Sol: Consider, total time period = T months

Given that, P's profit = Q's profit $\Rightarrow$ $I_1 T_1 = I_2 T_2$

$40000 \times T = 70000 \times (T - 6)$ $\Rightarrow$ $4T = 7T - 42$

$T = 14$ months $\Rightarrow$ $T = \frac{14}{12}$ years $= 1\frac{1}{6}$ years

$\therefore$ After $1\frac{1}{6}$ years from mthe beginning their profits are equal.

Q - 10 **Ameer and Sameer started a business. Ameer's investment was four times that of Sameer and the period of Sameer was thrice that of Ameer. If Sameer got Rs. 54000 as profit, then find the total profit.**

Sol: According to question,

			Ameer		Sameer

Investment ratio   = $I_1 : I_2$ ⟶   4   :   1

Time period ratio   = $T_1 : T_2$ ⟶   1   :   3

Profit ratio   = $P_1 : P_2 = 4 \times 1 : 1 \times 3 = 4 : 3$ ∵ $P_1 : P_2 = I_1T_1 : I_2T_2$

Sameer's profit = 3 parts ⟶ Rs. 54000

∴ Total profit = 7 parts ⟶ $? = \dfrac{54000 \times 7}{3} = $ Rs. 126000.

Q - 11 **Venu starts a business with Rs. 63000. After few months, Ram joined him with an investment of Rs. 72000. At the end of the year, the total profit was divided between them in the ratio of 3 : 2. After how many months did Ram joins with Venu?**

Sol: Consider, Ram joined after 'x' months, then he will be in the business for $(12 - x)$ months.

Profit ratio $3 : 2 = 63000 \times 12 : 72000 \times (12 - x)$ ∵ $P_1 : P_2 = I_1T_1 : I_2T_2$

$\dfrac{3}{2} = \dfrac{63000 \times 12}{72000(12 - x)}$ ⇒ $x = 5$ months

∴ Ram joined after 5 months in the business.

Q - 12 **Ankit and Rahul started a business by investing Rs. 25000 and Rs. 30000 respectively. After 9 months, Manish joined them with Rs. 35000. What is the share of Manish, if the total profit earned at the end of 3 years is Rs. 39000?**

Sol: Profit ratio $P_1 : P_2 : P_3 = I_1T_1 : I_2T_2 : I_3T_3$

$P_1 : P_2 : P_3 = 25000 \times 36 : 30000 \times 36 : 35000 \times 27 = 20 : 24 : 21$

Total profit = 65 parts ⟶   Rs. 39000   ⇒   1 part = Rs. 600

∴ Manish's share = 21 parts = 21 × 600 = Rs. 12600.

Q - 13 **P, Q and R invested a total capital of 70 lakhs in a business. P invested 10 lakhs more than R and 5 lakhs less than Q. What is the share of P in the total profit of 5.6 lakhs at the end of the year?**

Sol: Given that, total investment of P, Q and R = 70 lakhs

P = R + 10 lakhs = Q – 5 lakhs   ⇒   Q = P + 5 lakhs, R = P – 10 lakhs

P + Q + R = P + (P + 5 lakhs) + (P – 10 lakhs) = 70 lakhs

3P = 75 lakhs   ⇒   P = 25 lakhs

∴ Q = 25 lakhs + 5 lakhs = 30 lakhs, R = 25 lakhs – 10 lakhs = 15 lakhs.

If time periods are equal, then the profit ratio is equal to investment ratio.

$P_1 : P_2 : P_3 = 25$ lakhs : 30 lakhs : 15 lakhs $= 5 : 6 : 3$

Total profit = 14 parts $\longrightarrow$ 5.6 lakhs

$\therefore$ P's share = 5 parts $\longrightarrow$ $? = \dfrac{5.6 \ lakhs \times 5}{14} = 2$ lakhs.

Q - 14 A, B and C starts a business. Twice of A's capital, thrice of B's capital and four times of C's capital are equal. Find the share of B, out of total profit of Rs. 39000.

Sol: According to question,

$2A = 3B = 4C = k$

$A : B : C = \dfrac{K}{2} : \dfrac{K}{3} : \dfrac{K}{4} = \dfrac{1}{2} \times 12 : \dfrac{1}{3} \times 12 : \dfrac{1}{4} \times 12 = 6 : 4 : 3$

Total profit = 13 parts $\longrightarrow$ Rs. 39000

$\therefore$ B's share = 4 parts $\longrightarrow$ $? = \dfrac{39000 \times 4}{13} = $ Rs. 12000.

Q - 15 A started a business with Rs. 25000. After 4 months, B joined him with Rs. 45000. After some more months, C joined them with Rs. 75000. If A received Rs. 20000 out of the total profit of Rs. 74000, then after how many months did C joined from the start?

Sol: Consider, C joined after 'x' months in the business.

A   B   C

Profit ratio = $25000 \times 12 : 45000 \times 8 : 75000 \times (12 - x) = 20 : 24 : 60 - 5x$

Given that, A's share = Rs. 20000 out of Rs. 74000

A's share: $\dfrac{20}{20 + 24 + 60 - 5x} = \dfrac{20000}{74000}$ $\quad \Rightarrow \quad$ $x = 6$ months

$\therefore$ C joined after 6 months in the business.

Q - 16 Sachin and Ganguly start a business with Rs. 60000 each. At the end of 4 months Ganguly withdraws Rs. 10000 from his investment. What percent of the total profit should Ganguly receive at the end of the year?

Sol:

Sachin   Ganguly

Profit ratio = $60000 \times 12 : (60000 \times 4) + (50000 \times 8) = 9 : 8$

$\therefore$ Percentage of profit for Ganguly $= \dfrac{8}{9 + 8} \times 100\% = 47\dfrac{1}{17}\%$.

Q - 17   In a business, A invests $\frac{1}{5}$ of capital for $\frac{1}{4}$ of the time, B invests $\frac{1}{4}$ of capital for $\frac{1}{2}$ of the time and C invests, rest of the capital for the whole time. Out of a total profit of Rs. 11600, what is the share of C?

Sol: Consider total capital = x, total time = y

$$\qquad\quad \textbf{A} \qquad \textbf{B} \qquad \textbf{C}$$

Profit ratio $= \frac{x}{5} \times \frac{y}{4} : \frac{x}{4} \times \frac{y}{2} : (x - \frac{x}{5} - \frac{x}{4}) \times y = \frac{1}{5} : \frac{1}{2} : \frac{11}{5} = 2 : 5 : 22$

Total profit = 29 parts $\longrightarrow$ Rs. 11600

$\therefore$ C's share = 22 parts $\longrightarrow$ $? = \dfrac{11600 \times 22}{29} = $ Rs. 8800.

Q - 18   Rajesh and Srujan started a business by investing in the ratio of 5 : 7. Tarun joined them after 4 months investing an amount equal to that of Srujan's. At the end of the year, 18% profit was earned which was equal to Rs. 108000. What was the amount invested by Tarun?

Sol: Consider, total profit = 100%

Given that, 18% profit $\longrightarrow$ Rs. 108000

$\qquad\qquad\qquad$ 100% profit $\longrightarrow$ $? = \dfrac{108000 \times 100}{18} = $ Rs. 600000

Let the capital of Rajesh, Srujan and Tarun are $5x$, $7x$, and $7x$ respectively. Then,

$(5x \times 12) + (7x \times 12) + (7x \times 8) = 600000 \times 12$

$200x = 600000 \times 12$ $\Rightarrow$ $x = 36000$

$\therefore$ Tarun's investment $= 7x = 7 \times 36000 = $ Rs. 252000.

Q - 19   P, Q and R started a business with their investments in the ratio of 2 : 4 : 5. After 8 months, P invested twice amount more as before, Q invested one – fourth amount more as before, while R withdraws one – fifth of their investments. Find the ratio of their profits at the end of the year.

Sol: Consider P, Q and R investments are $2x$, $4x$ and $5x$ respectively.

$$\qquad\qquad\qquad \textbf{P} \qquad\qquad\qquad\qquad \textbf{Q} \qquad\qquad\qquad\qquad \textbf{R}$$

Profit ratio $= (2x \times 8) + (2x + 4x) \times 4 : (4x \times 8) + (4x + x) \times 4 : (5x \times 8) + (5x - x) \times 4$

$\qquad\qquad = 40x : 52x : 56x = 10 : 13 : 14.$

Q - 20   Four milkmen rented a pasture. A grazed 18 cows for 5 months, B grazed 15 cows for 8 months, C grazed 20 cows for 6 months and D grazed 24 cows for 4 months. If D's share of rent is Rs. 4000, then find the total rent of the pasture.

Sol: **A**   **B**   **C**   **D**

Rents ratio $= 18 \times 5 : 15 \times 8 : 20 \times 6 : 24 \times 4 = 15 : 20 : 20 : 16$

D's share of rent $= 16$ parts $\longrightarrow$ Rs. 4000

$\therefore$ Total rent $= 71$ parts $\longrightarrow$ $? = \dfrac{4000 \times 71}{16} = $ Rs. 17750.

Q - 21   **Balu, Sonu and Ramu started a business in partnership investing in the ratio of 4 : 5 : 7 respectively. At the end of the year, they earned a profit of Rs. 124800 which is 24% of their total investment. How much did Balu invest?**

Sol: Consider, total investment $= 100\%$

Given that, 24% of investment $\longrightarrow$ Rs. 124800

Total 100% of investment $\longrightarrow$ $? = \dfrac{124800 \times 100}{24} = $ Rs. 520000

Also, investment ratio of Balu, Sonu and Ramu $= 4 : 5 : 7$

Total investment $= 16$ parts $\longrightarrow$ Rs. 520000

$\therefore$ Balu's investment $= 4$ parts $\longrightarrow$ $? = \dfrac{520000 \times 4}{16} = $ Rs. 130000.

Q - 22   **A, B and C started a business with capitals in the ratio of 6 : 3 : 2. At the end of 12 years, they received the profits in the ratio of 3 : 5 : 6. If C stayed for 12 years, then how many years did A stayed in the business?**

Sol: Time period ratio $= \dfrac{P_1}{I_1} : \dfrac{P_2}{I_2} : \dfrac{P_3}{I_3}$ & Given that, $I_1 : I_2 : I_3 = 6 : 3 : 2$, $P_1 : P_2 : P_3 = 3 : 5 : 6$

Time period ratio $= \dfrac{3}{6} : \dfrac{5}{3} : \dfrac{6}{2} = \dfrac{3}{6} \times 6 : \dfrac{5}{3} \times 6 : \dfrac{6}{2} \times 6 = 3 : 10 : 18$

C's time period $= 18$ parts $\longrightarrow$ 12 years

$\therefore$ A's time period $= 3$ parts $\longrightarrow$ $? = \dfrac{12 \times 3}{18} = 2$ years.

Q - 23   **Girish started a business by investing Rs. 40000. He invested additional amount of Rs. 10000 every year. After two years his brother Harish joined him with an amount of Rs. 85000. Thereafter, Harish did not invest any additional amount. On completion of 4 years from the beginning they earned an amount of Rs. 195000. What will be Girish's share in the earnings?**

Sol: **Girish**   **Harish**

Profit ratio $= (40000 \times 1) + (50000 \times 1) + (60000 \times 1) + (70000 \times 1) : 85000 \times 2$

Profit ratio $= 220000 : 170000 = 22 : 17$

Total profit $= 39$ parts $\longrightarrow$ Rs. 195000

$\therefore$ Girish's share $= 22$ parts $\longrightarrow$ $? = \dfrac{195000 \times 22}{39} = $ Rs. 110000.

Q - 24 Vikram and Ajit invested some money in the ratio of 36 : 49 and their time periods are in the ratio of m : n. If the profits earned at the end of the year are in the ratio n : m, then find m : n.

Sol: Given that $I_1 : I_2 = 36 : 49$, $T_1 : T_2 = m : n$ and $P_1 : P_2 = n : m$

	Vikram	**Ajit**	

Profit ratio, n : m = 36 × m   :   49 × n   ∵ $P_1 : P_2 = I_1T_1 : I_2T_2$

$\dfrac{n}{m} = \dfrac{36m}{49n}$ ⇒ $\dfrac{m^2}{n^2} = \dfrac{49}{36}$ ⇒ $\dfrac{m}{n} = \dfrac{7}{6}$

∴ m : n = 7 : 6.

Q - 25 L invests Rs. 8000 more in a business than M but M has invested his capital for 8 months and L has invested his capital for 6 months. If the share of L is Rs. 50 more than that of M out of total profit of Rs. 2050, then find the capital invested by each of them.

Sol: Given that, total profit L + M = 2050 and L = M + 50

(M + 50) + M = 2050   ⇒   M = Rs. 1000

L's profit = 1000 + 50 = Rs. 1050

Consider, investment of M = x, then investment of L = x + 8000

$\dfrac{L's\ profit}{M's\ profit} = \dfrac{(x + 8000) \times 6}{x \times 8}$ ⇒ $\dfrac{1050}{1000} = \dfrac{(x + 8000) \times 6}{8x}$

$\dfrac{21}{20} = \dfrac{(x + 8000) \times 3}{4x}$ ⇒ $84x = 60x + 480000$ ⇒ $x = 20000$

Investment of M is Rs. 20000 & Investment of L is x + 8000 = 20000 + 8000 = Rs. 28000.

Q - 26 A and B hire a meadow for 12 months. A puts 280 cows for 5 months. How many can B put in for the remaining 7 months, if he pays $\dfrac{3}{4}$ as much again as A?

Sol: Rents ratio of A and B = $1 : 1 + \dfrac{3}{4} = 1 : \dfrac{7}{4} = 4 : 7$

Let, B puts 'x' cows

Rents ratio of A and B: $\dfrac{4}{7} = \dfrac{280 \times 5}{x \times 7}$ ⇒ $x = 350$

∴ B puts 350 cows for remaining 7 months.

Q - 27 | Varun, Arun and Karan enter into partnerships with shares are in the ratio of $\frac{7}{2} : \frac{4}{3} : \frac{6}{5}$. After 4 months, Varun increases his share by 50%. If the total profit at the end of the year is Rs. 21600, then what will be the share of Arun in the profit?

Sol: Given that investments ratio $= \frac{7}{2} : \frac{4}{3} : \frac{6}{5} = \frac{7}{2} \times 30 : \frac{4}{3} \times 30 : \frac{6}{5} \times 30 = 105 : 40 : 36$

Varun	**Arun**	**Karan**

Profit ratio $= (105 \times 4) + (105 \times \frac{150}{100}) \times 8 \quad : \quad 40 \times 12 \quad : \quad 36 \times 12$

Profit ratio $= 1680 : 480 : 432 = 35 : 10 : 9$

Total profit = 54 parts ⟶ Rs. 21600

∴ Arun's share = 10 parts ⟶ $? = \frac{21600 \times 10}{54} = $ Rs. 4000.

Q - 28 | Mohan and Santhosh invest in a business in the ratio 2 : 3. If 15% of the total profit goes to charity and Santhosh's share is Rs. 10200, then find the total profit.

Sol: Profit ratio = Investment ratio = 2 : 3 Total profit (100%)

Santhosh's share = 3 parts ⟶ Rs. 10200 Charity (15%) Mohan & Santhosh (85%)

Together share = 5 parts ⟶ $? = \frac{10200 \times 5}{3} = $ Rs. 17000

Together share of profit = 85% ⟶ Rs. 17000

∴ Total profit = 100% ⟶ $? = \frac{17000 \times 100}{85} = $ Rs. 20000.

Q - 29 | David and Warner enter into partnership with Rs. 12000 and Rs. 16000 respectively. After $\frac{1}{3}$ of time, David contributes an additional amount of Rs. 6000. Three months after the start Warner withdraws $\frac{1}{2}$ of his capital and also, then Watson joins the business with capital investment of Rs. 24000. Find the share of Watson, if the profit at the end of the year is Rs. 13068.

Sol:
David	**Warner**	**Watson**

Profit ratio $= (12000 \times 4) + (18000 \times 8) : (16000 \times 3) + (8000 \times 9) : 24000 \times 9$

Profit ratio $= 8 : 5 : 9$

Total Profit = 22 parts ⟶ Rs. 13068

∴ Watson's share = 9 parts ⟶ $? = \frac{13068 \times 9}{22} = $ Rs. 5346.

Q - 30 Deepak and Karthik starts a business. Deepak is an active partner and Karthik is a sleeping partner. Deepak invests Rs. 40000 and Karthik invests Rs. 50000. Deepak receives 20% of profit separately for managing the business and the remaining profit is divided in proportion of their respective investments. They earn a total profit of Rs. 90000, then how much money received by Deepak?

Sol: Working partner Deepak initial share = 20% of total profit = $\frac{20}{100} \times 90000$ = Rs. 18000

Remaining profit = 90000 – 18000 = Rs. 72000

Profit ratio of Deepak and Karthik = 40000 : 50000 = 4 : 5

Total remaining profit = 9 parts ⟶ Rs. 72000

Deepak's share = 4 parts ⟶ $? = \frac{72000 \times 4}{9}$ = Rs. 32000

∴ Deepak's total share of profit = 18000 + 32000 = Rs. 50000.

Q - 31 In a business, the total income of Vamsi and Krishna was in the ratio 4 : 5, at the end of the year. Vamsi was the working partner so that he will get Rs. 20000 per month as a salary. If Vamsi's salary is excluded the ratio of the shares of Vamsi and Krishna was 2 : 3, then what is Krishna's share of profit?

Sol: Let, Vamsi's income = 4x, Krishna's income = 5x

Vamsi's annual salary = 20000 × 12 = Rs. 240000

After exclusion of Vamsi's salary, income ratio = 2 : 3

$\frac{4x - 240000}{5x} = \frac{2}{3}$ ⟹ $12x - 720000 = 10x$ ⟹ $x = 3.6$ lakhs

∴ Krishna's share of profit = 5x = 5 × 3.6 lakhs = 18 lakhs.

Q - 32 A and B started a business with an investments of Rs. 20000 and Rs. 150000. From second month onwards, A increased his monthly investment by Rs. 6000, while B decreased his monthly investment by Rs. 6000. The total profit made by the business was Rs. 85000 for 2 years, then find the share of A.

Sol: According to question,

A's investment are 20000, 26000, 32000......, 158000

B's investments are 150000, 144000, 138000, ..., 12000

Total investments = Average × No. of observations

> Since, the gap between consecutive numbers is same, then
>
> Average $= \frac{First\ number\ +\ Last\ number}{2}$

Profit ratio of A and B = $\frac{20000 + 158000}{2} \times 24 : \frac{150000 + 12000}{2} \times 24 = 89 : 81$

Total profit = 170 parts $\longrightarrow$ Rs. 85000

∴ A's share = 89 parts $\longrightarrow$ $? = \frac{85000 \times 89}{170} = $ Rs. 44500.

Q - 33 | **A, B and C enter into business by investing in the ratio of 4 : 3 : 2. After one year, B invests another Rs. 320000 and after two years, C also invests another Rs. 320000. At the end of three years, profits are divided in the ratio of 12 : 17 : 10. Find the initial investment of B.**

Sol: Let the initial investments of A, B and C are $4x$, $3x$ and $2x$ respectively.

| A | B | C |

Profit ratio = $4x \times 36 : (3x \times 12) + (3x + 320000)24 : (2x \times 24) + (2x + 320000)12$

$12 : 17 : 10 = 144x : 108x + 7680000 : 72x + 3840000$

$\frac{A's\,profit}{B's\,profit} : \frac{12}{17} = \frac{144x}{108x + 7680000}$ $\Rightarrow$ $\frac{1}{17} = \frac{12x}{108x + 7680000}$

$204x = 108x + 7680000$ $\Rightarrow$ $x = $ Rs. 80000

∴ Initial investment of B = $3x = 3 \times 80000 = $ Rs. 240000.

Q - 34 | **A invests Rs. 400 at the end of every month, whereas B withdraws Rs. 400 at the end of every month. If their initial investments are Rs. 400 and Rs. 4800, then what is the ratio of their profits at the end of the year?**

Sol: According to question,

A's investments are 400, 800,…, 4400, 4800

B's investments are 4800, 4400,… , 800, 400

Total investments = Average × No. of observations

> Since, the gap between consecutive numbers is same, then
>
> Average = $\frac{First\ number + Last\ number}{2}$

∴ Profit ratio of A and B = $\frac{400 + 4800}{2} \times 12 : \frac{4800 + 400}{2} \times 12 = 1 : 1.$

Shortcut:

According to question,

A investments are 400, 800,…, 4400, 4800

B investments are 4800, 4400,… , 800, 400

From the above values, it is clear that A investments and B investments are equal.

∴ Profit ratio of A and B = 1 : 1.

Q - 35 **Ravi and Karan starts a business with an investments of Rs. 75000 and Rs. 90000 respectively. Abhinay joins them after 'n' months with an investment of Rs. 120000 and Karan leaves 'n' months before the end of the year. If they share the year end profit in the ratio of 15 : 12 : 16, then find the value of 'n'.**

Sol: **Ravi   Karan   Abhinay**

Profit ratio = $75000 \times 12 : 90000 \times (12 - n) : 120000 \times (12 - n) = 15 : 12 : 16$

$\dfrac{Ravi's\,profit}{Karan's\,profit} : \dfrac{15}{12} = \dfrac{75000 \times 12}{90000 \times (12-n)}$ $\Rightarrow$ $12 - n = 8$ $\Rightarrow$ $n = 4.$

Q - 36 **Sneha and Neha started a business with Rs. 24000 and Rs. 36000 respectively. At the end of the year, they gave Rs. 3840 to charity which is 15% of the total profit. Find the difference in the profit shares of both out of the remaining profit.**

Sol: Given that,   Total profit (100%)

To charity = 15% $\longrightarrow$ Rs. 3840   Charity (15%)   Sneha & Neha (85%)

Together share = 85% $\longrightarrow$ $? = \dfrac{3840 \times 85}{15} = $ Rs. 21760

Profit ratio of Sneha and Neha = $24000 \times 12 : 36000 \times 12 = 2 : 3$

Total together share = 5 parts $\longrightarrow$ Rs. 21760

$\therefore$ Difference of shares = 1 part $\longrightarrow$ $? = \dfrac{21760 \times 1}{5} = $ Rs. 4352.

Q - 37 **Mounika and Vahini together starts a business. Mounika invests Rs. 3000 at the beginning of every quarter and Vahini withdraws Rs. 2000 at the end of every quarter. If their initial investments were Rs. 2000 and Rs. 8000 respectively and they got a profit of Rs. 46000 at the end of the year, then find their respective shares.**

Sol: Profit ratio of Mounika and Vahini

$= (2000 \times 3) + (5000 \times 3) + (8000 \times 3) + (11000 \times 3) :$

$(8000 \times 3) + (6000 \times 3) + (4000 \times 3) + (2000 \times 3)$

Profit ratio = $78000 : 60000 = 13 : 10$

Total profit = 23 parts = Rs. 46000 $\Rightarrow$ 1 part = Rs. 2000

Mounika's share = 13 parts = $13 \times 2000 = $ Rs. 26000.

Vahini's share = 10 parts = $10 \times 2000 = $ Rs. 20000.

Q - 38 **A, B and C jointly starts a business. It was agreed that A would invest Rs. 80000 for 15 months, B invests Rs. 70000 for 21 months and C invests Rs. 65000 for 18 months. B is an active partner, so that he will receive 20% of total profit. If the total profit earned is Rs. 80000, then what is the share of B?**

Sol: B's working share = 20% of total profit = $\frac{20}{100} \times 80000$ = Rs. 16000

Remaining profit = 80000 − 16000 = Rs. 64000

Remaining profit is shared among the partners based on their profit ratio

Profit ratio of A, B and C = 80000 × 15 : 70000 × 21 : 65000 × 18 = 40 : 49 : 39

Total remaining profit = 128 parts ⟶ Rs. 64000

B's share = 49 parts ⟶ $? = \frac{64000 \times 49}{128}$ = Rs. 24500

∴ B's total profit share = 16000 + 24500 = Rs. 40500.

Q - 39 **Laxmi, Swathi and Priya invested amounts in the ratio of 4 : 6 : 7 in a business. After 4 months Laxmi invested additional 50% of her initial investment, Swathi invested additional 25% of her initial investment and Priya withdraws $14\frac{2}{7}$% of her initial investment. At the end of the year, a total profit of Rs. 28000 was earned, then find the share of Swathi.**

Sol: Consider, investments of Laxmi, Swathi and Priya are $4x$, $6x$ and $7x$ respectively.

Laxmi	**Swathi**	**Priya**

Profit ratio = $(4x \times 4) + (4x + 2x)8 : (6x \times 4) + (6x + \frac{3x}{2})8 : (7x \times 4) + (7x - x)8$

Profit ratio = $64x : 84x : 76x$ = 16 : 21 : 19

Since, Laxmi = 50% of $4x$ more, Swathi = 25% of $6x$ more & Priya = $14\frac{2}{7}$% of $7x$ less

Total profit = 56 parts ⟶ Rs. 28000

∴ Swathi's share = 21 parts ⟶ $? = \frac{28000 \times 21}{56}$ = Rs. 10500.

Q - 40 **P invests Rs. 75000 for 8 months, Q invests Rs. 50000 for 12 months and R invests Rs. 60000 for 9 months. R receives $\frac{1}{5}$ of the profit at the end of the year as commission for managing the business. If at the end of the year total amount that R got was Rs. 23640 less than P and Q together, then find the shares of P, Q and R respectively.**

Sol: Consider, total profit = Rs. x

R's commission = $\frac{x}{5}$, Remaining profit = $x - \frac{x}{5} = \frac{4x}{5}$

Profit ratio of P, Q and R = $75000 \times 8 : 50000 \times 12 : 60000 \times 9 = 10 : 10 : 9$

P's share = $\frac{10}{29}(\frac{4x}{5}) = \frac{40x}{145}$, Q's share = $\frac{10}{29}(\frac{4x}{5}) = \frac{40x}{145}$, R's share = $\frac{9}{29}(\frac{4x}{5}) = \frac{36x}{145}$

R's total share = $\frac{x}{5} + \frac{36x}{145} = \frac{65x}{145}$

Given that, R's share = (P + Q)'s share − Rs. 23640

$\frac{65x}{145} = \frac{40x}{145} + \frac{40x}{145} - 23640$ ⇒ $\frac{15x}{145} = 23640$ ⇒ $x = 228520$

∴ P's share = Q's share = $\frac{40x}{145} = \frac{4 \times 228520}{145}$ = Rs. 63040

R's share = $\frac{65x}{145} = \frac{65 \times 228520}{145}$ = Rs. 102440.

Q - 41 **A and B started a business with Rs. 26000 and Rs. 34000 respectively. They agreed to share the profit in the ratio of their investments. C joins the partnership with the condition that A, B and C will share profit equally and pays Rs. 180000 as premium for this, to be shared between A and B. In what ratio A and B divided this amount?**

Sol: A's total investment = 26000×12 = Rs. 312000

B's total investment = 34000×12 = Rs. 408000

B's investment is Rs. 96000 more than A, therefore A must receives Rs. 96000 more premium than B from C.

Consider B's premium = Rs. x, then A's premium = $x + 96000$

Given that, $(x + 96000) + x = 180000$ ⇒ x = Rs. 42000

B receives x = Rs. 42000

A receives $x + 96000 = 42000 + 96000$ = Rs. 138000

∴ Required ratio = $138000 : 42000 = 23 : 7$.

Alternate method:

Profits equal means investments of A and B must be equal.

	A	B
Investments of A and B = 26000×12, 34000×12 =	312000	408000
C's premium to A and B =	138000	42000
	450000	450000

∴ Required ratio = $138000 : 42000 = 23 : 7$.

Q - 42 In a business, the investments of Murali and Gopal are in the ratio of 4 : 3. At the end of the year, the profit was Rs. 140000. Both of them receives equal amount of salaries at the end of the year. The ratio of net profits of Murali and Gopal after excluding salaries are equal to 5 : 3. Find the salary received by each person.

Sol: Profit ratio = Investment ratio = 4 : 3

Total profit = 7 parts → Rs. 140000 ⟹ 1 part = Rs. 20000

Murali's share = 4 parts = 4 × 20000 = Rs. 80000

Gopal's share = 3 parts = 3 × 20000 = Rs. 60000

Consider their salary = Rs. x, then after excluding salary $\dfrac{80000 - x}{60000 - x} = \dfrac{5}{3}$

$240000 - 3x = 300000 - 5x$ ⟹ $x = $ Rs. 30000

∴ Salary received by each person = Rs. 30000.

Q - 43 Vasu started a business by investing Rs. 25000 in 2016. In 2017 he invested an additional amount of Rs. 10000 and Lokesh joined him with an investment of Rs. 35000. In 2018, Vasu invested another additional amount of Rs. 10000 and Prem joined them with an investment of Rs. 35000. What will be the share of Lokesh in the profit of Rs. 150000 at the end of 3 years from the start of the business in 2016?

Sol:

	Vasu	**Lokesh**	**Prem**

Profit ratio = (25000 × 1) + (35000 × 1) + (45000 × 1) : (35000 × 2) : (35000 × 1)

Profit ratio = 105000 : 70000 : 35000 = 3 : 2 : 1

Total profit = 6 parts ⟶ Rs. 150000

∴ Lokesh's share = 2 parts ⟶ $? = \dfrac{150000 \times 2}{6} = $ Rs. 50000.

Q - 44 Two partners invest Rs. 105000 and Rs. 135000 respectively in a business and agree that 70% of the profit is divided equally between them and the remaining profit is to be treated as interest on capital. If one partner gets Rs. 450 more than the other, then find the total profit made in the business.

Sol: Profit ratio of two partners = 105000 : 135000 = 7 : 9

Consider, total profit = Rs. x

Given that, 70% of total profit is divided equally

Remaining profit = 30% of total profit = $\dfrac{30x}{100}$

First partner share = $\dfrac{7}{16} \times \dfrac{30x}{100}$, Second partner share = $\dfrac{9}{16} \times \dfrac{30x}{100}$

Also, difference of shares: $= \dfrac{2}{16} \times \dfrac{30x}{100} = $ Rs. 450 ⟹ $x = 12000$

∴ Total profit $x = $ Rs. 12000.

Q - 45 P, Q and R are three partners in a business. P receives $\frac{3}{7}$ of the profit and Q and R share the remaining profit equally. P's income is increased by Rs. 945, when the profit rises from 12% to 15%. Find the capitals invested by P, Q and R.

Sol: According to question,

P's increased profit = 3% $\longrightarrow$ Rs. 945

P's total share = 100% $\longrightarrow$ $? = \frac{945 \times 100}{3} = $ Rs. 31500

P's share: $\frac{3}{7}$ $\longrightarrow$ Rs. 31500

(Q + R)'s share: $\frac{4}{7}$ $\longrightarrow$ $? = 31500 \times \frac{4}{7} \times \frac{7}{3} = $ Rs. 42000

Q's share = R's share = $\frac{42000}{2} = $ Rs. 21000 $\because$ Q = R

∴ P's share = Rs. 31500, Q's share = R's share = Rs. 21000.

Q - 46 A, B and C enter into partnership by making investments in the ratio 3 : 5 : 7. After a year, C invests another Rs. 337600, while A withdraws Rs. 45600. The ratio of investments then changes to 24 : 59 : 167. How much did A invest initially?

Sol: Consider, initial investments of A, B and C are $3x$, $5x$ and $7x$ respectively.

New investments ratio = $3x - 45600 : 5x : 7x - 337600 = 24 : 59 : 167$

$\frac{A}{B}: \frac{3x - 45600}{5x} = \frac{24}{59}$ $\Rightarrow$ $177x - 2690400 = 120x$

$57x = 2690400$ $\Rightarrow$ $x = 47200$

∴ Initial investment of A = $3x = 3 \times 47200 = $ Rs. 141600.

Q - 47 A, B and C invests 1.2 lakhs, 1.8 lakhs and 3 lakhs respectively in a business. A earns Rs. 2000 per month as salary and B is entitled to 20% commission of profits after deducting the salary given to A. If the total profit earned is Rs. 128000 at the end of the year, then what is income of B?

Sol: A's annual salary = 2000 × 12 = Rs. 24000

Remaining profit = 128000 – 24000 = Rs. 104000

B's commission = 20% of remaining profit = $\frac{20}{100} \times 104000 = $ Rs. 20800

Remaining profit = 104000 – 20800 = Rs. 83200

Profit ratio of A, B and C = 1.2 lakhs : 1.8 lakhs : 3 lakhs = 2 : 3 : 5

Total profit share = 10 parts $\longrightarrow$ Rs. 83200

B's share = 3 parts $\longrightarrow$ $? = \frac{83200 \times 3}{10} = $ Rs. 24960

∴ B's total income = 20800 + 24960 = Rs. 45760.

Q - 48 Aadhya started a business with 3 lakhs and after 3 months Pragnya joined her with 5 lakhs. Aadhya received Rs. 44000 as her annual profit share which included a salary of 19% of the annual profit. Find the share of Pragnya in annual profit.

Sol: Profit ratio of Aadhya and Pragnya = 3 lakhs × 12 : 5 lakhs × 9 = 4 : 5

Consider, total profit = Rs. x

Aadhya's salary = 19% of total profit = $\frac{19x}{100}$

Remaining profit = $x - \frac{19x}{100} = \frac{81x}{100}$

Aadhya's share = $\frac{4}{9} \times \frac{81x}{100} = \frac{324x}{900}$, Pragnya's share = $\frac{5}{9} \times \frac{81x}{100} = \frac{405x}{900}$

Aadhya's total profit: $\frac{19x}{100} + \frac{324x}{900}$ = Rs. 44000 $\Rightarrow$ $\frac{495x}{900}$ = 44000 $\Rightarrow x$ = Rs. 80000

∴ Pragnya's share = $\frac{405x}{900} = \frac{405}{900} \times 80000$ = Rs. 36000.

Q - 49 A, B, C and D started a business with investments in the ratio 2 : 3 : 5 : 7. As A and B are working partners, they get equal salaries. The ratio of A's and B's total annual income is 6 : 7. If the total annual profit is Rs.100000, then find the salary of A.

Sol: Profit ratio = Investment ratio = 2 : 3 : 5 : 7

Consider, salary of A and B = Rs. x each, then remaining profit = 100000 – 2x

A's share = $\frac{2}{17}(100000 - 2x)$, B's share = $\frac{3}{17}(100000 - 2x)$

Ratio of annual income of A and B, including their salaries = 6 : 7

$\frac{\frac{2}{17}(100000 - 2x) + x}{\frac{3}{17}(100000 - 2x) + x} = \frac{6}{7}$ $\Rightarrow$ $\frac{200000 - 4x + 17x}{300000 - 6x + 17x} = \frac{6}{7}$

$1400000 + 91x = 1800000 + 66x$ $\Rightarrow$ $x = 16000$

∴ Salary of both A and B is Rs. 16000 each.

Q - 50 X and Y entered into partnership with Rs. 1400 and Rs. 1800 respectively. After 4 months A withdraws $\frac{3}{7}$ of his stock but after 5 months more he puts back $\frac{2}{3}$ of what he had withdrawn. The profits at the end of the year are Rs. 13282, then find the share of A.

Sol: X investment for first 4 months = 1400 × 4 = Rs. 5600

X investment for next 5 months = 800 × 5 = Rs. 4000 ∵ withdrawn $\frac{3}{7} \times 1400 = 600$ less

X investment for last 3 months = 1200 × 3 = Rs. 3600 ∵ Invests $\frac{2}{3} \times 600 = 400$ more

Profit ratio of X and Y = 5600 + 4000 + 3600 : 1800 × 12 = 13200 : 21600 = 11 : 18

Total profit = 29 parts $\longrightarrow$ Rs. 13282

∴ A's share = 11 parts $\longrightarrow$? = $\frac{13282 \times 11}{29}$ = Rs. 5038.

ASSESSMENT TEST

1. Three partners A, B and C together invested Rs. 684000 in a business. At the end of the year, A got Rs. 85000, B got Rs. 102000 and C got Rs. 136000 as profit. How much amount did C invest?

2. Harini and Nalini started a business. Harini's investment was thrice that of Nalini and the time period of Harini was $1\frac{1}{2}$ times that of Nalini. If Nalini's profit was Rs. 26000, then what was the total profit earned?

3. Navya invested 10% less than the investment of Bhavya and Bhavya invested 20% more than the investment of Sravya. If the total investment of all three members Rs. 98400, then what is the investment of Navya?

4. A, B and C enter into partnership in a business. A got $\frac{3}{7}$ of the profit, B and C share the remaining profit equally. If A got Rs. 2500 more than B, then find the total profit.

5. P and Q enter into partnership in a business. P invests Rs. 84000 for 12 months and Q remains in the business for 7 months. If Q receives $\frac{5}{11}$ of the profit, then what is the contribution of Q?

6. Kunal and Rahul started a business with initial investments in the ratio of 7 : 10 and their annual profits were in the ratio of 2 : 5. If Kunal invested the money for 8 months, then for how long Rahul's investment was used?

7. M, N and O enter into a business. M invests some amount at the beginning, N invests thrice the amount of M after 4 months and O invests 4 times the amount of M after 7 months. If the total annual profit is Rs. 49000, then find the share of N.

8. In a business, P invests $\frac{1}{6}$ of capital for $\frac{1}{5}$ of the time, Q invests $\frac{1}{5}$ of capital for $\frac{1}{4}$ of the time, R invests $\frac{1}{10}$ of capital for $\frac{1}{2}$ of the time and S invest rest of the capital for $\frac{1}{8}$ of the time. Find the share of Q, if the total profit earned was Rs. 102300.

9. X, Y and Z enter into a business with a capital in which X's investment is Rs. 25000. If out of a total profit of Rs. 10000, X gets Rs. 2000 and Y gets Rs. 5000, then find Y's capital.

10. A and B invests in the ratio of 7 : 8 respectively in a business. After 6 months C joins them with an investment of the capital equal to that of B and B leaves the business after 9 months. Find the ratio of the profits at the end of the year.

11. P, Q and R rented a pasture. P grazed 48 cows for 3 months, Q grazed 42 cows for 4 months and R grazed 30 cows for 6 months. If the total rent paid by them is Rs. 12300, then what is the rent paid by Q?

12. Raj and Tej started a business jointly. Raj's investment was four times that of Tej's, but Raj stayed in the business for 6 months and Tej stayed for entire year. If Raj's share of profit is Rs. 42600, then find the total profit.

13. L, M and N starts a business. Thrice of L's capital, four times of M's capital and five times of N's capital are equal. Find the share of L, out of total profit of Rs. 94000.

14. Hari started a business with Rs. 30000. After 5 months Sudhir joined him with Rs. 60000 and at the beginning of sixth month, Hari added additional Rs. 10000. Find the ratio of profit at the end of the year.

15. Abhinay started a business with Rs. 15000. After 8 months, Balu joined him with Rs. 20000. After some more months, Chandu joined them with Rs. 25000. If at the end of two years Balu received Rs. 56000, out of the total profit of Rs. 171500, then when did Chandu joined in the business?

16. A, B and C start a business with an investments in the ratio 2 : 3 : 5. They got a profit of Rs. 455000. If B and C leaves after 8 months and 6 months respectively, then what is A's share of the total profit?

17. Mishra starts a business with Rs. 135000. After few months, Nitish joined him with an investment of Rs. 156000. At the end of the year, the total profit was divided between them in the ratio of 15 : 13. How long Nitish will be in the business?

18. M starts a business with Rs. 80000. N joined him after 6 months with an amount of Rs. 90000 and after another 6 months P joined with an amount of Rs. 100000. If at the end of two years from the starting, M earns a profit of Rs. 48000, then what is the total profit?

19. In a business, A and C invested amounts in the ratio of 3 : 4 and the ratio between amounts invested by A and B was 5 : 7. If they earn a total profit of Rs. 350000, then what is the share of A?

20. Ajay, Sravan and Ravi starts a business with an investments of 4 lakhs, 5 lakhs and 6 lakhs respectively. After 8 months Sravan leaves the business. 5 months before the end of the year Ravi withdraws 2 lakhs, then find the ratio of their profits at the end of the year.

21. Vinod and Jadeja starts a business with capitals Rs. 320000 and Rs. 480000 respectively. Vinod manages the business and draws a salary of Rs. 10500 per month. They earn 5 lakhs at the end of the year. What is Vinod's total income at the end of the year?

22. P and Q started a business with investments of Rs. 18000 and Rs. 25000 respectively. After 7 months Q withdraws Rs. 5000 from his investment and R joined the business with an investment of Rs. 15000. If the profit at the end of the year is Rs. 113200, then what is the share of P?

23. Madhu and Nandu started a business with their investments in the ratio of 5 : 7. Omkar joins the business after 4 months with an investment double that of Madhu's. If the total annual profit is Rs. 35000, then find the share of Omkar.

24. Prasad and Ramana invested some money in the ratio of 121 : 169 and their time periods are in the ratio of a : b. If the profits earned at the end of the year are in the ratio of b : a , then find a : b.

25. Venky started a business with an investment of 2 lakhs. After 8 months Praneeth joined the business with an investment of 3 lakhs. After how many years, from the beginning will the profit shares of both the members are equal?

26. Naveen and Kiran invest in a business in the ratio of 5 : 7. If 20% of the total profit goes to charity and Naveen's share is Rs. 20500, then find the total profit.

27. P and Q entered into a partnership just 8 months ago, but P had started the business 12 months ago. The capital invested by P was Rs. 48400. Find the capital invested by Q at the time of joining the business, if the profits are distributed in the ratio 121 : 76.

28. A invests Rs. 6250 less in a business than B, but B has invested his capital for 7 months and A has invested his capital for 9 months. If the share of A is Rs. 39 more than B out of the total profit of Rs. 2769, then find the capital invested by each of them.

29. L and M hire a meadow for 15 months. L puts 480 oxen for 7 months. How many can M put in for the remaining 8 months, if he pays $\frac{1}{3}$ as much again as L?

30. X and Y entered into a partnership by investing Rs. 24000 and Rs. 18000 respectively. After 4 months, X withdraws Rs. 4000, while Y invested Rs. 4000 more. After two more months, Z joins the business with a capital of Rs. 32000. How much is the share of Y exceeds that of Z, out of a total profit of Rs. 34800 after one year?

31. P, Q and R enter into partnership. P invests Rs. 7000 for the entire year, Q invests Rs. 8000 at first and increasing to Rs. 10000 after 6 months, while R invests Rs. 10000 at first but withdraws Rs. 4000 at the end of 8 months. Find the share of Q, if the profit at the end of the year is Rs. 31450.

32. A is a working partner and B is a sleeping partner in a business. A invests Rs. 42000 and B invests Rs.54000. A receives $14\frac{2}{3}$% of the profit for managing the business and the rest is divided in proportion to their capitals. What does each partner earned, if the total profit is Rs. 12000?

33. Praveen, Sathish and Revanth starts a business with capitals are in the ratio of 6 : 5 : 3. After 8 months, Sathish invests another Rs. 72000 and after one year, Revanth invests another Rs. 84000. At the end of two years, profits are divided in the ratio of 6 : 13 : 10. Find the initial investment of Revanth.

34. Anu invests Rs. 550 at the end of every month, whereas Bhanu withdraws Rs. 600 at the end of every month. If their initial investments are Rs. 550 and Rs. 7200, then what is the ratio of their profits at the end of the year?

35. Kamal and Vimal starts a business with an investments of Rs. 70000 and Rs. 90000 respectively. Anil joins them after 'x' months with an investment of Rs. 80000 and Vimal leaves 'x' months before the end of two years. If they share the profits in the ratio of 28 : 27 : 24, then find the value of 'x'.

36. Rama and Suma started a business with Rs. 27000 and Rs. 45000 respectively. At the end of the year, they gave Rs. 20600 to charity which is 25% of the total profit. Find the difference in the profit shares of both out of the remaining profit.

37. Karuna and Aruna together starts a business. Karuna invests Rs. 1500 at the beginning of every half-year and Aruna withdraws Rs. 2500 at the end of every half – year. If their initial investments were Rs. 3000 and Rs. 10000 respectively and they got a profit of Rs. 138000 after two years, then find their respective shares.

38. The salary of active partner is equal to 25% of the annual profit remaining after his salary is paid. If his salary is Rs. 20000, then find the annual profit.

39. Ramesh, Suresh and Mahesh started a business with their investments in the ratio of 2 : 4 : 5. After 8 months, Ramesh invested the same amount as before and Suresh withdraws half of his investment and Mahesh withdraws one – fifth of his investment. Find the ratio of their profits at the end of the year.

40. Kumar and Varma started a business by investing in the ratio of 4 : 5. Rahim joined them after 6 months investing an amount equal to that of Varma's. At the end of the year, 25% profit was earned which was equal to Rs. 86250. What was the amount invested by Rahim?

41. Sanju and Manju starts a business. Sanju is an active partner and Manju is a sleeping partner. Sanju invests Rs. 25000 and Manju invests Rs. 35000. Sanju receives 25% of profit separately for managing the business and the remaining profit is divided in proportion of their investments. They earn a total profit of Rs. 112000, then how much money received by Sanju?

42. P and Q started a business with Rs. 15000 and Rs. 20000 respectively. They agreed to share the profit in the ratio of their investments. R joins the partnership with the condition that P, Q and R will share profit equally and pays Rs. 80000 as premium for this, to be shared between P and Q. In what ratio P and Q divided this amount?

43. In a business the total income of Dhamu and Teja was in the ratio of 2 : 3, at the end of the year. Dhamu was the working partner, so that he will get Rs. 8000 per month as a salary. If Dhamu's salary was excluded the ratio of the shares of Dhamu and Teja was 3 : 5. What is Teja's share of profit?

44. In a business, the investments of Madan and Avinash are in the ratio of 3 : 5. At the end of the year, the total profit was Rs. 80000. Both of them receives equal amount of salaries at the end of the year. The ratio of net profits of Madan and Avinash after excluding salaries are equal to 1 : 2. Find the salary received by each person.

45. A, B and C are three partners in a business. A receives $\frac{4}{9}$ of the profit and B and C share the remaining profit equally. A's income is increased by Rs. 1424, when the profit rises from 13% to 17%. Find the capitals invested by A, B and C.

46. Two partners invests Rs. 126000 and Rs.150000 respectively in a business and agree that 65% of the profit is divided equally between them and the remaining profit is to be treated as interest on capital. If one partner gets Rs. 840 more than the other, then find the total profit made in the business.

47. P, Q and R invest 3.5 lakhs, 2.8 lakhs and 4.2 lakhs respectively in a business. Q earns Rs. 2500 per month as a salary and P is entitled to 25% commission of profits after deducting the salary given to Q. If the total profit earned is Rs. 150000, then what is the share of P?

48. Akmal started a business with Rs. 70000 and after 5 months Yuvi joined him with Rs. 60000. Yuvi received Rs. 65000 as his annual profit share which included a salary of 28% of the annual profit. Find the share of Akmal in the annual profit.

49. A, B, C and D started a business with investments in the ratio of 4 : 3 : 2 : 6. As B and C are working partners they get equal salaries. The ratio of B's and C's total annual income is 8 : 7. If the total annual profit is Rs. 125000, then find the salary of B.

50. M and N starts a business, M invests double that of N. M withdraws $\frac{1}{4}$ of his stock at the end of 3 months, but at the end of 8 months he puts back $\frac{1}{3}$ of what he has taken out and at the end of the 7 months N withdraws $\frac{1}{5}$ of his stock. If M receives Rs. 8250 profit at the end of the year, what does N receives?

KEY

1. Rs. 288000
2. Rs. 143000
3. Rs. 32400
4. Rs. 17500
5. Rs. 120000
6. 14 months
7. Rs. 21000
8. Rs. 25200
9. Rs. 62500
10. 7 : 6 : 4
11. Rs. 4200
12. Rs. 63900
13. Rs. 40000
14. 43 : 42
15. After 12 months
16. Rs. 140000
17. 9 months
18. Rs. 129000
19. Rs. 125000
20. 24 : 20 : 31
21. Rs. 275600
22. Rs. 43200
23. Rs. 12500
24. 13 : 11
25. 2 years
26. Rs. 61500
27. Rs. 45600
28. A = Rs. 25000, B = Rs. 31250
29. 560 Oxen
30. Rs. 2800
31. Rs. 11475
32. A = Rs. 6240, B = Rs. 5760

33. Rs. 18000

34. 143 : 156

35. 6 months

36. Rs. 15450

37. K = Rs. 63000, A = Rs. 75000

38. Rs. 100000

39. 4 : 5 : 7

40. Rs. 150000

41. Rs. 63000

42. 7 : 1

43. 14.4 lakhs

44. Rs. 10000

45. A = Rs. 35600, B = Rs. 22250, C = Rs. 22250

46. Rs. 27600

47. Rs. 60000

48. Rs. 60000

49. Rs. 25000

50. Rs. 4500

MIXTURES AND ALLIGATIONS

MIXTURE: If two or more than two products are mixed in any ratio, the resultant product is called as mixture.

MEAN PRICE:

✓   The cost price of a unit quantity of the mixture is called as mean price.

✓   Mean Price is always between highest price and lowest price of the product.

ALLIGATION RULE:

       Alligation is the rule to find the ratio of quantities when two (or) more than two different varieties of products are mixed.

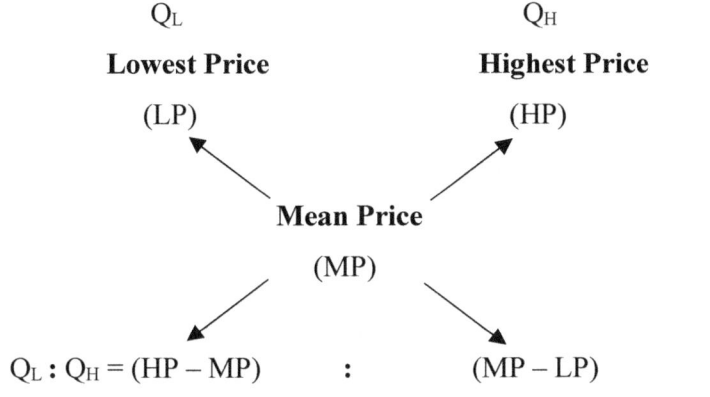

Q_L : Q_H = (HP – MP)   :   (MP – LP)

Where, Q_L = Quantity of lowest price & Q_H = Quantity of highest price

> **Note:** We can also apply this alligation rule, while solving the problems on averages, percentages, profit and loss, time and distance etc.

Some important points to be remember:

✓   If a vessel initially contains 'x' units of one liquid and 'y' units of liquid is taken out and is filled with 'y' units of another liquid, the same process is repeated 'n' times. Then,

$$\text{Final quantity of original liquid} = \text{Initial quantity} \times \left[1 - \frac{Replaced\ quantity}{Total\ quantity}\right]^n$$

$$\text{Final quantity of original liquid} = x\left[1 - \frac{y}{x}\right]^n \text{ units.}$$

✓   A jar has whisky and water in the ratio of a : b and second jar has whisky and water in the ratio c : d. If both the mixtures are emptied into a third jar, then the ratio of whisky and water in third jar is $\left(\frac{a}{a+b} + \frac{c}{c+d}\right) : \left(\frac{b}{a+b} + \frac{d}{c+d}\right)$.

SOLVED EXAMPLES

Q - 1   **In what ratio two qualities of wheat costing Rs. 65 per kg and Rs. 80 per kg be mixed in order to get a new mixture costing Rs. 72 per kg?**

Sol: According to alligation rule,

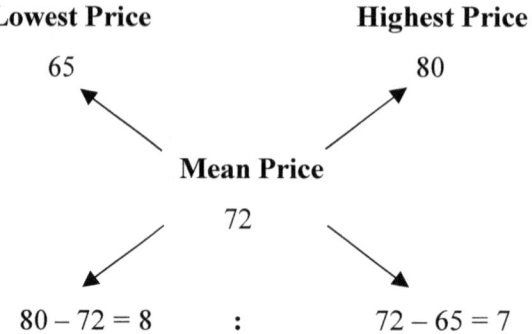

∴ Required ratio = 8 : 7.

Q - 2   **In what ratio two qualities of rice costing Rs. 42.5 per kg and Rs. 59.75 per kg be mixed in order to get a new mixture costing Rs. 51.5?**

Sol: According to alligation rule,

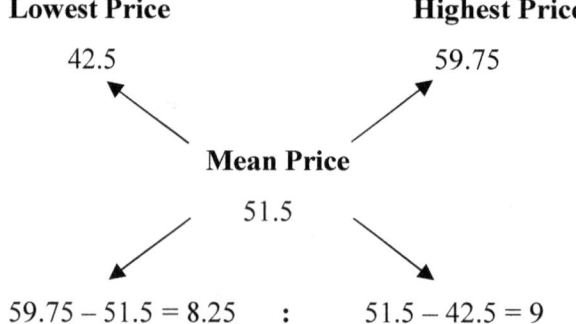

∴ Required ratio = 8.25 : 9 = 11 : 12.

Q - 3   **How many kilograms of sugar costing Rs. 36 per kg should be mixed with 26 kg of high quality sugar costing Rs. 54 per kg, so that the mixture worth is Rs. 48 per kg?**

Sol: Let us consider, low quality of sugar = 'x' kg

> Total cost = Quantity × Price

Total cost of mixture = Total cost of low quality + Total cost of high quality

$$(x + 26) \, 48 = x \times 36 + 26 \times 54$$

$$48x + 1248 = 36x + 1404 \quad \Rightarrow \quad 12x = 156 \quad \Rightarrow \quad x = 13$$

∴ Low quality of sugar = 13 kg.

Shortcut:

According to alligation rule,

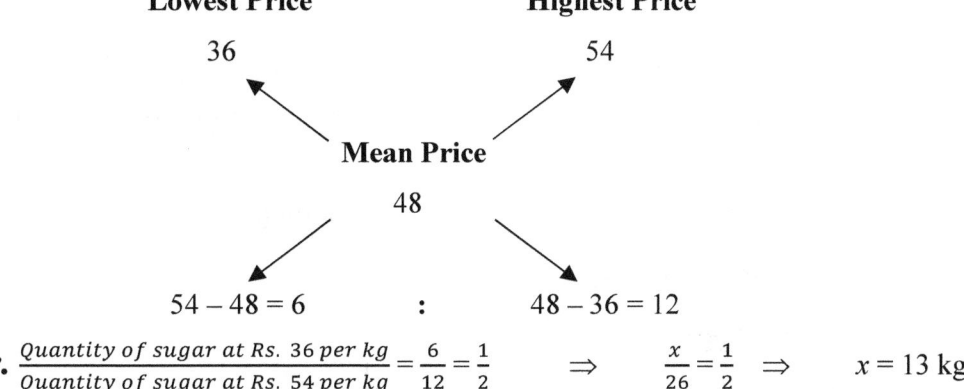

Lowest Price	Highest Price
36	54

Mean Price

48

54 − 48 = 6   :   48 − 36 = 12

$$\therefore \frac{Quantity\ of\ sugar\ at\ Rs.\ 36\ per\ kg}{Quantity\ of\ sugar\ at\ Rs.\ 54\ per\ kg} = \frac{6}{12} = \frac{1}{2} \quad \Rightarrow \quad \frac{x}{26} = \frac{1}{2} \quad \Rightarrow \quad x = 13 \text{ kg.}$$

Q - 4   **450 grams of sugar solution has 30% sugar in it. How much sugar should be added to make it 40% in the solution?**

Sol: Given that, total sugar solution = 450 grams

Sugar = 30% of 450 = $\frac{30}{100} \times 450 = 135$ grams

Consider 'x' grams of sugar is added to make it 40% in the solution.

Sugar concentration = $\frac{Quantity\ of\ sugar}{Total\ quantity} \times 100\%$

$40 = \frac{135 + x}{450 + x} \times 100 \qquad \Rightarrow \qquad \frac{2}{5} = \frac{135 + x}{450 + x}$

$900 + 2x = 675 + 5x \qquad \Rightarrow \qquad 3x = 225 \qquad \Rightarrow \qquad x = 75$ grams

∴ 75 grams of sugar is added to make it 40% in the solution.

Alternate method:

Total sugar solution (100%)

Sugar	Water
30%	70%
$\frac{30}{100} \times 450 = 135$	$\frac{70}{100} \times 450 = 315$

According to question, we are adding sugar to the solution, so the quantity of water is constant, even after adding sugar.

After adding sugar

Water quantity = 60% ⟶ 315 grams

Total quantity = 100% ⟶ ? = $\frac{315 \times 100}{60} = 525$ Sugar (40%)   Water (60%)

Total quantity increased by (525 - 450) = 75 grams

∴ 75 grams of sugar is added (since water constant).

Q - 5 **In a class there are 480 students. All the boys contributed Rs. 210 each and all the girls contributed Rs. 170 each and the average contribution of the class is Rs. 185. Find the number of girls.**

Sol: Consider, no. of girls = x, no. of boys = $480 - x$ $\because$ Total = 480

Total contribution of class = Contribution of boys + Contribution of girls

$185 \times 480 = 210 (480 - x) + 170x$

$8880 = 10080 - 21x + 17x$ $\Rightarrow$ $4x = 1200$ $\Rightarrow$ $x = 300$

$\therefore$ No. of girls are 300.

Alternate method:

According to alligation rule,

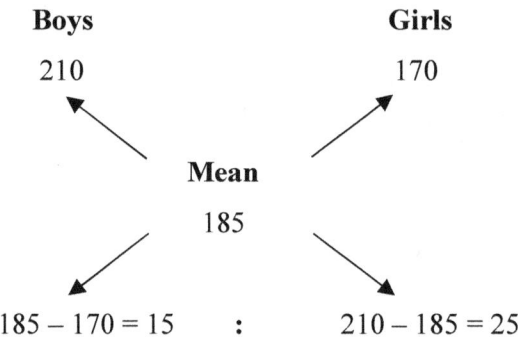

Boys   **Girls**

210   170

Mean

185

$185 - 170 = 15$: $210 - 185 = 25$

Boys : Girls = 15 : 25 = 3 : 5

Total strength = 8 parts $\longrightarrow$ 480

$\therefore$ No. of girls = 5 parts $\longrightarrow$ $? = \frac{480 \times 5}{8} = 300.$

Q - 6 **In an examination out of 320 students, 80% of the boys and 90% of the girls are passed. How many boys appeared in the exam, if the total pass percentage is 84%?**

Sol: Consider, no. of boys appeared = x, no. of girls = $320 - x$ $\because$ Total strength = 320

Total passed students = Total passed boys + Total passed girls

84% of 320 = 80% of x + 90% of $(320 - x)$

$\frac{84}{100} \times 320 = \frac{80}{100} \times x + \frac{90}{100} \times (320 - x)$

$2688 = 8x + 2880 - 9x$ $\Rightarrow$ $x = 192$

$\therefore$ No. of boys appeared in the exam is 192.

Alternate method:

According to alligation rule,

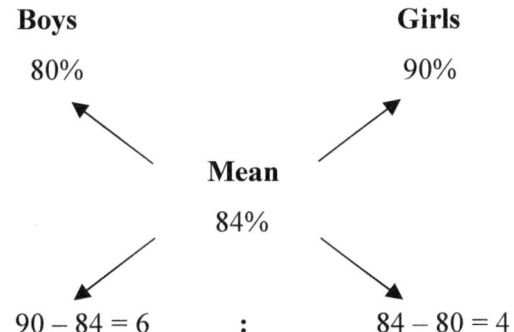

Boys   **Girls**

80%   90%

Mean

84%

90 − 84 = 6   :   84 − 80 = 4

Boys : Girls = 6 : 4 = 3 : 2

Total strength = 5 parts ⟶ 320

∴ No. of boys = 3 parts ⟶ $? = \frac{320 \times 3}{5} = 192$.

Q - 7   **A shopkeeper has 84 kg of wheat. Part of which he sell at 7% profit and the rest at 21% profit. He gains 15% on the whole. Find the quantity sold at 21% profit.**

Sol: According to alligation rule,

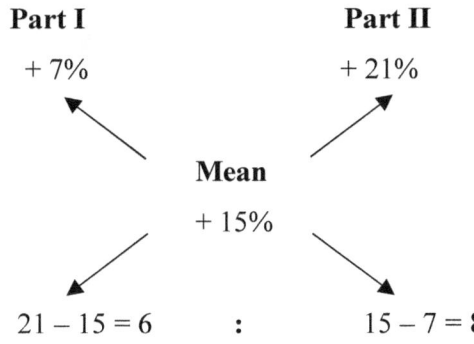

Part I   **Part II**

+ 7%   + 21%

Mean

+ 15%

> **Note:** Use +ve sign for profit and −ve sign for loss.

21 − 15 = 6   :   15 − 7 = 8

∴ Ratio of quantities = 6 : 8 = 3 : 4.

Given that, total quantity = 84 kg

        Total quantity = 7 parts ⟶ 84 kg

∴ Quantity sold at 21% profit = 4 parts ⟶ $? = \frac{84 \times 4}{7} = 48$ kg.

Q - 8   **A trader has 70 kg of sugar. Part of which he sell at 8% profit and the rest at 6% loss. He got 3% profit on the whole. Find the quantity sold at 8% profit and 6% loss.**

Sol: According to alligation rule,

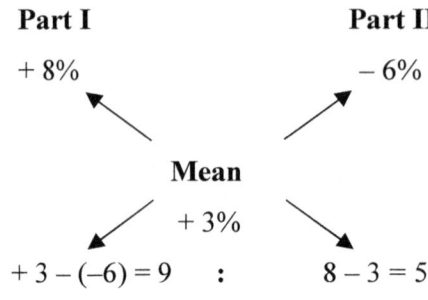

Part I   **Part II**

+ 8%   – 6%

Mean

+ 3%

$+ 3 - (-6) = 9$: $8 - 3 = 5$

Note: Use +ve sign for profit and –ve sign for loss.

Ratio of quantities = 9 : 5

Total quantity = 14 parts   →   70 kg   ⇒   1 part = 5 kg

∴ Quantity sold at 8% profit = 9 parts = 9 × 5 = 45 kg.

Quantity sold at 6% loss = 5 parts = 5 × 5 = 25 kg.

Q - 9   **How many liters of water is to be added to 90 liters of a mixture having milk and water in the ratio 4 : 1, to make it a mixture with milk concentration as 60%?**

Sol: Given that, milk : water = 4 : 1

Total mixture = 5 parts   →   90 L   ⇒   1 part = 18 L

Quantity of milk = 4 parts = 4 × 18 = 72 L

Quantity of water = 1 part = 1 × 18 = 18 L

Consider, 'x' liters of water is added.

According to question,

After adding water to the initial solution milk concentration is 60%.

So water concentration is 40%.

Water concentration $= \dfrac{Quantity\ of\ water}{Total\ quantity} \times 100\%$

$40 = \dfrac{18 + x}{90 + x} \times 100$ ⇒ $\dfrac{2}{5} = \dfrac{18 + x}{90 + x}$

$180 + 2x = 90 + 5x$ ⇒ $3x = 90$ ⇒ $x = 30$ L

∴ 30 L of water is added to 90 liters of mixture.

Shortcut:

Milk : Water = 4 : 1

Initial mixture = 5 parts → 90 L ⇒ 1 part = 18 L

Initial milk = 4 parts = 4 × 18 = 72 L & Initial water = 1 part = 1 × 18 = 18 L

According to question,

We are adding water, so quantity of milk is constant even after adding water.

Milk concentration = 60% ⟶ 72 L

Total concentration = 100% ⟶ $? = \frac{72 \times 100}{60} = 120$ L

Total quantity increased by (120 – 90) be 30 L

∴ 30 L of water is added (since, milk is constant).

Q - 10 How much water to be added to 60 liters of milk worth Rs. 7.8 per liter. So that the value of mixture may be Rs. 4.5 per liter?

Sol: Consider, 'x' L of water is added and water is available at free of cost.

According to alligation rule,

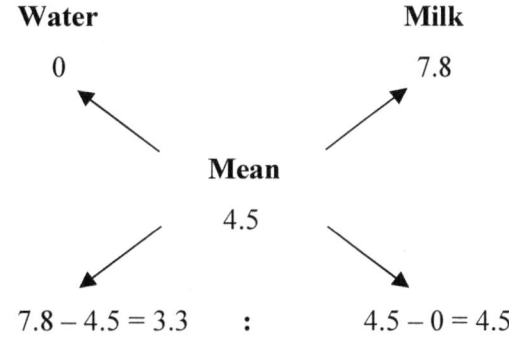

Water		**Milk**
0		7.8

Mean

4.5

7.8 – 4.5 = 3.3 : 4.5 – 0 = 4.5

Water **:** Milk = 3.3 : 4.5 = 11 : 15

Given that, total milk = 15 parts → 60 L ⇒ 1 part = 4 L

∴ Total water = 11 parts = 11 × 4 = 44 L.

Q - 11 912 mL of a mixture contains milk and water in the ratio 11 : 5. How much water is to be added to get a new mixture containing milk and water in the ratio 11 : 6?

Sol: Given that, milk: water = 11 : 5

Total mixture = 16 parts → 912 mL ⇒ 1 part = 57 mL

Milk = 11 parts = 11 × 57 = 627 mL & Water = 5 parts = 5 × 57 = 285 mL

After adding water, M **:** W = 11 : 6

Consider, 'x' mL of water is added

$\frac{M}{W} : \frac{627}{285 + x} = \frac{11}{6}$ ⇒ 627 × 6 = 11 (285 + x) ⇒ x = 57

∴ 57 mL of water is added to the initial mixture.

Shortcut:

Given that, M : W = 11 : 5

Total mixture = 16 parts   →   912 mL   ⇒   1 part = 57 mL

		Milk		**Water**	
Before adding water	→	11	:	5	⎤ 1 part increases
After adding water	→	11	:	6	⎦

∴ Amount of water added = 1 part = 57 mL.

Q - 12 **Lead and tin are in the ratio of 5 : 4 in 360 grams of an alloy. How many grams of tin is added to make the ratio as 4 : 5?**

Sol: Given that, Lead : Tin = 5 : 4

Total alloy = 9 parts   →   360 grams   ⇒   1 part = 40 grams

Lead = 5 parts = 5 × 40 = 200 grams, Tin = 4 parts = 4 × 40 = 160 grams

Consider, 'x' grams of Tin is added.

After adding Tin, L : T = 4 : 5

$\dfrac{\text{Lead}}{\text{Tin}} = \dfrac{200}{160 + x}$ ⇒ $\dfrac{4}{5} = \dfrac{200}{160 + x}$ ⇒ $x = 90$ grams

∴ 90 grams of Tin is added.

Alternate method:

Given that, L : T = 5 : 4

Total alloy = 9 parts   →   360   ⇒   1 part = 40 grams

Lead = 5 × 40 = 200 grams, Tin = 4 × 40 = 160 grams

According to question, we are adding Tin, so quantity of Lead is constant.

Lead = 4 parts   →   200 grams   ⇒   1 part = 50 grams

Tin = 5 parts = 5 × 50 = 250 grams

Here, Tin is increased by (250 – 160) 90 grams

∴ 90 grams of Tin is added.

Q - 13 **How many kilograms of rice costing Rs. 50 per kg must be mixed with 70 kg of rice costing Rs. 68 per kg, so that 20% gain is obtained by selling the mixture at Rs. 72 per kg?**

Sol: Consider, Cost price of mixture = 100%

SP of mixture = 120%   ⟶   Rs. 72   ∵ Gain = 20%

CP of mixture = 100%   ⟶ $? = \dfrac{72 \times 100}{120} = $ Rs. 60

According to alligation rule,

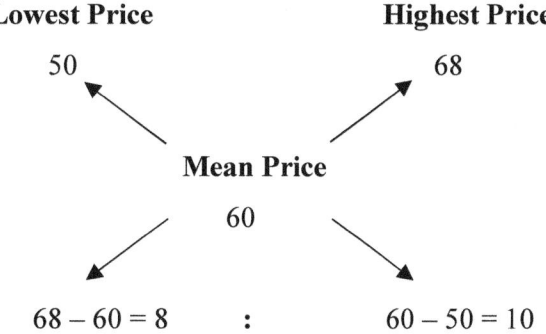

Ratio of quantities $Q_L : Q_H = 8 : 10 = 4 : 5$

$$\frac{Q_L}{Q_H} = \frac{4}{5} \qquad \Rightarrow \qquad \frac{Q_L}{70} = \frac{4}{5} \qquad \Rightarrow \qquad Q_L = 56 \text{ kg}$$

∴ Quantity of rice at Rs. 50 per kg is 56 kg.

Q - 14 **Tea powder at Rs. 144 per kg and Rs. 186 per kg are mixed with a third variety in the ratio 2 : 1 : 3. If the mixture worth is Rs. 165 per kg, then find the price of third variety per kg.**

Sol: Consider, price of third variety be Rs. x per kg.

Cost of 1st variety + 2nd variety + 3rd variety = Total mixture cost

$2 \times 144 + 1 \times 186 + 3 \times x = (2 + 1 + 3) \times 165 \qquad \Rightarrow \qquad x = \text{Rs. } 172$

∴ Price of third variety be Rs. 172 per kg.

Q - 15 **A mixture of 15 kg wheat flour costing Rs. 24 per kg and 10 kg of corn flour costing Rs. 16 per kg is sold for Rs. 24 per kg. What is the profit made by selling 70 kg of mixture?**

Sol: Consider, CP of mixture be Rs. x per kg.

According to alligation rule,

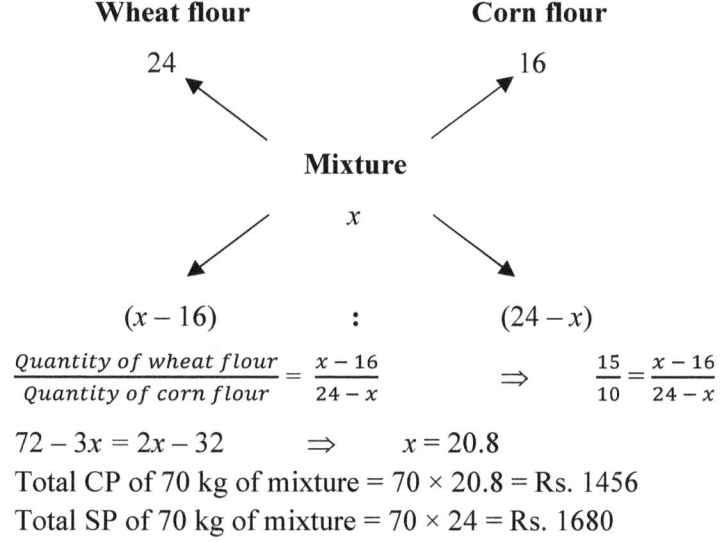

$$\frac{\text{Quantity of wheat flour}}{\text{Quantity of corn flour}} = \frac{x-16}{24-x} \qquad \Rightarrow \qquad \frac{15}{10} = \frac{x-16}{24-x}$$

$72 - 3x = 2x - 32 \qquad \Rightarrow \qquad x = 20.8$

Total CP of 70 kg of mixture = $70 \times 20.8 = \text{Rs. } 1456$

Total SP of 70 kg of mixture = $70 \times 24 = \text{Rs. } 1680$

∴ Profit = SP − CP = 1680 − 1456 = Rs. 224.

Q - 16 A mixture contains 75% acid and the remaining is water. What part of the mixture that should be removed and replaced by same amount of water to make the ratio of acid and water is 3 : 2?

Sol: Initially, the ratio of acid and water = 75 : 25 = 3 : 1

Let 'x' liters of mixture is replaced by 'x' liters of water.

In 'x' L of mixture, acid = $\frac{3x}{4}$ L and water = $\frac{x}{4}$ L

$$\frac{Acid}{Water} = \frac{3 - \frac{3x}{4}}{1 - \frac{x}{4} + x} = \frac{3}{2} \qquad \Rightarrow \qquad \frac{12 - 3x}{4 + 3x} = \frac{3}{2}$$

$$24 - 6x = 12 + 9x \qquad \Rightarrow \qquad x = \frac{4}{5}$$

∴ Part of mixture replaced = $\frac{4}{5} \times \frac{1}{4} = \frac{1}{5}$.

Q - 17 The cost of pure milk is Rs. 50 per liter. After adding water, the milk man sells the mixture for Rs. 52 per liter and thereby makes a profit of 30%. In what ratio does he mix the milk and water respectively?

Sol: Consider, CP of mixture = 100%

Given that, SP of mixture = Rs. 52 per liter and gain = 30%

SP of mixture = 130% ⟶ 52

CP of mixture = 100% ⟶ ? = $\frac{52 \times 100}{130}$ = Rs. 40

According to alligation rule,

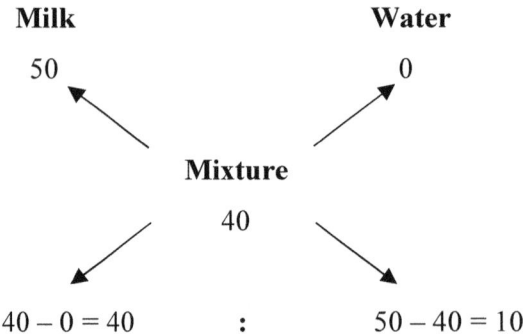

∴ Ratio of milk and water = 40 : 10 = 4 : 1.

Q - 18 A petrol pump owner mixed leaded and unleaded petrol in such a way that the mixture contains 15% unleaded petrol. What quantity of leaded petrol should be added to 1 liter mixture, so that the percentage of unleaded petrol becomes 10%?

Sol: Given that, total petrol mixture = 1 liter = 1000 mL

Unleaded petrol = $\frac{15}{100} \times 1000$ = 150 mL

Leaded petrol = $\frac{85}{100} \times 1000$ = 850 mL

Let 'x' mL of leaded petrol is added to make unleaded petrol as 10% in the mixture.

Unleaded petrol percentage $= \dfrac{Quantity\ of\ unleaded\ petrol}{Total\ quantity} \times 100\%$

$10 = \dfrac{150}{1000 + x} \times 100 \qquad \Rightarrow \qquad 1000 + x = 1500 \qquad \Rightarrow \qquad x = 500\ \text{mL}$

∴ 500 mL of leaded petrol is added to the initial mixture.

Alternate method:

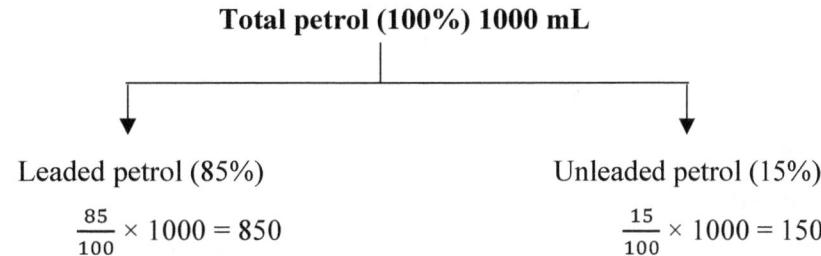

Total petrol (100%) 1000 mL

Leaded petrol (85%)   Unleaded petrol (15%)

$\dfrac{85}{100} \times 1000 = 850$ $\dfrac{15}{100} \times 1000 = 150$

According to question,

We are adding leaded petrol. So, the quantity of unleaded petrol is constant.

Unleaded petrol = 10% $\longrightarrow$ 150 mL   **After adding leaded petrol**

Total petrol = 100% $\longrightarrow$ $? = \dfrac{150 \times 100}{10} = 1500\ \text{mL}$

Total quantity increased by $(1500 - 1000) = 500\ \text{mL}$ Leaded (90%)   Unleaded (10%)

∴ 500 mL of leaded petrol is added to the mixture.

Q - 19 **Nikitha's expenditure and savings are in the ratio of 4 : 3. Her income increases by 12% and expenditure increases by 15%. Find the percentage increase in the savings.**

Sol: Consider, savings are increases by x%.

According to alligation rule,

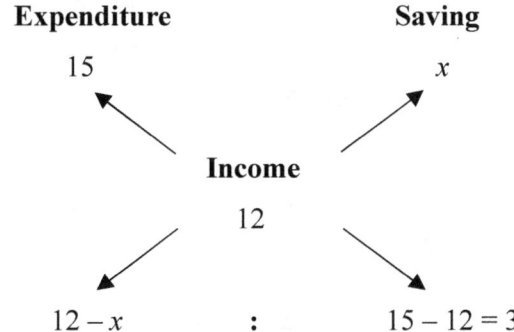

Expenditure   **Saving**

15   x

Income

12

$12 - x$: $15 - 12 = 3$

Given that, Expenditure : Savings = 4 : 3

$\Rightarrow \qquad \dfrac{12 - x}{3} = \dfrac{4}{3} \qquad \Rightarrow \qquad x = 8\%$

∴ Savings are increases by 8%.

Q - 20 **In a zoo, there are rabbits and pigeons. If number of heads are 150 and number of legs are 460. Find the number of rabbits.**

Sol: Consider, no. of rabbits $= x$ and no. of pigeons $= y$.

According to question, $x + y = 150$ .... (1)

$\qquad 4x + 2y = 460$ .... (2) $\qquad$ ∵ No. of legs for rabbit $= 4$

From (1), $y = 150 - x$ .... substitute in (2) $\qquad$ No. of legs for pigeon $= 2$

$\qquad 4x + 2(150 - x) = 460$ ⇒ $4x + 300 - 2x = 460$ ⇒ $x = 80$

∴ No. of rabbits are 80.

Shortcut:

According to alligation rule,

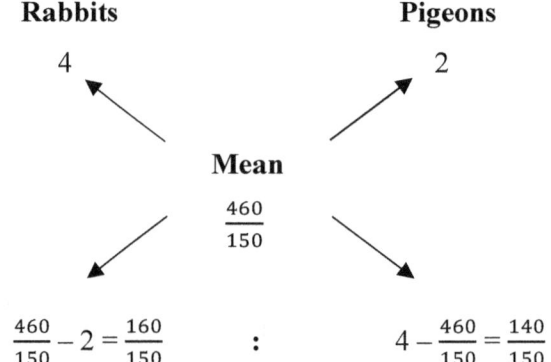

Rabbits $\qquad\qquad$ **Pigeons**

4 $\qquad\qquad\qquad\qquad$ 2

Mean

$\dfrac{460}{150}$

$\dfrac{460}{150} - 2 = \dfrac{160}{150}$ $\qquad : \qquad$ $4 - \dfrac{460}{150} = \dfrac{140}{150}$

Ratio of rabbits and pigeons $= 160 : 140 = 8 : 7$

Total no. of heads $= 15$ parts → 150 ⇒ 1 part $= 10$

∴ No. of rabbits $= 8$ parts $= 8 \times 10 = 80$.

Q - 21 **If three jars of equal capacity are filled with a mixture of wine and water. The ratios of wine and water are 2 : 3, 1 : 2 and 5 : 1 in the three jars respectively. If all the three jars are emptied into a single large jar, then find the proportion of wine and water in the mixture.**

Sol: Given that, ratios of wine and water in 3 jars are 2 : 3, 1 : 2 and 5 : 1

Total wine in 3 jars $= \dfrac{2}{5} + \dfrac{1}{3} + \dfrac{5}{6} = \dfrac{47}{30}$

Total water in 3 jars $= \dfrac{3}{5} + \dfrac{2}{3} + \dfrac{1}{6} = \dfrac{43}{30}$

∴ Required ratio $= \dfrac{47}{30} : \dfrac{43}{30} = 47 : 43$.

Q - 22 Two vessels of capacity 24 liters and 36 liters, contain mixture of acid and water in the ratio of 9 : 11 and 13 : 12 respectively. The contents of all the two vessels are poured into single vessel. Find the ratio of acid and water in the resultant mixture.

Sol: Given that, capacity of two vessels are 24 L and 36 L

Also, acid and water ratios are 9 : 11 and 13 : 12

Total acid in 2 vessels $= 24 \times \dfrac{9}{20} + 36 \times \dfrac{13}{25} = \dfrac{2952}{100}$

Total water in 2 vessels $= 24 \times \dfrac{11}{20} + 36 \times \dfrac{12}{25} = \dfrac{3048}{100}$

$\therefore$ Required ratio $= \dfrac{2952}{100} : \dfrac{3048}{100} = 123 : 127$.

Q - 23 Three jugs of capacity 5 liters, 6 liters and 7 liters, contain mixture of spirit and water in the ratio 4 : 1, 3 : 7 and 7 : 8 respectively. The contents of all three jugs are poured into single vessel. Find the ratio of spirit and water in the resultant mixture.

Sol: Given that, capacity of 3 jugs are 5 L, 6 L and 7 L

Also, spirit and water ratio are 4 : 1, 3 : 7, 7 : 8

Total spirit in 3 jugs $= 5 \times \dfrac{4}{5} + 6 \times \dfrac{3}{10} + 7 \times \dfrac{7}{15} = \dfrac{272}{30}$

Total water in 3 jugs $= 5 \times \dfrac{1}{5} + 6 \times \dfrac{7}{10} + 7 \times \dfrac{8}{15} = \dfrac{268}{30}$

$\therefore$ Required ratio $= \dfrac{272}{30} : \dfrac{268}{30} = 68 : 67$.

Q - 24 Fresh fruit contains 80% water and dry fruit contains 36% water. How much dry fruit is obtained from 400 kg of fresh fruit?

Sol:

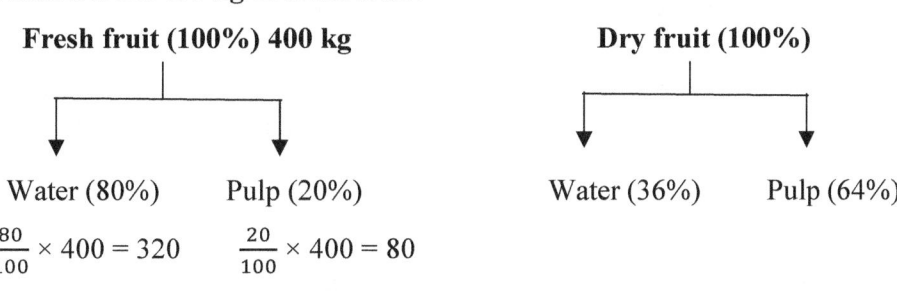

Fresh fruit (100%) 400 kg   **Dry fruit (100%)**

Water (80%)   Pulp (20%)   Water (36%)   Pulp (64%)

$\dfrac{80}{100} \times 400 = 320$ $\dfrac{20}{100} \times 400 = 80$

> **Note:** Quantity of pulp is same for both fresh fruit and dry fruit.

Dry fruit pulp = 64% $\longrightarrow$ 80 kg

Total dry fruit = 100% $\longrightarrow$? $= \dfrac{80 \times 100}{64} = 125$ kg

$\therefore$ 125 kg of dry fruit is obtained from 400 kg of fresh fruit.

Q - 25 **A person mixes 3 kiloliters of milk at a rate of Rs. 800 per kiloliter with 4 kiloliters at a rate of Rs. 720 per kiloliter. How many kiloliters of water should be added to make the average value of the mixture is Rs. 640 per kiloliter?**

Sol: Consider, 'x' kilolitres of water is added and also water is available at free of cost.

$$\boxed{\text{Total cost} = \text{Quantity} \times \text{Price}}$$

Total cost of mixture = Cost of 1st variety + Cost of 2nd variety + Cost of water

$$(3 + 4 + x)\,640 = 3 \times 800 + 4 \times 720 + x \times 0$$

$$640x = 800 \qquad \Rightarrow \qquad x = 1.25$$

∴ 1.25 kiloliters of water is added to make the average value of mixture is Rs. 640.

Q - 26 **In what ratio must water be mixed with milk to gain $12\frac{1}{2}\%$ on selling the mixture at cost price?**

Sol: Let, CP of 1 liter milk is Rs. 1000

SP of 1 liter of mixture = Rs. 1000 and gain = $12\frac{1}{2}\% = \frac{25}{2}\%$

CP of mixture = 100% and SP of mixture = $112\frac{1}{2}\% = \frac{225}{2}\%$

SP of 1 liter of mixture = $\frac{225}{2}\%$ $\longrightarrow$ 1000

CP of 1 liter of mixture = 100% $\longrightarrow$ $? = 100 \times 1000 \times \frac{2}{225} = \frac{8000}{9}$

According to alligation rule,

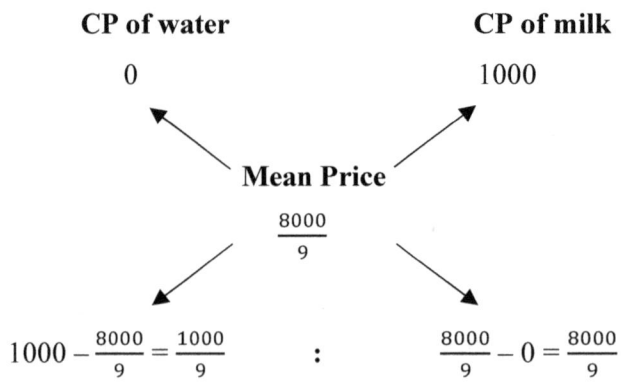

∴ Ratio of water and milk = $\frac{1000}{9} : \frac{8000}{9} = 1 : 8.$

Shortcut:

As the gain is $12\frac{1}{2}\%$, so the same amount of water is mixed with milk.

∴ Required ratio of water and milk = $12\frac{1}{2} : 100 = 1 : 8.$

Q - 27 A dishonest milk man professes to sell his milk at cost price but he mixes it with water and thereby gains 20%. Find the percentage of water in the mixture.

Sol: Consider, CP of 1 liter milk = Rs. 1000

SP of 1 liter of mixture = 1000 and gain = 20%

$\Rightarrow$ SP of mixture = 120% $\longrightarrow$ 1000 $\because$ CP = 100%

CP of mixture = 100% $\longrightarrow$ $? = \frac{1000 \times 100}{120} = \frac{2500}{3}$

According to alligation rule,

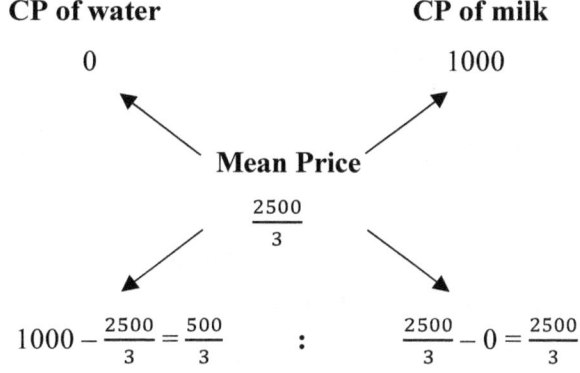

CP of water   **CP of milk**

0   1000

Mean Price

$\frac{2500}{3}$

$1000 - \frac{2500}{3} = \frac{500}{3}$: $\frac{2500}{3} - 0 = \frac{2500}{3}$

Ratio of water and milk $= \frac{500}{3} : \frac{2500}{3} = 1 : 5$.

$\therefore$ Percentage of water in the mixture $= \frac{Water}{Total} \times 100\% = \frac{1}{1+5} \times 100\% = 16\frac{2}{3}\%$.

Shortcut:

As the gain is 20%, so the same amount of water is mixed with milk. Then

Ratio of water and milk = 20 : 100 = 1 : 5

$\therefore$ Percentage of water in the mixture $= \frac{Water}{Total} \times 100\% = \frac{1}{1+5} \times 100\% = 16\frac{2}{3}\%$.

Q - 28 A man has Rs. 8000 with him. Part of which he lent out at 15% p.a. and the rest at 25% p.a., so that he will get a total interest of Rs. 4650 from both the amounts after 3 years. Find the amount lent at 25% p.a.

Sol: Given that, total amount = Rs. 8000 and total interest = Rs. 4650

Average interest p.a. $= \frac{Total\ interest}{No.of\ years} = \frac{4650}{3} = $ Rs. 1550

Mean rate of interest p.a. $= \frac{1550}{8000} \times 100\% = \frac{155}{8}\%$

According to alligation rule,

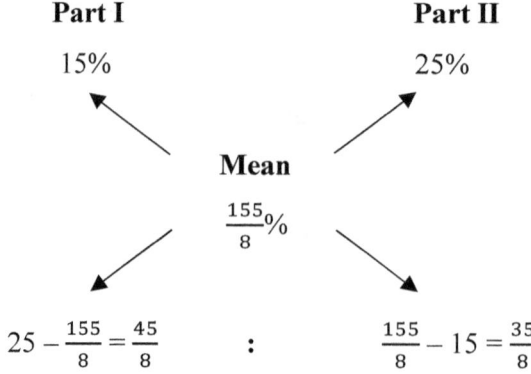

$$25 - \frac{155}{8} = \frac{45}{8} \qquad : \qquad \frac{155}{8} - 15 = \frac{35}{8}$$

∴ Required ratio $= \frac{45}{8} : \frac{35}{8} = 9 : 7$

Total amount = 16 parts → 8000 ⇒ 1 part = 500

∴ Amount lent at 25% p.a. = 7 parts = 7 × 500 = Rs. 3500.

Q - 29 | **Two alloys A and B contain gold and silver in the ratio of 3 : 1 and 5 : 3 respectively. In what ratio will they are mixed such that there will be 70% gold?**

Sol: Given that, alloy A (G : S) = 3 : 1, alloy B (G : S) = 5 : 3

Mixture = Gold : Silver = 70 : 30 = 7 : 3

According to alligation rule,

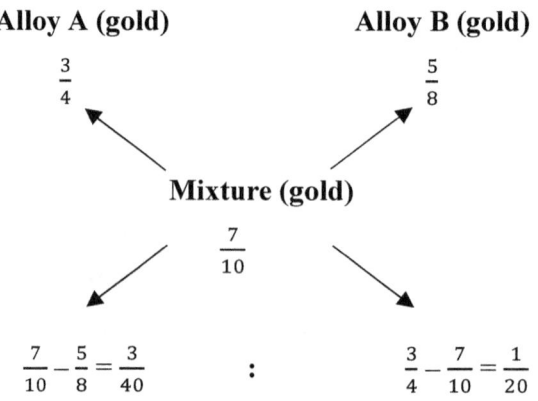

$$\frac{7}{10} - \frac{5}{8} = \frac{3}{40} \qquad : \qquad \frac{3}{4} - \frac{7}{10} = \frac{1}{20}$$

∴ Two alloys are mixed in the ratio of $\frac{3}{40} : \frac{1}{20} = 3 : 2$.

> **Note:** We can take the contents of either gold (or) silver from both alloys, we will get the same result.

(or)

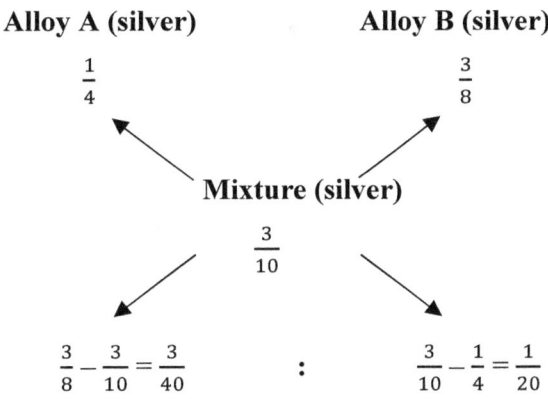

Alloy A (silver)   **Alloy B (silver)**

$$\frac{1}{4} \qquad\qquad \frac{3}{8}$$

Mixture (silver)

$$\frac{3}{10}$$

$$\frac{3}{8} - \frac{3}{10} = \frac{3}{40} \qquad : \qquad \frac{3}{10} - \frac{1}{4} = \frac{1}{20}$$

$\therefore$ Two alloys are mixed in the ratio of $\dfrac{3}{40} : \dfrac{1}{20} = 3 : 2$.

Q - 30   **Ankit has 75 notes. Out of these some are of Rs. 100 and other are Rs. 50. If Ankit has Rs. 5050, then find the total number of notes of Rs. 50.**

Sol:

> Total amount = No. of notes × Note denomination

Given that, total amount = Rs. 5050 and total no. of notes = 75

Average note denomination $= \dfrac{5050}{75} = \dfrac{202}{3}$

According to alligation rule,

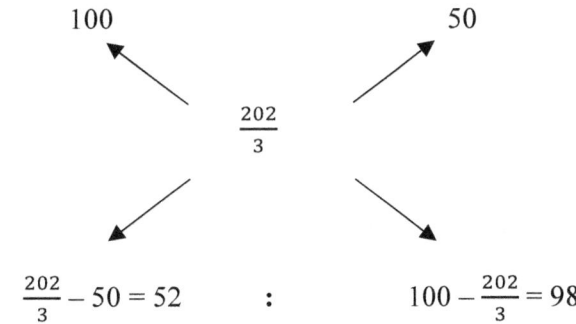

$$100 \qquad\qquad\qquad 50$$

$$\frac{202}{3}$$

$$\frac{202}{3} - 50 = 52 \qquad : \qquad 100 - \frac{202}{3} = 98$$

No. of notes ratio = 52 : 98 = 26 : 49

Total no. of notes = 75 parts $\rightarrow$ 75 notes $\Rightarrow$ 1 part = 1 note

$\therefore$ No. of Rs. 50 notes = 49 parts = 49 notes.

Q - 31   **The concentration of milk in three vessels X, Y and Z are 54%, 45% and 39% respectively. If 3 liters from X, 5 liters from Y and 7 liters from Z are mixed. Find the concentration of milk in the resultant solution.**

Sol: Milk concentration $= \dfrac{Milk\ quantity}{Total\ quantity} \times 100\%$

Milk concentration $= \dfrac{3 \times \frac{54}{100} + 5 \times \frac{45}{100} + 7 \times \frac{39}{100}}{3 + 5 + 7} \times 100\% = 44\%$.

Q - 32 A man covers a distance of 150 km in 7 hours. He covers some distance with a speed of 15 km/hr by cycle and the remaining distance with a speed of 30 km/hr by bike. Find the distance travelled by bike.

Sol: Given that, total distance = 150 km and total time = 7 hours.

Average speed $= \dfrac{Total\ distance}{Total\ time} = \dfrac{150}{7}$ km/hr

According to alligation rule,

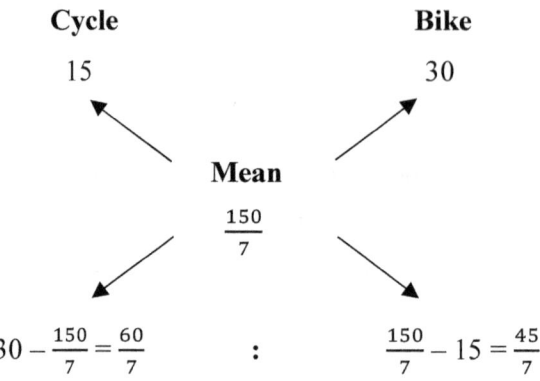

$$30 - \dfrac{150}{7} = \dfrac{60}{7} \qquad : \qquad \dfrac{150}{7} - 15 = \dfrac{45}{7}$$

∴ Time ratio $= \dfrac{60}{7} : \dfrac{45}{7} = 4 : 3$

Total time = 7 parts   →   7 hours   ⇒   1 part = 1 hour

Time taken by riding on bike = 3 parts = 3 hours

∴ Distance travelled by bike = S × T = 30 × 3 = 90 km.

Q - 33 In a laboratory, two bottles contain mixture of acid and water in the ratio 3 : 7 in the first bottle and 8 : 5 in the second bottle. Find the ratio in which the contents of these two bottles are mixed such that the new mixture has acid and water in the ratio 4 : 5.

Sol: According to alligation rule,

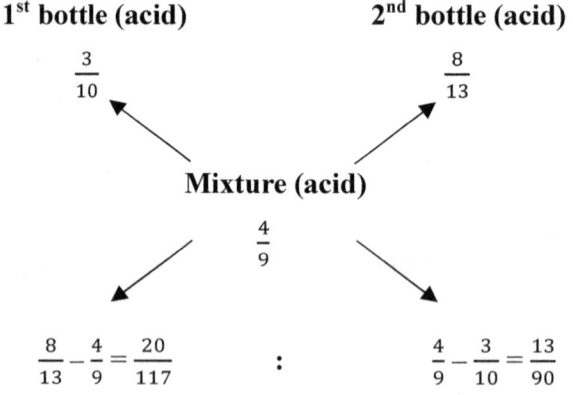

$$\dfrac{8}{13} - \dfrac{4}{9} = \dfrac{20}{117} \qquad : \qquad \dfrac{4}{9} - \dfrac{3}{10} = \dfrac{13}{90}$$

∴ Contents of two bottles are mixed in the ratio of $\dfrac{20}{117} : \dfrac{13}{90} = 200 : 169$.

Q - 34 There are two alloys X and Y which contains lead and tin in the ratio of 4 : 7 and 9 : 13 respectively. Equal quantities of these alloys are melted to form a third alloy Z. Find the ratio of lead and tin in the alloy Z.

Sol: Given that, equal quantities of both the alloys are taken to form third alloy Z.

∴ Ratio of lead and tin in alloy Z $= \frac{4}{11} + \frac{9}{22} : \frac{7}{11} + \frac{13}{22} = 17 : 27$.

Q - 35 A person travels 930 km in 15 hours in two stages. One part of the journey, he travels by bus at a rate of 50 km/hr and other part of the journey, he travels by car at a rate of 70 km/hr. How much distance did he travel by car?

Sol: Given that, total distance = 930 km and total time = 15 hours.

Average speed $= \frac{Total\ distance}{Total\ time} = \frac{930}{15} = 62$ km/hr

According to alligation rule,

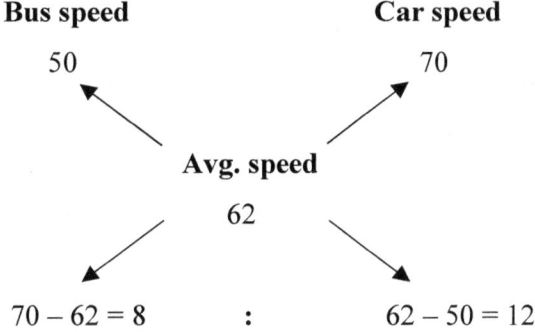

Time ratio = 8 : 12 = 2 : 3

Total time = 5 parts → 15 hours ⇒ 1 part = 3 hours

Time taken for car = 3 parts = 3 × 3 = 9 hours

∴ Distance travelled by car = S × T = 70 × 9 = 630 km.

Q - 36 An 15 liters cylinder contains a mixture of oxygen and nitrogen, the volume of oxygen is 25% of the total volume. A few liters of the mixture is replaced and an equal amount of nitrogen is added. Then, the same amount of the mixture as before is released and replaced by nitrogen for the second time. As a result, the oxygen content becomes 9% of the total volume. How many liters of mixture is released each time?

Sol:

$$\text{Final quantity} = \text{Initial quantity} \left[1 - \frac{replaced\ quantity}{total\ quantity}\right]^n$$

Where, n = no. of operations

Consider 'x' liters of mixture is released each time.

$$9 = 25\left[1 - \frac{x}{15}\right]^2 \quad \Rightarrow \quad 1 - \frac{x}{15} = \sqrt{\frac{9}{25}} \quad \Rightarrow \quad 1 - \frac{x}{15} = \frac{3}{5} \quad \Rightarrow \quad x = 6$$

∴ 6 L of mixture is released each time.

Q - 37 **A milk man has two vessels of milk. The first vessel contains 20% water and the rest milk. The second vessel contains 55% water. How much milk should he mix from each of the vessels, so as to get 14 liters of milk such that the ratio of water and milk is 2 : 3?**

Sol: Given that, ratio of water and milk = 2 : 3

Water in the mixture = 40% and milk = 60%

According to alligation rule,

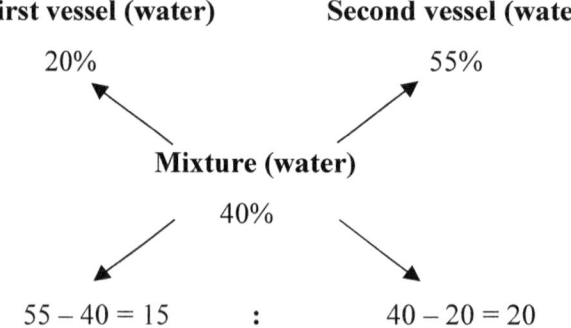

Ratio of quantities mix from both the vessels = 15 : 20 = 3 : 4

Total milk = 7 parts → 14 L ⇒ 1 part = 2 L

∴ Milk taken from 1ˢᵗ vessel = 3 × 2 = 6 L.

Milk taken from 2ⁿᵈ vessel = 4 × 2 = 8 L.

Q - 38 A glass full of wine contains 48% alcohol. A part of this wine is replaced by another containing 16% alcohol and now the percentage of alcohol was found to be 28%. Find the quantity of wine replaced.

Sol: According to alligation rule,

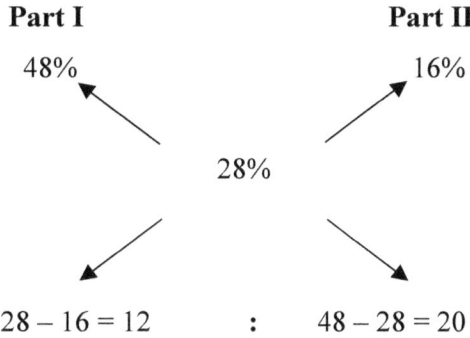

Required ratio = 12 : 20 = 3 : 5

∴ Quantity of wine replaced = $\frac{5}{3+5} = \frac{5}{8}$.

Q - 39 In a bag there are 25 paisa and 50 paisa coins. The total amount in the bag is Rs. 28.5 and the total number of coins are 75. Then, find the number of 50 paisa coins.

Sol: Given that, total amount = Rs. 28.5 = 2850 paisa

Total no. of coins = 75

Total amount = No. of coins × Coin denomination

Average coin denomination = $\frac{2850}{75}$ = 38 paisa

According to alligation rule,

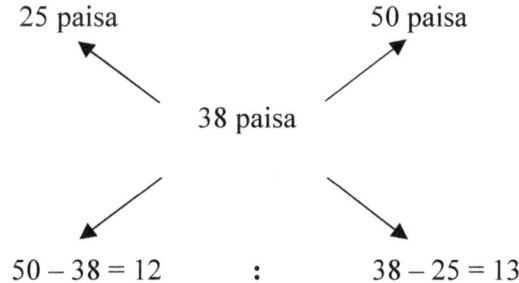

No. of coins ratio = 12 : 13

Total no. of coins = 25 parts   →   75   ⇒   1 part = 3 coins

∴ No. of 50 paisa coins = 13 parts = 13 × 3 = 39 coins.

Q - 40 The average marks of the students in four sections A, B, C and D together is 70%. The average marks of the students of A, B, C and D individually are 50%, 65%, 75% and 80% respectively. If the average marks of the students of section A and B together is 57% and that of the students of B and C is 70%. Find the ratio of number of students in sections A and D.

Sol: Given that, average marks of all 4 sections = 70%

Average marks of B and C is also 70%, so average marks of A and D together must be 70%

According to alligation rule,

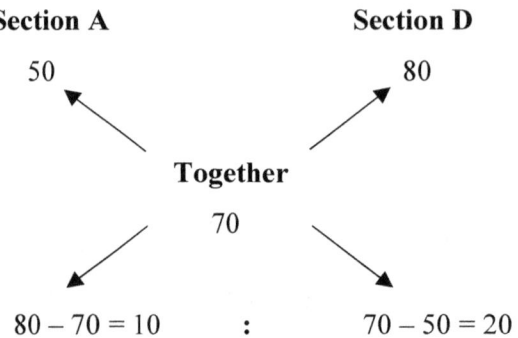

∴ Ratio of no. of students in A and D = 10 : 20 = 1 : 2.

Q - 41 In a municipal parking there are some two wheelers and rest are four wheelers. If wheels are counted, there are total 480 wheels but the watchman of parking told that there are only 155 vehicles. If no vehicle has a stepney, then find the number of four wheelers.

Sol: Given that, total wheels = 480 and total vehicles = 155

Average no. of wheels per vehicle $= \frac{Total\ wheels}{Total\ vehicles} = \frac{480}{155} = \frac{96}{31}$

According to alligation rule,

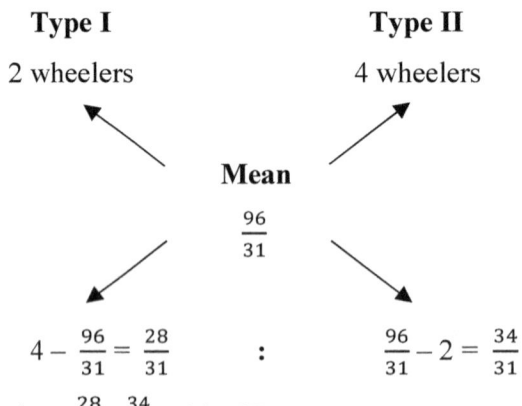

Required ratio $= \frac{28}{31} : \frac{34}{31} = 14 : 17$

Total no. of vehicles = 31 parts → 155 ⇒ 1 part = 5 vehicles

∴ Total no. of four wheelers = 17 parts = 17 × 5 = 85 vehicles.

Q - 42 A jar is filled with spirit and water. 60% of spirit and 40% of water is taken out of the jar. It is found that the jar is vacated by 48% and has 210 L mixture. Find the quantity of spirit and water in the mixture.

Sol: Total quantity in the jar = 210 L

Given that, 60% of spirit is taken out, then remaining spirit = 40%

40% of water is taken out, then remaining water = 60%

Jar is vacated by 48%, then remaining quantity left = 52%

According to alligation rule,

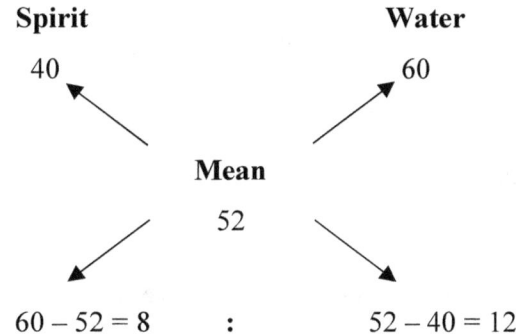

| **Spirit** | | **Water** |
| 40 | | 60 |

Mean

52

$60 - 52 = 8$: $52 - 40 = 12$

Ratio of spirit and water in the mixture = 8 : 12 = 2 : 3

Total quantity = 5 parts $\rightarrow$ 210 L $\Rightarrow$ 1 part = 42 L

∴ Quantity of spirit = 2 × 42 = 84 L & Quantity of water = 3 × 42 = 126 L.

Q - 43 A vessel contains a mixture of two liquids milk and water in the ratio of 5 : 2, when 14 L of mixture are withdrawn off and the vessel is filled with water, then the ratio of milk and water becomes 3 : 4. Find the initial quantity of milk in the vessel.

Sol: Consider, initially quantity of milk = 5x L and water = 2x L.

According to question,

$$\frac{New\ milk}{New\ water} : \frac{5x - 14 \times \frac{5}{7}}{2x - 14 \times \frac{2}{7} + 14} = \frac{3}{4} \quad \Rightarrow \quad \frac{5x - 10}{2x + 10} = \frac{3}{4}$$

$$20x - 40 = 6x + 30 \quad \Rightarrow \quad x = 5$$

∴ Initially quantity of milk = 5x = 5 × 5 = 25 L.

Q - 44 From a container, full of pure milk. 10% is replaced by water and this process is repeated three times. At the end of third operation, the quantity of pure milk is reduces to how much percentage?

Sol: Consider, initial quantity of milk = 100 L

According to question, every time 10% is removed and it is replaced by water.

Total 100 L (Pure milk)

	Milk	**Water**

After 1ˢᵗ operation: $100 - \frac{10}{100} \times 100 = 90$ $100 - 90 = 10$

After 2ⁿᵈ operation: $90 - \frac{10}{100} \times 90 = 81$ $100 - 81 = 19$

After 3ʳᵈ operation: $81 - \frac{10}{100} \times 81 = 72.9$ $100 - 72.9 = 27.1$

∴ After 3ʳᵈ operation, pure milk is reduces to 72.9%.

Shortcut:

Consider, initial quantity of pure milk = 100%

Final milk quantity left = Initial quantity $\left[1 - \frac{Replaced\ quantity}{Total\ quantity}\right]^n$

Where, n = no. of operations

∴ Final milk quantity left = $100 \left[1 - \frac{10}{100}\right]^3 = 72.9\%$.

Q - 45   **9 liters of milk is drawn from a container full of milk and is then filled with water. This operation is performed three more times. The ratio of the quantity of milk left in the container and that of water is 81 : 175. How much milk did the container hold initially?**

Sol: Consider, total quantity of milk in the container = 'T' L

$\frac{Final\ milk\ quantity\ left}{Initial\ quantity} = \left[1 - \frac{replaced\ quantity}{Total\ quantity}\right]^n$

Where, n = no. of operations.

Given that, final milk and water ratio = 81 : 175,

no. of operations = 4 and replaced quantity = 9 L

$\frac{81}{256} = \left[1 - \frac{9}{T}\right]^4$ ⇒ $1 - \frac{9}{T} = \frac{3}{4}$ ⇒   T = 36 L

∴ Total quantity of milk in the container initially = 36 L.

Q - 46   **A mixture contains wine and water in the ratio 4 : 3 and another mixture contains them in the ratio 5 : 6. How many liters of water must be mixed with 5 L of the former, so that the resulting mixture may contain equal quantities of wine and water?**

Sol: According to alligation rule,

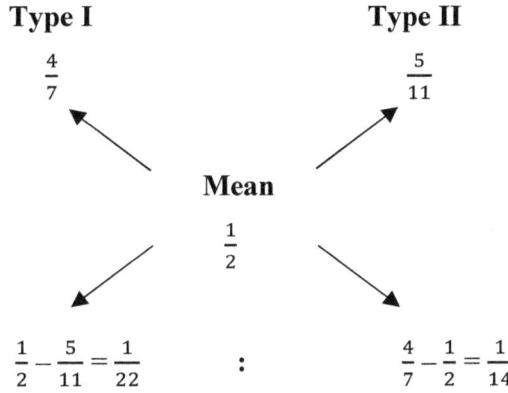

Required ratio $= \dfrac{1}{22} : \dfrac{1}{14} = 7 : 11$

Given that, former liquid = 7 parts $\rightarrow$ 5 L $\Rightarrow$ 1 part $= \dfrac{5}{7}$ L

∴ Quantity of water mixed = 11 parts $= 11 \times \dfrac{5}{7} = 7\dfrac{6}{7}$ L.

Q - 47 **A cask is full of acid. One-fourth of it is taken out and then an equal amount of water is poured into the cask to fill it. This operation is performed three times. Find the final ratio of acid and water in the cask.**

Sol: Consider, initial acid = 100 L

	Acid	**Water**
After 1st operation:	$100 - \dfrac{100}{4} = 75$	$100 - 75 = 25$
After 2nd operation:	$75 - \dfrac{75}{4} = \dfrac{225}{4}$	$100 - \dfrac{225}{4} = \dfrac{175}{4}$
After 3rd operation:	$\dfrac{225}{4} - \dfrac{225}{4} \times \dfrac{1}{4} = \dfrac{675}{16}$	$100 - \dfrac{675}{16} = \dfrac{925}{16}$

∴ Ratio of acid and water $= \dfrac{675}{16} : \dfrac{925}{16} = 27 : 37$

Shortcut:

Final quantity of acid = Initial quantity $\left[1 - \dfrac{Replaced\ quantity}{Total\ quantity}\right]^n$

Where, n = no. of operations

Consider, initial acid = 100 L

Given that, replaced quantity $= \dfrac{1}{4}$ of total $= \dfrac{1}{4} \times 100 = 25$ and n = 3

Final quantity of acid left $= 100 \left[1 - \dfrac{25}{100}\right]^3 = 100 \times \dfrac{27}{64} = \dfrac{2700}{64}$

Final quantity of water left = Total − Final acid $= 100 - \dfrac{2700}{64} = \dfrac{3700}{64}$

∴ Ratio of acid and water $= \dfrac{2700}{64} : \dfrac{3700}{64} = 27 : 37.$

Q - 48 The ratio of petrol and kerosene in the container is 5 : 4. When 12 liters of the mixture is taken out and is replaced by the kerosene, the ratio becomes 4 : 5. Find the total quantity of the mixture in the container.

Sol: Consider, initially quantity of petrol and kerosene are $5x$ and $4x$ respectively.

According to question,

$$\frac{New\ Petrol\ quantity}{New\ Kerosene\ quantity} : \frac{5x - 12 \times \frac{5}{9}}{4x - 12 \times \frac{4}{9} + 12} = \frac{4}{5}$$

$$\left(5x - \frac{20}{3}\right) \times 5 = \left(4x - \frac{16}{3} + 12\right) \times 4 \qquad \Rightarrow \qquad 75x - 100 = 48x + 80$$

$$\Rightarrow \qquad x = \frac{180}{27} = \frac{20}{3}$$

∴ Total quality of mixture in the container $= 5x + 4x = 9x = 9 \times \frac{20}{3} = 60$ L.

Q - 49 If the price of three types of oil are Rs. 78, Rs. 92 and Rs. 114 per liter, then find the ratio in which these types of oil should be mixed, so that the resultant mixture cost is Rs. 88 per liter.

Sol: Consider, 3 varieties of oil are x, y and z.

According to alligation rule,

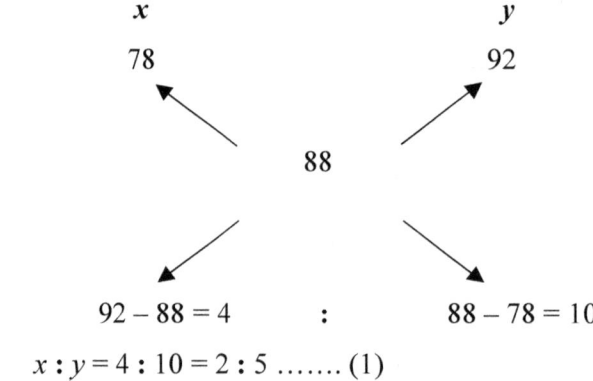

$$92 - 88 = 4 \qquad : \qquad 88 - 78 = 10$$

$$x : y = 4 : 10 = 2 : 5 \ \dots\dots\ (1)$$

According to alligation rule,

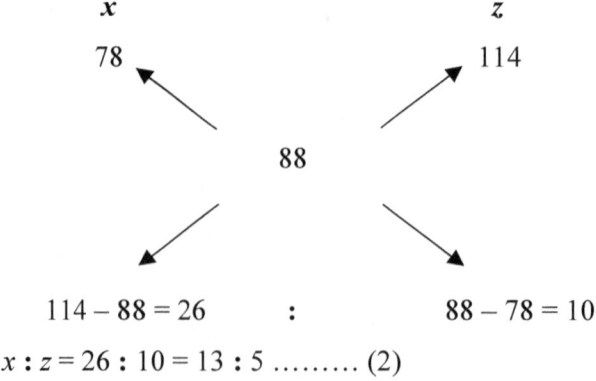

$$114 - 88 = 26 \qquad : \qquad 88 - 78 = 10$$

$$x : z = 26 : 10 = 13 : 5 \ \dots\dots\ (2)$$

The value of '*x*' must be equal in both ratios. So, multiply first ratio with 13 and second ratio with 2.

$$x : y = (2 : 5) \times 13 = 26 : 65$$
$$x : z = (13 : 5) \times 2 = 26 : 10$$

∴ Three varieties of oil are mixed in the ratio $x : y : z = 26 : 65 : 10$.

Q - 50 | **A butler stole wine from a butt of sherry which contained 85% of the spirit and he replaced it by wine containing only 34% spirit. Then, the butt was of 68% strength only. How much of the butt did he steal?**

Sol: According to alligation rule,

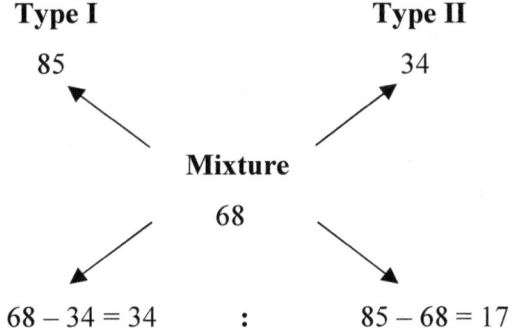

Required ratio = 34 : 17 = 2 : 1

Finally there is $\frac{2}{3}$ of butt

∴ Butler stole $\frac{1}{3}$ of butt.

ASSESSMENT TEST

1. How many kilograms of tea powder costing Rs. 52 per kg should be mixed with 63 kg of high quality tea powder costing Rs. 74 per kg, so that the mixture worth is Rs. 66 per kg?

2. 520 grams of salt solution has 25% salt in it. How much salt should be added to make it 48% in the solution?

3. In an examination out of 510 students, 70% of boys and 85% of girls are passed. How many girls appeared in an exam, if the total pass percentage is 81%?

4. Sugar at Rs. 64 per kg and Rs. 75 per kg are mixed with a third variety in the ratio 1 : 2 : 2. If the mixture worth is Rs. 66 per kg, then find the price of third variety per kg.

5. A merchant has 95 kg of rice. Part of which he sell at 12% profit and the rest at 7% loss. He gains 1% on the whole. Find the quantity sold at 7% loss.

6. How many liters of water is to be added to 96 liters of a mixture having milk and water in the ratio 5 : 3, to make it a mixture with milk concentration as 48%?

7. How much water to be added to 51 liters of milk worth Rs. 4.6 per liter. So that the value of mixture may be Rs. 3.4 per liter?

8. 884 ml of a mixture contains milk and water to ratio of 9 : 4. How much water is to be added to get a new mixture containing milk and water in the ratio 9 : 5?

9. A vessel contains 240 liters of solution in which 60% is milk. How many liters of water must be added, so that the water concentration is increased to 55%?

10. Silver and copper are in the ratio of 7 : 4 in 550 grams of an alloy. How many grams of copper is added to make the ratio as 4 : 7?

11. How many kilograms of sugar costing Rs. 30 per kg must be mixed with 72 kg of sugar costing Rs. 45 per kg, so that 25% gain is obtained by selling the mixture at Rs. 50 per kg?

12. A mixture of 14 kg wheat flour costing Rs. 45 per kg and 12 kg of corn flour Rs. 30 per kg is sold for Rs. 50 per kg. What is the profit made by selling 65 kg of mixture?

13. A mixture contains 70% acid and the remaining is water. What part of the mixture that should be removed and replaced by same amount of water to make the ratio of acid and water is 7 : 5?

14. The cost of pure milk is Rs. 60 per liter. After adding water, the milk man sells the mixture for Rs. 42 per liter and thereby makes a profit of 20%. In what ratio does he mix the milk and water respectively?

15. Sham's expenditure and savings are in the ratio of 3 : 5. His income increases by 9% and expenditure increases by 14%. Find the percentage increase in the savings.

16. In a zoo, there are rabbits and pigeons. If number of heads are 250 and number of legs are 740, then find the number of pigeons.

17. If three vessels of equal capacity are filled with a mixture of alcohol and water. The ratios of alcohol and water are 3 : 2, 1 : 4 and 2 : 1 in three vessels respectively. If all the three vessels are emptied into a single large vessel, then find the proportion of alcohol and water in the mixture.

18. Three jugs of capacities 8 L, 9 L, and 10 L contain mixture of spirit and water in the ratio 8 : 5, 7 : 19 and 22 : 17 respectively. The contents of all the three jugs are poured into single vessel. Find the ratio of spirit and water in the resultant mixture.

19. Fresh fruit contains 75% water and dry fruit contains 30% water. How much dry fruit is obtained from 200 kg of fresh fruit?

20. Vikas bought 200 kg of rice at a rate of Rs. 25 per kg. He sold 80 kg at a profit of 20%. At what rate per kg should he sell the remaining to get a profit of 17% on the whole?

21. A person mixes 7 kiloliters of milk at a rate of Rs. 500 per kiloliter with 8 kiloliters at a rate of Rs. 450 per kiloliter. How many kiloliters of water should be added to make the average value of the mixture is Rs.400 per kiloliter?

22. The weights of Roger and Nadal are in the ratio of 5 : 4. Roger's weight is increases by 15% and the total weight of Roger and Nadal together becomes 75.6 kg, with an increase of 19%. By what percent did the weight of Nadal increase?

23. In what ratio must water be mixed with the milk to gain $14\frac{2}{7}$% on selling the mixture at cost price?

24. A dishonest milk man professes to sell his milk at cost price but he mixes it with water and there by gains $16\frac{2}{3}$%. Find the percentage of water in the mixture.

25. There are three types of milk. Vishaka, Amul and Nestle. The ratio of fat to the non-fat contents in milk is 2 : 3, 3 : 4, and 4 : 5 respectively. If all three types of milk is mixed in equal quantity, then what is the ratio of fat to non-fat contents in the mixture?

26. A man has Rs. 9000 with him. Part of which he lent out at 20% p.a and the rest at 30% p.a. So that he will get a total interest of Rs. 11700 from both the amounts after 5 years. Find the amount lent at 20% p.a.

27. A milk man purchased 20 liters of milk and mixed 7 liters of water in it. If the price per liter of the mixture becomes Rs. 24, then find the cost price of milk per liter.

28. 30 kg of a certain variety of wheat at Rs. 48 per kg is mixed with 42 kg of another variety of wheat and the mixture is sold at the average price of Rs.55 per kg. If there be no profit or loss due to the new sale price, then find the price of the second variety wheat.

29. A vessel is filled with liquid. 8 parts of which are water and 12 parts are milk. How much of the mixture is withdrawn and replaced with water so that, the mixture may be half water and half milk?

30. Two alloys A and B contains copper and silver in the ratio of 2 : 5 and 5 : 8 respectively. In what ratio will they are mixed such that there will be 30% copper?

31. The concentration of milk in three vessels A, B and C are 75%, 60% and 45% respectively. If 2 liters from A, 5 liters from B and 8 liters from C are mixed, then find the concentration of milk in resultant solution.

32. Ravi has 90 notes. Out of these some are of Rs. 50 and other are of Rs. 20. If Ravi has Rs. 2760, then find the total number of 20 rupees notes.

33. A man covers a distance of 70 km in 9 hours. He covers some distance with a speed of 10 km/hr by cycle and the remaining distance with a speed of 6kmph on foot. Find the distance travelled on foot.

34. In a laboratory, two bottles contain mixture of acid and water in the ratio 2 : 7 in first bottle and 7 : 5 in second bottle. Find the ratio in which the contents of these two bottles are mixed such that the new mixture has acid and water in the ratio 4 : 11.

35. There are two alloys A and B which contains copper and silver in the ratio of 3 : 5 and 7 : 13 respectively. Equal quantities of these alloys are melted to form a third alloy C. Find the ratio of copper and silver in the alloy C.

36. A vessel contains a mixture of two liquids acid and water in the ratio of 7 : 5, when 12 liters of mixture are withdrawn off and the vessel is filled with water, then the ratio of acid and water becomes 7 : 11. Find the initial quantity of acid in the vessel.

37. From a container, full of pure milk. 30% is replaced by water and this process is repeated 3 times. At the end of third operation, the quantity of pure milk is reduces to how much percentage?

38. 5 liters of milk is drawn from a container full of milk and is then filled with water. This operation is performed two more times. The ratio of the quantity of milk left in the container and that of water 64 : 61. How much milk did the container hold initially?

39. A mixture contains wine and water in the ratio 5 : 4 and another mixture contains them in the ratio 6 : 7. How many liters of water must be mixed with 7 liters of the former, so that the resulting mixture may contain equal quantities of wine and water?

40. A person travels 660 km in 11 hours in two stages. One part of the journey, he travels by bus at a rate of 54 km/hr and other part of the journey, he travels by train at a rate of 65 km/hr. How much distance did he travel by bus?

41. A can contains a mixture of whisky and water in the ratio 5 : 3. If 16 liters of the mixture is taken out and 16 liters of water is poured into the can, then the ratio becomes 5 : 7. How many liters of whisky was contained in the can?

42. A cask is full of alcohol. Two-fifth of it is taken out and then an equal amount of water is poured into the cask to fill it. This operation is performed four times. Find the final ratio of alcohol and water in the cask.

43. An 28 liters cylinder contains a mixture of oxygen and nitrogen, the volume of oxygen is 49% of total volume. A few liters of the mixture is released and an equal amount of nitrogen is added. Then, the same amount of mixture as before is released and replaced by nitrogen for the second time. As a result, the oxygen content becomes 16% of the total volume. How many liters of mixture is released each time?

44. A milk man has two vessels of milk. The first vessel contains 15% water and the rest milk. The second vessel contains 45% water. How much milk should he mix from each of the vessels so as to get 32 liters of milk such that the ratio of water and milk is 3 : 5?

45. A bottle full of wine contains 51% alcohol. A part of this wine is replaced by another containing 24% alcohol and now the percentage of alcohol was found to be 36%. Find the quantity of wine replaced.

46. In a bag there are 50 paisa and 1 rupee coins. The total amount in the bag is Rs. 90 and the total number of coins are 125. Find the number of 1 rupee coins.

47. A bottle is filled with whisky and water. 55% of whisky and 35% of water is taken out of the bottle. It is found that the bottle is vacated by 46% and has 340 liters mixture. Find the quantity of whisky and water in the mixture.

48. In a mixture of milk and water, there is only 32% water. After replacing the mixture with 9 liters of pure milk, then the percentage of milk in the mixture becomes 72%. Then, find the quantity of mixture.

49. If the price of three types of wheat are Rs. 56, Rs. 70 and Rs. 84 per kg, then find the ratio in which these types of wheat should be mixed, so that the resultant mixture cost is Rs. 62 per kg.

50. A butler stole wine from a butt of sherry which contained 70% of spirit and he replaced it by wine containing only 26% spirit. Then, the butt was of 46% strength only. How much of the butt did he steal?

KEY

1. 36 kg
2. 230 grams
3. 374
4. Rs. 58
5. 55 kg
6. 29 L
7. 18 L
8. 68 mL
9. 80 L
10. 412.5 grams
11. 36 kg
12. Rs. 775
13. $\frac{1}{2}$
14. 7 : 5
15. 6% increases
16. 130
17. 22 : 23
18. 1013 : 1093
19. $71\frac{3}{7}$ kg
20. Rs. 28.75 per kg
21. $2\frac{3}{4}$ kiloliters
22. 24%
23. 1 : 7
24. $14\frac{2}{7}$%
25. 401 : 544
26. Rs. 3600
27. Rs. 32.4
28. Rs. 60
29. $\frac{1}{6}$
30. 77 : 13

31. 44%

32. 58

33. 30 km

34. 57 : 8

35. 29 : 51

36. 21 L

37. 34.3%

38. 25 L

39. $10\frac{1}{9}$ L

40. 270 km

41. 30 L

42. 81 : 544

43. 12 L

44. 8 L, 24 L

45. $\frac{5}{9}$

46. 55

47. 187 L, 153 L

48. 72 L

49. 44 : 33 : 12

50. $\frac{6}{11}$

CPSIA information can be obtained
at www.ICGtesting.com
Printed in the USA
LVHW060034031120
670493LV00020B/477